D0437806

Also by Leonard Mlodinow

Subliminal: How Your Unconscious Mind Rules Your Behavior

War of the Worldviews (with Deepak Chopra)

The Grand Design (with Stephen Hawking)

The Drunkard's Walk: How Randomness Rules Our Lives

Feynman's Rainbow: A Search for Beauty in Physics and in Life

Euclid's Window: The Story of Geometry from Parallel Lines to Hyperspace

For Children (with Matt Costello)

The Last Dinosaur

Titanic Cat

The Upright Thinkers

The Upright Thinkers

The Human Journey from Living in Trees to Understanding the Cosmos

Leonard Mlodinow

Pantheon Books, New York

Library of Congress Cataloging-in-Publication Data
Mlodinow, Leonard, [date]
The upright thinkers : the human journey from living in trees to understanding the cosmos / Leonard Mlodinow.
pages cm
Includes index.
ISBN 978-0-307-90823-0 (hardcover : alk. paper).
ISBN 978-0-307-90824-7 (eBook).
1. Science—History. I. Title.
Q126.M56 2015 509—dc23 2014040067

www.pantheonbooks.com

Maps by Mapping Specialists

Jacket design by Kelly Blair

Printed in the United States of America
First Edition
2 4 6 8 9 7 5 3 1

To Simon Mlodinow

Contents

Part I: The Upright Thinkers

1. *Our Drive to Know* 3
   A starving man's hunger for knowledge . . . The human odyssey of discovery

2. *Curiosity* 10
   Lizards don't ask questions . . . From Handy Man to Wise, Wise Man . . . What infants ask but chimps don't

3. *Culture* 24
   Humanity's first church . . . Knowledge, ideas, and values go viral . . . Human and primate culture

4. *Civilization* 39
   From the savanna to the city . . . How the charms and headaches of neighbors led to the new arts of writing and arithmetic . . . The invention of law, from peasant (*Don't vomit in streams*) to planet (*Don't stray from your orbit*)

5. *Reason* 61
   Bad crops and angry gods . . . A new framework for looking at the world . . . The mystery of change and the tyranny of common sense . . . Aristotle, the one-man Wikipedia

Part II: The Sciences

6. *A New Way to Reason* 85
   Trusting your eyes over your ancestors . . . Castrated boars and universal laws of motion . . . The tactless Professor Galileo

7. *The Mechanical Universe* 114
   The good, the bad, and the ugly: Isaac Newton . . . The bet that turned Newton from alchemy to authoring the greatest scientific treatise ever written . . . The force of Newtonian thinking

8. *What Things Are Made Of* 148
From embalming to alchemy . . . The similarities between burning and breathing . . . Lavoisier loses his head . . . Mendeleev and his periodic table

9. *The Animate World* 183
Cells and the complexity of life . . . A recipe for making mice and the revolution of the microscope . . . Tragedy, illness, and Darwin's secret research

Part III: Beyond the Human Senses

10. *The Limits of Human Experience* 215
The billion billion tiny universes in a drop of water . . . Cracks in the Newtonian worldview . . . Accepting an unseeable reality . . . Planck and Einstein invent the quantum

11. *The Invisible Realm* 247
The insights of a dreamer . . . The crazy ideas of a pale and modest young man . . . The early quantum laws, "awful nonsense, bordering on fraud"

12. *The Quantum Revolution* 264
Heisenberg's new physics . . . The bizarre reality of the quantum universe . . . The empowering and humbling legacy of a new science

Epilogue 293
The advance of human understanding as a succession of fantasies . . . The importance of critical and innovative thinking . . . Where we are and where we are going

Acknowledgments 299

Notes 301

Index 321

Part I

The Upright Thinkers

The most beautiful and deepest experience a man can have is the sense of the mysterious. It is the underlying principle of religion as well as of all serious endeavor in art and science. He who never had this experience seems to me, if not dead, then at least blind.

—Albert Einstein, *My Credo*, 1932

1

Our Drive to Know

My father once told me of an emaciated fellow inmate in the Buchenwald concentration camp who had been educated in mathematics. You can tell something about people from what comes to mind when they hear the term "pi." To the "mathematician" it was the ratio of the circumference of a circle to its diameter. Had I asked my father, who had but a seventh-grade education, he would have said it was a circle of crust filled with apples. One day, despite that gulf between them, the mathematician inmate gave my father a math puzzle to solve. My father thought about it for a few days but could not master it. When he saw the inmate again, he asked him for the solution. The man wouldn't say, telling my father he must discover it for himself. Sometime later, my father again spoke to the man, but the man held on to his secret as if it were a hunk of gold. My father tried to ignore his curiosity, but he couldn't. Amid the stench and death around him, he became obsessed with knowing the answer. Eventually the other inmate offered my father a deal—he would reveal the puzzle's solution if my father would hand over his crust of bread. I don't know what my father weighed at the time, but when the American forces liberated him, he weighed eighty-five pounds. Still, my father's need to know was so powerful that he parted with his bread in exchange for the answer.

I was in my late teens when my father recounted that episode, and it made a huge impact on me. My father's family was gone, his possessions confiscated, his body starved, withered, and beaten. The Nazis had stripped him of everything palpable, yet his drive to think and reason and know survived. He was imprisoned, but his mind was free to roam, and it did. I realized then that the search for knowledge is the most human of all our desires, and that, different as our circumstances were, my own passion for understanding the world was driven by the same instinct as my father's.

As I went on to study science in college and after, my father would question me not so much about the technicalities of what I was learning, but about the underlying meaning—where the theories came from, why I felt they were beautiful, and what they said about us as human beings. This book, written decades later, is my attempt, finally, to answer those questions.

* * *

A few million years ago, we humans began to stand upright, altering our muscles and skeletons so that we could walk in an erect posture, which freed our hands to probe and manipulate the objects around us and extended the range of our gaze so that we could explore the far distance. But as we raised our stance, so too did our minds rise above those of other animals, allowing us to explore the world not just through eyesight but with our thoughts. We stand upright, but above all, we are thinkers.

The nobility of the human race lies in our drive to know, and our uniqueness as a species is reflected in the success we've achieved, after millennia of effort, in deciphering the puzzle that is nature. An ancient, given a microwave oven to heat his auroch meat, might have theorized that inside it was an army of hardworking, pea-size gods who built miniature bonfires under the food, then miraculously disappeared when the door was opened. But just as miraculous is the truth—that a handful of simple and inviolable abstract laws account for everything in our universe, from the workings of that microwave to the natural wonders of the world around us.

As our understanding of the natural world evolved, we progressed from perceiving the tides as being governed by a goddess to understanding them as the result of the gravitational pull of the moon, and we graduated from thinking of the stars as gods floating in the heavens to identifying them as nuclear furnaces that send photons our way. Today we understand the inner workings of our sun, a hundred million miles away, and the structure of an atom more than a billion times smaller than ourselves. That we have been able to decode these and other natural phenomena is not just a marvel. It also makes a gripping tale, and an epic one.

Some time ago, I spent a season on the writing staff of the television series *Star Trek: The Next Generation.* At my first story meeting there, at a table populated by all the show's other writers and producers, I pitched an idea for an episode that excited me because it involved the real astrophysics of solar wind. All eyes were focused on me, the new guy, the *physicist* in their midst, as I enthusiastically detailed my idea, and the science behind it. When I was done—the pitch had taken less than a minute—I looked with great pride and satisfaction at my boss, a gruff, middle-aged producer who had once been an NYPD homicide detective. He stared at me for a moment, his face strangely unreadable, and then he said with great force, "Shut up, you f—king egghead!"

When I got over my embarrassment, I realized that what he was so succinctly telling me was that they had hired me for my storytelling abilities, not to conduct an extension school class on the physics of stars. His point was well taken, and I have let it guide my writing ever since. (His other memorable suggestion: if you ever sense that you are going to be fired, turn down the heat on your swimming pool.)

In the wrong hands, science can be famously boring. But the story of what we know and how we know it isn't boring at all. It is supremely exciting. Full of episodes of discovery that are no less compelling than a *Star Trek* episode or our first trip to the moon, it is peopled by characters as passionate and quirky as those we know from art and music and literature, seekers whose insatiable curiosity took our species from its origins on the African savanna to the society we live in today.

How did they do that? How did we go from a species that had barely learned to walk upright and lived off whatever nuts and berries and roots we could harvest with our bare hands to one that flies airplanes, sends messages instantly around the globe, and re-creates in enormous laboratories the conditions of the early universe? That is the story I want to tell, for to know it is to understand your heritage as a human being.

* * *

It has become a cliché to say that today the world is flat. But if the distances and differences between countries are effectively diminishing, the differences between today and tomorrow are growing. When the first cities were built, around 4000 B.C., the fastest way to travel long distances was by camel caravan, which averaged just a few miles per hour. Somewhere between a thousand and two thousand years later, the chariot was invented, raising the maximum speed to about 20 miles per hour. Not until the nineteenth century did the steam locomotive finally allow faster travel, reaching speeds up to 100 miles per hour by the century's end. But although it took humans two million years to graduate from running at 10 miles per hour to racing across the country at 100 miles per hour, it took only fifty years to achieve the next factor of ten, with the creation of an airplane that could fly 1,000 miles an hour. And by the 1980s, humans were traveling at more than 17,000 miles per hour in the space shuttle.

The evolution of other technologies exhibits a similar acceleration. Take communications. As late as the nineteenth century, the Reuters news service was still using carrier pigeons to fly stock prices between cities. Then, by the mid-nineteenth century, the telegraph had become widely available, and in the twentieth century, the telephone. But while it took eighty-one years for the landline telephone to achieve a market penetration of 75 percent, the cell phone achieved that in twenty-eight years, and the smartphone in thirteen. In recent years, first email, and then texting, largely supplanted phone calls as a tool for communication, while the phone is increasingly employed not for phone calls but as a pocket computer.

"The world of today," said economist Kenneth Boulding, "is as different from the world in which I was born as that world was from Julius Caesar's." Boulding was born in 1910 and died in 1993. The changes he witnessed—and many others that have occurred since—were the products of science, and the technology it spawns. Such changes are a greater part of human life than ever before, and our success at work and in society is increasingly predicated on our ability to both assimilate innovation and to create innovations ourselves. For today, even those of us who don't work in science or technology face challenges that require us to innovate if we are to remain competitive. And so the nature of discovery is an important topic for us all.

To gain perspective on where we are today, and to have a hope of understanding where we are going, one must know where we came from. The greatest triumphs of human intellectual history—writing and mathematics, natural philosophy, and the various sciences—are usually presented in isolation, as if each had nothing to do with the others. But that approach emphasizes the trees and not the forest. It neglects, by its very nature, the unity of human knowledge. The development of modern science, for example—often heralded as the work of "isolated geniuses" such as Galileo and Newton—did not spring from a social or cultural vacuum. It had its roots in the approach to knowledge invented by the ancient Greeks; it grew from the big questions posed by religion; it developed hand in hand with a new approach to art; it was colored by the lessons of alchemy; and it would have been impossible without social progress ranging from the grand development of the great universities of Europe to mundane inventions such as that of the postal systems that grew to connect nearby cities and countries. The Greek enlightenment, similarly, sprouted from the astounding intellectual inventions of earlier peoples in lands such as Mesopotamia and Egypt.

As a result of such influences and connections, the account of how humans came to understand the cosmos does not consist of isolated vignettes. It forms, like the best fiction, a coherent narrative, a unified whole whose parts have numerous interconnections, and which begins at the dawn of humanity. In what follows, I offer a selective, guided tour of that odyssey of discovery.

Our tour begins with the development of the modern human mind and features the critical eras and turning points at which that mind leapt to new ways of looking at the world. Along the way, I will also portray some of the fascinating characters whose unique personal qualities and modes of thought played an important role in those innovations.

Like many tales, this is a drama in three parts. Part I, spanning millions of years, traces the evolution of the human brain and its propensity to ask "Why?" Our *why*'s propelled us to our earliest spiritual inquiries and led, eventually, to the development of writing and mathematics and the very concept of laws—the necessary tools of science. Ultimately those *whys* led to the invention of philosophy, the insight that the material world operates according to rhyme and reason that can, in principle, be understood.

The next phase of our journey explores the birth of the hard sciences. It is a story of revolutionaries who had the gift of seeing the world differently, and the patience, grit, brilliance, and courage to continue striving through all the years and even decades it sometimes took to develop their ideas. These pioneers—thinkers such as Galileo, Newton, Lavoisier, and Darwin—fought long and hard against the established doctrine of their day, so their stories are inevitably tales of personal struggle, sometimes with stakes as high as life itself.

Finally, as in many good stories, our tale takes an unexpected turn just when its heroes have reason to think they're close to the end of their journey. For no sooner did humanity come to believe it had deciphered all the laws of nature than, in a strange twist of the plot, such thinkers as Einstein, Bohr, and Heisenberg discovered a new realm of existence, an invisible one, in which those laws had to be rewritten. That "other" world—with its otherworldly laws—plays out on a scale too small to be apprehended directly: the microcosmos of the atom, ruled by the laws of quantum physics. It is those laws that are responsible for the enormous and still accelerating changes that we are experiencing in society today, for it is our understanding of the quantum that enabled the invention of computers, cell phones, televisions, lasers, the Internet, medical imaging, genetic mapping, and most of the new technologies that have revolutionized modern life.

While Part I of the book covers millions of years, and Part II covers hundreds, Part III spans mere decades, reflecting the exponential acceleration in the accumulation of human knowledge—and the newness of our forays into this strange new world.

* * *

The odyssey of human discovery stretches over many eras, but the themes of our quest to understand the world never vary, as they arise from our own human nature. One theme is familiar to anyone who works in a field dedicated to innovation and discovery: the difficulty in conceiving of a world, or an idea, that is any different from the world or ideas that we already know.

In the 1950s, Isaac Asimov, one of the greatest and most creative science fiction writers of all time, wrote the *Foundation* trilogy, a series of novels set many thousands of years in the future. In those books, the men commute to work in offices every day, and the women stay home. Within only a few decades, that vision of the distant future was already a thing of the past. I bring this up because it illustrates an almost universal limitation of human thought: our creativity is constrained by conventional thinking that arises from beliefs we can't shake, or never even think of questioning.

The flip side of the difficulty of conceiving change is the difficulty of accepting it, and that is another recurring theme of our story. We human beings can find change overwhelming. Change makes demands on our minds, takes us beyond our comfort zones, shatters our mental habits.

It produces confusion and disorientation. It requires that we let go of old ways of thinking, and the letting go is not our choice but is imposed upon us. What's more, the changes resulting from scientific progress often upend belief systems to which large numbers of people—and possibly their careers and their livelihoods—are attached. As a result, new ideas in science are often met with resistance, anger, and ridicule.

Science is the soul of modern technology, the root of modern civilization. It underpins many of the political, religious, and ethical issues of our day, and the ideas that underlie it are transforming society at an ever faster pace. But just as science plays a key role in shaping the patterns of human thought, it is also true that the patterns of human thinking have played a key role in shaping our theories of science. For science is, as Einstein remarked, "as subjective and psychologically conditioned as any other branch of human endeavor." This book is an effort to describe the development of science in that spirit—as an intellectual as well as a culturally determined enterprise, whose ideas can best be understood by an examination of the personal, psychological, historical, and social situations that molded them. To look at science that way sheds light not just on the enterprise itself, but on the nature of creativity and innovation, and, more broadly, on the human condition.

2

Curiosity

To understand the roots of science, we must look back to the roots of the human species itself. We humans are uniquely endowed with both the capacity and the desire to understand ourselves and our world. That is the greatest of the gifts that set us apart from other animals, and it is why it is we who study mice and guinea pigs and not they who study us. The urge to know, to ponder, and to create, exercised through thousands of millennia, has provided us with the tools to survive, to build for ourselves a unique ecological niche. Using the power of our intellect rather than our physique, we have shaped our environment to our needs, rather than allowing our environment to shape—or defeat—us. Over millions of years, the force and creativity of our minds have triumphed over the obstacles that challenged the strength and agility of our bodies.

When he was young, my son Nicolai used to like to capture small lizards to keep as pets—something you can do when you live in Southern California. We noticed that when we approached the animals, they would first freeze, and then, as we reached down for them, they would flee. We eventually figured out that if we brought along a large box, we could invert it over a lizard before it scurried off, then slide a piece of cardboard under it to complete the capture. Personally, if I'm walking down a dark, deserted street and see anything suspicious going on, I don't freeze; I immediately cross to the other side. So it's pretty safe to assume that if I were approached by two giant predators heading toward me, staring eagerly and carrying a gargantuan box, I'd assume the worst and instantly race away. Lizards, though, don't question their situation. They act purely by instinct. That instinct no doubt served them well over the many millions of years that preceded Nicolai and the box, but it failed them here.

We humans may not be the ultimate in physical specimens, but we do have the ability to supplement instinct with reason and—most important for our purposes—to ask questions about our environment. Those are the prerequisites of scientific thought, and they are crucial characteristics of our species. And so that is where our adventure begins: with the development of the human brain, with its unique gifts.

We call ourselves the "human" species, but the word "human" actually refers not only to us—*Homo sapiens sapiens*—but to an entire genus called *Homo*. This genus includes other species such as *Homo habilis* and *Homo erectus*, but those relatives are all long gone. In the single-elimination tournament called evolution, all the other human species proved inadequate. Only we, due to the power of our minds, have met all survival challenges (thus far).

Not too long ago, the man who was then the president of Iran was quoted as saying that Jews descended from monkeys and pigs. It is always heartening when a fundamentalist of any religion professes a belief in evolution, so I hesitate to criticize, but actually Jews—and all other humans, too—are descended not from monkeys and pigs, but from apes and rats, or at least ratlike creatures. Called *Protungulatum donnae* in the scientific literature, our great-great-great-and-so-on-grandmother—the progenitor of our ancestral primates, and of all mammals like us—seems to have been a cute, furry-tailed species whose members weighed in at no more than half a pound.

Artist's conception of *Protungulatum*

Scientists believe that these tiny animals first scurried across their habitat around sixty-six million years ago, soon after a six-mile-wide asteroid slammed into the earth. That catastrophic collision threw enough debris into the atmosphere to choke off the sun's rays for an extended period of time—and generated enough greenhouse gases to send temperatures soaring once the dust had settled. The double whammy of darkness followed by heat killed off about 75 percent of all plant and animal species, but for us it was fortunate: it created an ecological niche in which animals that give birth to live young could survive and thrive without being gobbled up by ravenous dinosaurs and other predators. Over the subsequent tens of millions of years, as species after species emerged and then faded into extinction, one branch of the *Protungulatum* family tree evolved into our ape and monkey ancestors, and then it branched further, producing our closest living relatives, the chimpanzees and bonobos (pygmy chimpanzees) and, finally, you, the reader of this book, and your fellow humans.

Today most people are comfortable with the fact that Grandma had a tail and ate insects. I go further than mere acceptance: I am excited and fascinated by our ancestry, and the story of our survival and cultural evolution. I think the fact that our ancient grandparents were rats and apes is one of the coolest facts about nature: on our amazing planet, a rat plus sixty-six million years yielded scientists who study the rat, and thus discover their own roots. Along the way, we developed culture and history and religion and science, and we replaced our ancestors' nests of twigs with gleaming skyscrapers of concrete and steel.

The speed of this intellectual development has been increasing dramatically. Nature required about sixty million years to produce the "ape" from which all humans descended; the rest of our physical evolution occurred in a few million; our cultural evolution has taken only about ten thousand. It is as if, in the words of psychologist Julian Jaynes, "all life evolved to a certain point, and then in ourselves turned at a right angle and simply exploded in a different direction."

Animal brains first evolved for the most primal of reasons: to better enable motion. The ability to move—to find food and shelter, and to escape from enemies—is of course one of the most fundamental characteristics of animals. Looking backward along the road of evolution to animals such as nematodes, earthworms, and mollusks, we find that the earliest brainlike precursors function to control motion by exciting muscles in the right order. But movement does little good without the ability to perceive the environment, and so even simple animals have a way to sense what is around them—cells that react to certain chemicals, for example, or to photons of light, by sending electrical impulses to the nerves governing motion control. By the time *Protungulatum donnae* appeared, these chemical- and photosensitive cells had evolved into senses of smell and sight, and the nexus of nerves that controlled muscle movement had become a brain.

No one knows exactly how our ancestors' brains were organized into functional components, but even in the modern human brain, far more than half the neurons are devoted to motor control and the five senses. That part of our brain that sets us apart from "lower" animals, on the other hand, is relatively small, and was late in coming.

One of the first nearly human creatures roamed the earth just three to four million years ago. It was unknown to us until one blistering day in 1974, when an anthropologist named Donald Johanson, from Berkeley's Institute of Human Origins, stumbled across a tiny fragment of an arm bone sticking out of the scorched terrain in a parched ravine in remote northern Ethiopia. Johanson and a student soon dug up more bones—thigh and rib bones, vertebrae, even a partial jawbone. All told, they found nearly half the skeleton of a female. She had a woman's pelvis, a small skull, short legs, and long, dangling arms. Not someone you'd ask to the prom, but this 3.2-million-year-old lady is thought to be a link to our past, a transitional species, and possibly the ancestor from which our entire genus evolved.

Johanson named the new species *Australopithecus afarensis,* which means "southern ape of Afar," Afar being the region of Ethiopia in which he made his discovery. He also named the bones: Lucy, after the Beatles song "Lucy in the Sky with Diamonds," which was playing on the camp radio as Johanson and his team celebrated. Andy Warhol said everyone gets fifteen minutes of fame, and after millions of years, this woman finally got hers. Or to be more accurate, half of her did; her other half was never found.

It is surprising how much anthropologists can discern from half a skeleton. Lucy's large teeth, with jaws adapted for crushing, suggest that she had a vegetarian diet consisting of tough roots, seeds, and fruit with hard outer coverings. Her skeletal structure indicates that she had a huge belly, which would have been necessary to hold the great length of intestine she'd have needed in order to digest the quantity of vegetable matter she required to survive. Most important, the structure of her spine and knees indicates that she walked more or less upright, and a bone from a member of her species that Johanson and his colleagues discovered nearby in 2011 reveals a humanlike foot, with arches suited to walking, as opposed to gripping branches. Lucy's species had evolved from a life up in the trees to a life on the ground, enabling its members to forage through the mixed ecology of forest and grassland and to exploit new ground-based food sources such as protein-rich roots and tubers. It was a style of living that many believe spawned the entire *Homo* genus.

Imagine living in a house and having your mother live next door, her mother next door to her, and so on. Our human heritage is not actually that linear, but, complexities aside, it is interesting to imagine driving along that street, traveling backward in time, and passing generation after generation of ancestors. If you did that, you'd have to drive almost four thousand miles to reach the home of Lucy, who, as a three-foot-seven, sixty-five-pound hairy "woman," would look more like a chimp to you than a relative. About halfway there, you'd have passed the home of an ancestor who was 100,000 generations removed from Lucy, the first species similar enough to people today—in skeleton and, scientists theorize, in mind—to be classed in the genus *Homo.* Scientists have dubbed that two-million-year-old species of human *Homo habilis*, or "Handy Man."

Homo habilis lived on the immense African savannas at a time when, due to climatic change, the forests were receding. Those grassy plains were not an easy environment, for they served as home to an enormous number of terrifying predators. The less dangerous of the predators competed with *Homo habilis* for their dinner; the more dangerous tried to *make them* dinner. One way the Handy Men survived was through their wits—they had a new, larger brain, the size of a small grapefruit. On the fruit-salad scale of brain heft, that is smaller than our cantaloupe but twice the volume of Lucy's orange.*

When comparing different species, we know from experience that there is usually a rough correlation between intellectual capability and

*For those who prefer precision to fruit, I should add that Handy Man's brain was half the size of our own.

average brain weight relative to body size. Hence, from his brain size we can conclude that Handy Man was an intellectual improvement over Lucy and her kind. Luckily, we can gauge brain size and shape in humans and other primates even if their species is long extinct, because their brains fit snugly into their skulls, which means that if we find a primate skull, we in essence have a cast of the brain that once occupied it.

Lest I appear to be advocating hat size as a proxy for intelligence testing, I should add the disclaimer that when scientists speak of being able to gauge intelligence by comparing brain size, they are talking only about making comparisons between the average brain sizes of *different species*. Brain size varies considerably *among* the individuals of a species, but within a species brain size is *not* directly related to intelligence. For example, modern human brains average about three pounds. Yet the brain of the English poet Lord Byron weighed in at about five pounds, while that of the French writer and Nobel Prize winner Anatole France weighed just a bit more than two, and Einstein's brain weighed about 2.7. And then there is the case of a man named Daniel Lyons, who died in 1907, at age forty-one. He had normal body weight and normal intelligence, but when his brain was weighed at his autopsy, it was found to tip the scale at a mere 680 grams, about a pound and a half. The moral of the story is that, within a species, the architecture of the brain—the nature of the connections among neurons and groups of neurons—is far more important than the size of the brain.

Lucy's brain was only slightly larger than that of a chimp. More important, the *shape* of her skull indicates that the increased brain power was concentrated in regions of the brain that deal with sensory processing, while the frontal, temporal, and parietal lobes—the regions of the brain in which abstract reasoning and language are seated—remained relatively undeveloped. Lucy was a step toward the *Homo* genus, but not yet there. That changed with Handy Man.

Like Lucy, Handy Man stood upright, freeing his hands to carry things, but unlike Lucy, Handy Man used that freedom to experiment with his environment. And so it happened that roughly two million years ago, a *Homo habilis* Einstein, or a Madame Curie, or—perhaps more likely—several ancient geniuses working independently of one another, made humankind's first momentous discovery: if you smash one stone into another at an oblique angle, you can flake off a sharp, knife-edged shard of rock. Learning to bash one rock against another

Homo habilis

doesn't sound like the beginning of a social and cultural revolution. Certainly, producing a shard of stone pales in comparison with inventing the lightbulb, the Internet, or chocolate chip cookies. But it was our first baby step toward the realization that we could learn about and transform nature to improve our existence, and that we could rely on our brains to bestow us with powers that complemented and often exceeded those of our bodies.

To a creature who has never seen a tool of any sort, a kind of jumbo artificial tooth you could grasp in your hand and use for cutting and chopping is a life-changing invention, and it did help to change, completely, the way humans live. Lucy and her kind were vegetarians; microscopic studies of the wear and tear on *Homo habilis* teeth, and butchering marks on bones found near their skeletons, indicate that the Handy Men had used their stone cutters to add meat to their diet.

Being a vegetarian had exposed Lucy and her species to seasonal shortages of plant food. Having a mixed diet helped *Homo habilis* bridge those shortages. And because meat is a more concentrated form of nutrients than vegetable matter, meat eaters required a smaller quantity of food than vegetarians. On the other hand, you don't have to chase down and slaughter a head of broccoli, while procuring animal matter can be quite difficult if you don't have lethal weapons, which Handy Men lacked. As a result, the Handy Men obtained most of their meat from carcasses left behind by predators like saber-toothed tigers, which, with their powerful front paws and butcher-knife teeth, killed prey that were often far larger than what they could consume themselves. But even scavenging can be difficult if, like the Handy Men, you have to compete with other species. So next time you fret about the half-hour wait for a seat in your favorite restaurant, try to remember that, to obtain their food, your ancestors had to battle wandering packs of fierce hyenas.

In their struggle to procure food, Handy Men's sharp stones would have made the task of ripping stray flesh off the bone faster and easier, helping to level the playing field with animals who'd been born with the equivalent tools. So once these implements appeared, they became enormously popular, and they remained the human tool of choice for nearly two million years. In fact, it was just such a scattering of stones found alongside fossils of *Homo habilis* that inspired the name "Handy Man," bestowed on the species by Louis Leakey and his colleagues in the early 1960s. Since then, the stone choppers have been found to be so abundant at excavation sites that you often have to tread carefully if you don't want to step on one.

* * *

It is a long journey from sharp stones to liver transplants, but, as reflected by his tool use, *Homo habilis*'s mind had already advanced beyond the capabilities of any of our extant primate relatives. For example, even after years of training by primate researchers, bonobos fail to become proficient in the use of simple stone tools of the type employed by *Homo habilis*. Recent neuroimaging studies suggest that this ability to design, plan, and use tools arose from the evolutionary development of a specialized "tool use" network in our left brain. Sadly, there are rare cases in which humans with damage to that network do no better than the bonobos: they can identify tools, but, like me before coffee, they cannot figure out how to employ even simple devices such as a toothbrush or comb.

Despite the improved cognitive power, this more-than-two-million-year-old species of human—*Homo habilis*—is but a shadow of the modern human: still relatively small-brained and also physically small, with long arms and a face only a zookeeper could love. But after its appearance, it didn't take long—on geological time scales—for several other *Homo* species to emerge. The most important of those—the one that most experts agree was the direct ancestor of our own species—was *Homo erectus*, or "Erect Man," which originated in Africa about 1.8 million years ago. Skeletal remains reveal Erect Man as a species that bore a far greater resemblance to our present-day species than did Handy Man, not just more erect, but bigger and taller—nearly five feet in stature—with long limbs and a much larger skull, which allowed for expansion of the frontal, temporal, and parietal lobes of the brain.

That new larger skull had implications for the birthing process. One thing carmakers don't have to contend with when redesigning one of their models is how to get the new Hondas to squeeze out the tailpipe of the older Hondas. But nature does have to worry about such things, and so in the case of *Homo erectus*, the head redesign caused some issues: *Homo erectus* females had to be larger than their predecessors in order to birth their big-headed, large-brained babies. As a result, while the *Homo habilis* female was only 60 percent as large as her male counterpart, the average *Homo erectus* woman weighed 85 percent as much as her mate.

The new brains were worth the cost, for Erect Man marked another abrupt and magnificent shift in human evolution. They saw the world, and approached its challenges, differently than their predecessors had. In particular, they were the first humans to have the imaginative and planning skills to create complex stone and wooden tools—carefully crafted hand axes, knives, and cleavers that required *other tools* for their construction. Today we credit our brains with giving us the ability to create science and technology, art and literature, but our brain's ability to conceive of complex tools was far more important to our species—it gave us an edge that aided our very survival.

With their advanced tools, Erect Men could hunt and not just scavenge, increasing the availability of meat in their diet. If the veal recipes in modern cookbooks began by saying, "Hunt down and slaughter a calf," most people would stick to recipes in books like *The Joys of Eggplant*. But in the history of human evolution, the new ability to hunt was a giant leap forward, allowing greater consumption of protein and less dependence on the high volumes of plant food previously necessary for survival. Erect Man was probably also the first species to learn that rubbing materials together creates heat, and to discover that heat creates fire. With fire, Erect Man could do what no other animal is capable of: stay warm in climates that would otherwise have been too cold to sustain its life.

I find it somehow comforting that, while I do my hunting at the butcher counter, and my idea of tool use is to call a carpenter, I hail from folks who were quite adept at the practical—even if they did have protruding foreheads and teeth that could gnaw through a yardstick. More important, these new achievements of the mind enabled *Homo erectus* to branch out from Africa to Europe and Asia, and to persist as a species for well over a million years.

* * *

If advances in our intelligence allowed us to make complex hunting and butchering tools, they also created a new, pressing need—for chasing down and cornering a large, fast animal on the savanna is best done by a team of hunters. And so, long before we humans formed all-star basketball and soccer teams, there was evolutionary pressure for our genus to evolve enough social intelligence and planning skill to cohere and band together to score antelopes and gazelles. Erect Man's new lifestyle therefore favored the survival of those who could best communicate and plan. Here again we see the origins of modern human nature rooted in the African savanna.

Somewhere toward the end of Erect Man's reign, perhaps half a million years ago, *Homo erectus* evolved into a new form, *Homo sapiens*, with even greater cerebral power. Those early, or "archaic," *Homo sapiens* were still not beings we would recognize as present-day humans: they had more robust bodies and larger and thicker skulls but brains that were still not quite as large as ours today. Anatomically, modern humans are classified as a subspecies of *Homo sapiens* that probably didn't emerge from early *Homo sapiens* until around 200,000 B.C.

We almost didn't make it: an amazing recent analysis of DNA by genetic anthropologists indicates that sometime around 140,000 years ago, a catastrophic event—probably related to climate change—decimated the ranks of modern humans, all of whom then lived in Africa. During that period, the entire population of our subspecies plummeted to just several hundred—making us what we would today call an "endangered species," like the mountain gorilla or the blue whale. Isaac Newton, Albert Einstein, and everyone else you've ever heard of, and the billions of us who live in the world today, are all descendants of those mere hundreds who survived.

That close call was perhaps an indication that the new subspecies with the larger brain still wasn't quite smart enough to make it in the long run. But then we underwent another transformation, one that gave us astonishing new mental powers. It does not seem to have been due to a change in our physical anatomy, not even in the anatomy of our brain. Instead it seems to have been a reworking of how our brain operates. However it happened, it was that metamorphosis that enabled our species to produce scientists, artists, theologians, and, more generally, people who think the way we do.

Anthropologists refer to that final mental transformation as the development of "modern human behavior." By "modern behavior" they don't mean shopping or gulping intoxicating beverages while watching sports competitions; they mean the exercise of complex symbolic thought, the kind of mental activity that would eventually lead to our current human culture. There is some controversy about just when that happened, but the most generally accepted date puts the transition at around 40,000 B.C.

Today we call our subspecies *Homo sapiens sapiens*, or "Wise, Wise Man." (Your own species ends up with a name like that when you get to choose it yourself.) But all the changes that led to our large brains didn't come cheap. From the point of view of energy consumption, the modern human brain is the second-most-expensive organ in the body, after the heart.

Rather than endowing us with brains that have such high operating costs, nature could have bestowed upon us more powerful muscles, which, compared with the brain, consume just one-tenth the calories per unit mass. Yet nature chose not to make our species the most physically fit. We humans are not particularly strong, nor are we the most agile. Our nearest relatives, the chimpanzee and the bonobo, savaged their way into their ecological niche through their ability to pull with a force exceeding twelve hundred pounds, and by having teeth so sharp and rugged they tear with ease through hard nutshells. I, on the other hand, have trouble with popcorn.

Rather than impressive muscle mass, humans have oversize craniums, making us inefficient users of food energy—our brains, which account for only about 2 percent of our body weight, consume about 20 percent of our calorie intake. So while other animals are adept at surviving the harshness of the jungle or the savanna, we seem more suited to sitting in a café, sipping mochas. That sitting, though, is not to be underestimated. For as we sit, we think, and we question.

In 1918, German psychologist Wolfgang Köhler published a book, destined to become a classic, called *The Mentality of Apes*. It was an account of experiments he had conducted on chimpanzees while he was director of a Prussian Academy of Sciences outpost on Tenerife, in the Canary Islands. Köhler was interested in understanding the way chimpanzees solve problems, such as how to procure food that has been placed out of reach, and his experiments reveal much about the mental gifts we share with other primates. But if one contrasts chimpanzee behavior with our own, his book also reveals much about the human talents that help compensate for our physical shortcomings.

One of Köhler's experiments was especially telling. He nailed a banana to the ceiling and noted that the chimpanzees could learn to stack boxes in order to climb up and get it, but they seemed to have no knowledge of the forces involved. For example, they would sometimes try to stack a box on its edge, or, if stones were placed on the floor so that the boxes toppled over, they would not think to remove the stones.

In an updated version of the experiment, chimps and human children, aged three to five, were taught to stack L-shaped blocks to obtain a reward. Then weighted blocks were surreptitiously substituted for the originals; these toppled over when the chimps and children tried to stack them. The chimps persisted for a while, employing trial and error in failed attempts to earn the reward—but they did not pause to examine the off-balance blocks. The human children also failed the revised task (it wasn't actually achievable), but they did not simply give up. They examined the blocks in an attempt to determine what the problem was. From an early age, we humans seek answers; we seek a theoretical understanding of our environment; we ask the question "Why?"

Anyone who has experience with young children knows their love of the *why* question. In the 1920s, psychologist Frank Lorimer made it official: he observed a four-year-old boy over a period of four days and scribbled down all the *why*s the child asked during that time. There were forty of them, questions such as *Why does the watering pot have two handles? Why do we have eyebrows?* And my favorite, *Why don't you have a beard, Mother?* Human children all around the world ask their first questions at an early age, while they are still babbling and don't yet speak grammatical language. The act of questioning is so important to our species that we have a universal indicator for it: all languages, whether tonal or nontonal, employ a similar rising intonation for questions. Certain religious traditions see questioning as the highest form of apprehension, and in both science and industry, the ability to ask the right questions is probably the greatest talent one can have. Chimpanzees and bonobos, on the other hand, can learn to use rudimentary signing to communicate with their trainers, and even to answer questions, but they never ask them. They are physically powerful, but they are not thinkers.

* * *

If we humans are born with a drive to understand our environment, we also seem to be born with—or at least acquire at a very early age—a gut feeling regarding how the laws of physics operate. We seem to innately understand that all events are caused by other events, and to have a rudimentary intuition for the laws that, after millennia of effort, were eventually uncovered by Isaac Newton.

In the Infant Cognition Laboratory at the University of Illinois, scientists have spent the past thirty years studying the physical intuition of babies by sitting mothers and their children at a small stage or table and observing how the infants react to staged events. The scientific question at hand: With regard to the physical world, what do these infants know, and when did they know it? What they have discovered is that possessing a certain feeling for the workings of physics seems to be an essential aspect of what it means to be human, even in infancy.

In one series of studies, six-month-olds sat in front of a horizontal track that was attached to an inclined ramp. At the bottom of the ramp researchers had placed a toy bug mounted on wheels. At the top of the ramp was a cylinder. Once the cylinder was let loose, the infants watched excitedly as it rolled downward, crashed into the bug, and sent it rolling halfway along the horizontal track for a distance of a couple of feet. Next came the part that excited the *researchers:* if they reproduced the setup with a different-size cylinder atop the ramp, would the infants predict that, upon collision, the bug would be sent a distance that was proportional to the cylinder's size?

The first question that came to mind when I heard of this experiment was: How does one know *what* an infant is predicting? Personally, I have trouble understanding what my kids are thinking, and, being in their teens and twenties, they are all capable of speech. Did I have any insight at all, back when they were limited to smiles, grimaces, and drooling? The truth is, if you hang around babies enough, you do start attributing thoughts to them based on their facial expressions, but it is difficult to confirm scientifically whether your intuition is correct. If you see a baby's face crinkle up like a prune, is that due to severe gas pains or dismay because the radio just said the stock market tumbled five hundred points? I know my own expression would look the same either way, and with babies, the look is all we have to go by. When it comes to determining what a baby is predicting, though, psychologists have an app for that. They show the baby some chain of events and then measure how long the infant gazes at the scene. If events don't unfold in the way the baby expected, the baby will stare, and the more surprising the occurrence, the longer the stare.

In the ramp experiment, psychologists arranged for half the infants to watch a second collision in which the rolling cylinder was larger than before, while the others watched a second collision in which the cylinder was smaller. In *both* cases, however, the tricky researchers had arranged artificially that the bug would be sent rolling *farther* than in the first collision—all the way, in fact, to the end of the track. The infants who watched the larger cylinder send the bug farther had no exceptional reaction to the events. But sure enough, the infants who saw the bug go farther when struck by a *smaller* cylinder stared at the bug for a prolonged period, giving the impression that they'd be scratching their heads if only they knew how.

To know that big impacts will send bugs rolling farther than small impacts does not quite make you a peer of Isaac Newton, but, as this experiment illustrates, humans do seem to have a certain built-in understanding of the physical world, a sophisticated intuitive feeling for the environment that complements our built-in curiosity and is far more developed in humans than in other species.

Over millions of years, our species evolved and progressed, gaining a more powerful brain, striving as individuals to learn what we could about the world. The development of the modern human mind was a necessary development if we were to understand nature, but it was not sufficient. And so the next chapter in our story is the tale of how we began to ask questions about our surroundings and to band together intellectually to answer them. It is the tale of the development of human culture.

3

Culture

Those of us who look in the mirror each morning see something that few other animals ever recognize: ourselves. Some of us smile at our image and blow ourselves a kiss; others rush to cover the disaster with makeup or to shave, lest we appear unkempt. Either way, as animals go, the human reaction is an odd one. We have it because somewhere along the path of evolution we humans became self-aware. Even more important, we began to have a clear understanding that the face we see in reflections will in time grow wrinkles, sprout hair in embarrassing places, and, worst of all, cease to exist. That is, we had our first intimations of mortality.

Our brain is our mental hardware, and it was for purposes of survival that we developed one with the capacity to think symbolically and to question and reason. But hardware, once you have it, can be put to many uses, and as our *Homo sapiens sapiens* imagination leapt forward, the realization that we will all die helped turn our brains toward existential questions such as "Who is in charge of the cosmos?" This is not a scientific question per se, yet the road to questions like "What is an atom?" began with such queries, as well as more personal ones, like "Who am I?" and "Can I alter my environment to suit me?" It is when we humans rose above our animal origins and began to make these inquiries that we took our next step forward as a species whose trademark is to think and question.

The change in human thought processes that led us to consider these issues probably simmered for tens of thousands of years, beginning around the time—probably forty thousand years ago or thereabouts—when our subspecies began to manifest what we think of as modern behaviors. But it didn't boil over until about twelve thousand years ago, around the end of the last ice age. Scientists call the two million years leading up to that period the Paleolithic era, and the following seven or eight thousand years the Neolithic era. The names come from the Greek words *palaio*, meaning "old," *neo*, meaning "new," and *lithos*, meaning "stone"—in other words, the Old Stone Age (Paleolithic) and the New Stone Age (Neolithic), both of which were characterized by the use of stone tools. Though we call the sweeping change that took us from the Old to the New Stone Age the "Neolithic revolution," it wasn't about stone tools. It was about the way we think, the questions we ask, and the issues of existence that we consider important.

* * *

Paleolithic humans migrated often, and, like my teenagers, they followed the food. The women gathered plants, seeds, and eggs, while the men generally hunted and scavenged. These nomads moved seasonally—or even daily—keeping few possessions, chasing the flow of nature's bounty, enduring her hardships, and living always at her mercy. Even so, the abundance of the land was sufficient to support only about one person per square mile, so for most of the Paleolithic era, people lived in small wandering groups, usually numbering fewer than a hundred individuals. The term "Neolithic revolution" was coined in the 1920s to describe the transition from that lifestyle to a new existence in which humans began to settle into small villages consisting of one or two dozen dwellings, and to shift from gathering food to producing it.

With that shift came a movement toward actively shaping the environment rather than merely reacting to it. Instead of simply living off the bounty nature laid before them, the people living in these small settlements now collected materials with no intrinsic worth in their raw form and remade them into items of value. For example, they built homes from wood, mud brick, and stone; forged tools from naturally occurring metallic copper; wove twigs into baskets; twisted fibers gleaned from flax and other plants and animals into threads and then wove those threads into clothing that was lighter, more porous, and more easily cleaned than the animal hides people had formerly worn; and formed and fired clay into pots and pitchers that could be used for cooking or storing surplus food products.

At face value, the invention of objects like clay pitchers seems to represent nothing deeper than the realization that it is hard to carry water around in your pocket. And indeed, until recently many archaeologists thought that the Neolithic revolution was merely an adaptation aimed at making life easier. Climate change at the end of the last ice age, ten to twelve thousand years ago, resulted in the extinction of many large animals and altered the migration patterns of others. This, it was assumed, had put pressure on the human food supply. Some also speculated that the number of humans had finally grown to the point that hunting and gathering could no longer support the population. Settled life and the development of complex tools and other implements were, in this view, a reaction to these circumstances.

But there are problems with this theory. For one, malnutrition and disease leave their signature on bones and teeth. During the 1980s, however, research on skeletal remains from the period preceding the Neolithic revolution revealed no such damage, which suggests that people of that era were not suffering nutritional deprivation. In fact, paleontological evidence suggests that the early farmers had more spinal issues, worse teeth, and more anemia and vitamin deficiencies—and died younger—than the populations of human foragers who preceded them. What's more, the adoption of agriculture seems to have been gradual, not the result of any widespread climatic catastrophe. Besides, many of the first settlements showed no sign of having domesticated plants or animals.

We tend to think of humanity's original foraging lifestyle as a harsh struggle for survival, like a reality show in which starving contestants live in the jungle and are driven to eat winged insects and bat guano. Wouldn't life be better if the foragers could get tools and seeds from the Home Depot and plant rutabagas? Not necessarily, for, based on studies of the few remaining hunters and gatherers who lived in untouched, unspoiled parts of Australia and Africa as late as the 1960s, it seems that the nomadic societies of thousands of years ago may have had a "material plenty."

Typically, nomadic life consists of settling temporarily and remaining in place until food resources within a convenient range of camp are exhausted. When that happens, the foragers move on. Since all possessions must be carried, nomadic peoples value small goods over larger items, are content with few material goods, and in general have little sense of property or ownership. These aspects of nomadic culture made them appear poor and wanting to the Western anthropologists who first began to study them in the nineteenth century. But nomads do not, as a rule, face a daunting struggle for food or, more generally, to survive. In fact, studies of the San people (also called Bushmen) in Africa revealed that their food-gathering activities were more efficient than those of farmers in pre–World War II Europe, and broader research on hunter-gatherer groups ranging from the nineteenth to the mid-twentieth centuries shows that the average nomad worked just two to four hours each day. Even in the scorched Dobe area of Africa, with a yearly rainfall of just six to ten inches, food resources were found to be "both varied and abundant." Primitive farming, by contrast, requires long hours of back-breaking work—farmers must move stones and rock, clear brush, and break up hard land using only the most rudimentary of tools.

Such considerations suggest that the old theories of the reason for human settlement don't tell the whole story. Instead, many now believe that the Neolithic revolution was not, first and foremost, a revolution inspired by practical considerations, but rather a mental and *cultural* revolution, fueled by the growth of human spirituality. That viewpoint is supported by perhaps the most startling archaeological discovery of modern times, a remarkable piece of evidence suggesting that the new human approach to nature didn't *follow* the development of a settled life-style but rather *preceded* it. That discovery is a grand monument called Göbekli Tepe, Turkish for what it looked like before it was excavated: "hill with a potbelly."

* * *

Göbekli Tepe is located on the summit of a hill in what is now the Urfa Province in southeastern Turkey. It is a magnificent structure, built 11,500 years ago—7,000 years before the Great Pyramid—through the herculean efforts not of Neolithic settlers but of hunter-gatherers who hadn't yet abandoned the nomadic lifestyle. The most astounding thing about it, though, is the use for which it was constructed. Predating the Hebrew Bible by about 10,000 years, Göbekli Tepe seems to have been a religious sanctuary.

The pillars at Göbekli Tepe were arranged in circles as large as sixty-five feet across. Each circle had two additional, T-shaped pillars in its center, apparently humanoid figures with oblong heads and long, narrow bodies. The tallest of them stand eighteen feet high. Its construction required transporting massive stones, some weighing as much as sixteen tons. Yet it was accomplished prior to the invention of metal tools, prior to the invention of the wheel, and before people learned to domesticate animals as beasts of burden. What's more, unlike the religious edifices of later times, Göbekli Tepe was built before people lived in cities that could have provided a large and centrally organized source of labor. As *National Geographic* put it, "discovering that hunter-gatherers had constructed Göbekli Tepe was like finding that someone had built a 747 in a basement with an X-Acto knife."

The first scientists to stumble across the monument were anthropologists from the University of Chicago and Istanbul University conducting a survey of the region in the 1960s. They spotted some broken slabs of limestone peeking up through the dirt but dismissed them as remnants of an abandoned Byzantine cemetery. From the anthropology community came a huge yawn. Three decades passed. Then, in 1994, a local farmer ran his plow into the top of what would prove to be an enormous buried pillar. Klaus Schmidt, an archaeologist working in the area who had read the University of Chicago report, decided to have a look. "Within a minute of first seeing it, I knew I had two choices," he said. "Go away and tell nobody, or spend the rest of my life working here." He did the latter, working at the site until his death in 2014.

Since Göbekli Tepe predated the invention of writing, there are no sacred texts scattered about whose decoding could shed light on the rituals practiced at the site. Instead, the conclusion that Göbekli Tepe was a place of worship is based on comparisons with later religious sites and practices. For example, carved on the pillars at Göbekli Tepe are various animals, but unlike the cave paintings of the Paleolithic era, they are not likenesses of the game on which Göbekli Tepe's builders subsisted, nor do they represent any icons related to hunting or to the actions of daily life. Instead the carvings depict menacing creatures such as lions, snakes, wild boars, scorpions, and a jackal-like beast with an exposed rib cage. They are thought to be symbolic or mythical characters, the types of animals later associated with worship.

Those ancients who visited Göbekli Tepe did so out of great commitment, for it was built in the middle of nowhere. In fact, no one has uncovered evidence that anyone *ever* lived in the area—no water sources, houses, or hearths. What archaeologists found instead were the bones of thousands of gazelles and aurochs that seem to have been

The ruins of Göbekli Tepe

brought in as food from faraway hunts. To come to Göbekli Tepe was to make a pilgrimage, and the evidence indicates that it attracted nomadic hunter-gatherers from as far as sixty miles away.

Göbekli Tepe "shows sociocultural changes come first, agriculture comes later," says Stanford University archaeologist Ian Hodder. The emergence of group-based religious ritual, in other words, appears to have been an important reason humans started to settle as religious centers drew nomads into tight orbits, and eventually villages were established, based on shared beliefs and systems of meaning. Göbekli Tepe was constructed in an age in which saber-toothed tigers* still prowled the Asian landscape, and our last non–*Homo sapiens* human relative, the three-foot-tall hobbitlike hunter and toolmaker *Homo floresiensis*, had only centuries earlier become extinct. And yet its ancient builders, it seems, had graduated from asking practical questions about life to asking spiritual ones. "You can make a good case," says Hodder, that Göbekli Tepe "is the real origin of complex Neolithic societies."

Other animals solve simple problems to get food; other animals use simple tools. But one activity that has never been observed, even in a

*The technical term is saber-toothed cat.

rudimentary form, in any animal other than the human is the quest to understand its own existence. So when late Paleolithic and early Neolithic peoples turned their focus away from mere survival and toward "nonessential" truths about themselves and their surroundings, it was one of the most meaningful steps in the history of the human intellect. If Göbekli Tepe is humanity's first church—or at least the first we know of—it deserves a hallowed place in the history of religion, but it deserves one in the history of science, too, for it reflected a leap in our existential consciousness, an era in which people began to expend great effort to answer grand questions about the cosmos.

* * *

Nature had required millions of years to evolve a human mind capable of asking existential questions, but once that happened, it took an infinitesimal fraction of that time for our species to evolve cultures that would remake the way we live and think. Neolithic peoples began to settle into small villages and then, as their backbreaking work eventually increased food production, into larger ones, with population density ballooning from a mere one person per square mile to a hundred people.

The most impressive of the new mammoth Neolithic villages was Çatalhöyük, built around 7500 B.C. on the plains of central Turkey, just a few hundred miles west of Göbekli Tepe. Analysis of animal and plant remains there suggests that the inhabitants hunted wild cattle, pigs, and horses and gathered wild tubers, grasses, acorns, and pistachios but engaged in little domestic agriculture. Even more surprising, the tools and implements found in the dwellings indicate that the inhabitants

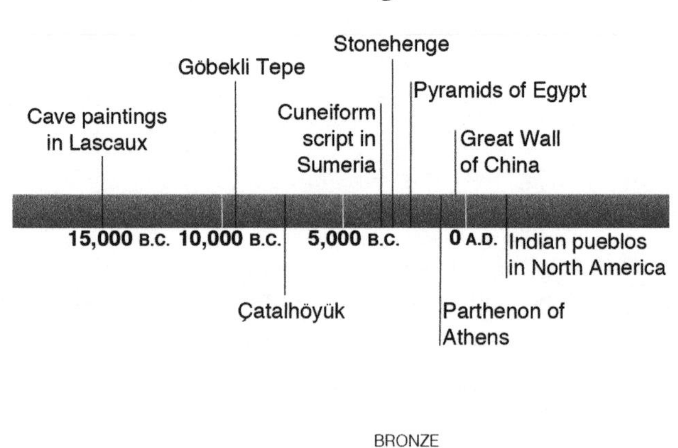

built and maintained their own houses and made their own art. There seems to have been no division of labor at all. That wouldn't sound unusual for a small settlement of nomads, but Çatalhöyük was home to up to eight thousand people—roughly two thousand families—all, in one archaeologist's words, going about and "doing their own thing."

For that reason, archaeologists don't consider Çatalhöyük and similar Neolithic villages cities, or even towns. The first of those would not come for several millennia. The difference between a village and a city is not just a matter of size. It hinges on the social relationships within the population, as those relationships bear on means of production and distribution. In cities, there is a division of labor, meaning that individuals and families can rely on others for certain goods and services. By centralizing the distribution of various goods and services that everyone requires, the city frees individuals and families from having to do everything for themselves, which in turn allows some of them to engage in specialized activities. For example, if the city becomes a center where agricultural surpluses harvested by farmers who live in the surrounding countryside can be distributed to the inhabitants, people who would otherwise have been focused on gathering (or farming) food will be free to practice professions; they can become craftsmen, or priests. But in Çatalhöyük, even though the inhabitants lived in neighboring houses, the artifacts indicate that individual families engaged in the practical activities of life more or less independently of one another.

If each extended family had to be self-sufficient—if you couldn't get your meat from a butcher, your pipes repaired by a plumber, and your water-damaged phone replaced by bringing it to the nearest Apple Store and pretending you hadn't dropped it in the toilet—then why bother to settle side by side and form a village? What bonded and united the people of settlements like Çatalhöyük seems to have been the same glue that drew Neolithic humans to Göbekli Tepe: the beginnings of a common culture and shared spiritual beliefs.

The contemplation of human mortality came to be a feature of these emerging cultures. In Çatalhöyük, for example, we see evidence of a new culture of death and dying, one that differed drastically from that of the nomads. Nomads, on their long journeys over hills and across raging rivers, cannot afford to carry the sick or infirm. And so it is common for nomadic tribes on the move to leave behind their aged who are too weak to follow. The settlers of Çatalhöyük and other forgotten villages of the Near East had a practice that was quite the opposite. Their extended family units often remained physically close, not only in life but in death: in Çatalhöyük, they buried the dead under the floors of their homes. Infants were sometimes interred beneath the threshold at the entrance to a room. Beneath one large building alone, an excavation team discovered seventy bodies. In some cases, a year after burial, the inhabitants would open a grave and use a knife to cut off the head of the deceased, to be used for ceremonial purposes.

If the settlers at Çatalhöyük worried about mortality, they also had a new feeling of human superiority. In most hunter-gatherer societies, animals are treated with great respect, as if hunter and prey are partners. Hunters don't seek to control their prey but instead form a kind of friendship with the animals that will be yielding their lives to the hunter. At Çatalhöyük, however, wall murals depict people teasing and baiting bulls, wild boars, and bears. People are no longer viewed as partnering with animals but instead are seen as dominating them, using them in much the same way they came to use twigs to make their baskets.

The new attitude would eventually lead to the domestication of animals. Over the next two thousand years, sheep and goats were tamed, then cattle and pigs. At first, there was selective hunting—wild herds were culled to achieve an age and gender balance, and people sought to protect them from natural predators. Over time, though, humans took responsibility for all aspects of the animals' lives. As domesticated animals no longer had to fend for themselves, they responded by evolving new physical attributes as well as tamer behavior, smaller brains, and lower intelligence. Plants, too, were brought under human control—wheat, barley, lentils, and peas, among others—and became the concern not of gatherers but of gardeners.

The invention of agriculture and the domestication of animals catalyzed new intellectual leaps related to maximizing the efficiency of those enterprises. Humans were now motivated to learn about and exploit the rules and regularities of nature. It became useful to know how animals bred, and what helped plants grow. This was the beginnings of what would become science, but in the absence of the scientific method or any appreciation of the advantages of logical reasoning, magic and religious ideas blended with and often superseded empirical observations and theories, with a goal that is more practical than that of pure science today: to help humans exert their power over the workings of nature.

As humans began to ask new questions about nature, the great expansion of Neolithic settlements provided a new way to answer them. For the quest to know was no longer necessarily an enterprise of individuals or small groups; it could now draw upon contributions from a great many minds. And so, although these humans had largely given up the practice of hunting and gathering food, they now joined forces in the hunting and gathering of ideas and knowledge.

* * *

When I was a graduate student, the problem I chose for my Ph.D. dissertation was the challenge of developing a new method for finding approximate solutions to the unsolvable quantum equations that describe the behavior of hydrogen atoms in the strong magnetic field outside neutron stars—the densest and tiniest stars known to exist in the universe. I have no idea why I chose that problem, and apparently neither did my thesis adviser, who quickly lost interest in it. I then spent an entire year developing various new approximation techniques that, one after another, proved no better at solving my problem than did existing methods, and hence not worthy of earning me a degree. Then one day I was talking to a postdoctoral researcher across the hall from my office. He was working on a novel approach to understanding the behavior of elementary particles called quarks, which come in three "colors." (The term, when applied to quarks, has nothing to do with the everyday definition of "color.") The idea was to imagine (mathematically) a world in which there are an infinite number of colors, rather than three. As we talked about the quarks, which had no relation at all to the work I was doing, a new idea was born: What if I solved *my* problem by pretending that we lived not in a three-dimensional world, but in a world of infinite dimensions?

If that sounds like a strange, off-the-wall idea, it was. But as we churned through the math, we found that, oddly, though I could not solve my problem as it arose in the real world, I could if I rephrased it in infinite dimensions. Once I had the solution, "all" I had to do to graduate was figure out how the answer should be modified to account for the fact that we actually live in three-dimensional space.

This method proved powerful—I could now do calculations on the back of an envelope and achieve results that were more accurate than those from the complex computer calculations others were using.

After a year of fruitless effort, I ended up doing the bulk of what would become my Ph.D. dissertation on the "large N expansion" in just a few weeks, and over the following year that postdoc and I turned out a series of papers applying the idea to other situations and atoms. Eventually a Nobel Prize–winning chemist named Dudley Herschbach read about our method in a journal with the exciting name *Physics Today*. He renamed the technique "dimensional scaling" and started to apply it in his own field. Within a decade there was even an academic conference entirely devoted to it. I tell this story not because it shows that one can choose a lousy problem, waste a year on dead ends, and still come away with an interesting discovery, but rather to illustrate that the human struggle to know and to innovate is not a series of isolated personal struggles but a cooperative venture, a social activity that requires for its success that humans live in settlements that offer minds a plentitude of other minds with which to interact.

Those other minds can be found in both the present and the past. Myths abound about isolated geniuses revolutionizing our understanding of the world or performing miraculous feats of invention in the realm of technology, but they are invariably fiction. For example, James Watt, who developed the concept of horsepower and for whom the unit of power, the watt, is named, is said to have hatched the idea for the steam engine from a sudden inspiration he had while watching steam spewing from a teapot. In reality, Watt concocted the idea for his device while repairing an earlier version of the invention that had already been in use for some fifty years by the time he got his hands on it. Similarly, Isaac Newton did not invent physics after sitting alone in a field, watching an apple fall; he spent years gathering information, compiled by others, regarding the orbits of planets. And if he hadn't been inspired by a chance visit from astronomer Edmond Halley (of comet fame), who inquired about a mathematical issue that had intrigued him, Newton would never have written the *Principia*, which contains his famous laws of motion and is the reason he is revered today. Einstein, too, could not have completed his theory of relativity had he not been able to hunt down old mathematical theories describing the nature of curved space, aided by his mathematician friend Marcel Grossmann. None of these great thinkers could have achieved their grand accomplishments in a vacuum; they relied on other humans and on prior human knowledge, and they were shaped and fed by the cultures in which they were immersed. And it's not just science and technology that build on the work of prior practitioners: the arts do, too. T. S. Eliot even went so far as to say, "Immature poets imitate; mature poets steal . . . and good poets make it into something better, or at least something different."

"Culture" is defined as behavior, knowledge, ideas, and values that you acquire from those who live around you, and it is different in different places. We modern humans act according to the culture in which we are raised, and we also acquire much of our knowledge through culture, which is true for us far more than it is for other species. In fact, recent research suggests that humans are even evolutionarily adapted to teach other humans.

It's not that other species don't exhibit culture. They do. For example, researchers studying distinct groups of chimps found that, just as people around the world have a pretty good record of identifying as American a person who travels abroad and then seeks restaurants that serve milkshakes and cheeseburgers, so too could they observe a group of chimps and identify, from their repertoire of behaviors alone, their place of origin. All told, the scientists identified thirty-eight traditions that vary among those chimp communities. Chimps at Kibale, in Uganda, in Gombe, Nigeria, and in Mahale, Tanzania, prance about in heavy rain, dragging branches and slapping the ground. Chimps in the Tai forest of Côte d'Ivoire and in Bossou, Guinea, crack open *Coula* nuts by smashing them with a flat stone on a piece of wood. Other groups of chimpanzees have been reported to have culturally transmitted usage of medicinal plants. In all these cases, the cultural activity is not instinctive or rediscovered in each generation, but rather something that juveniles learn by mimicking their mothers.

The best-documented example of the discovery and cultural transmission of knowledge among animals comes from the small island of Kojima, in the Japanese archipelago. In the early 1950s, the animal keepers there would feed macaque monkeys each day by throwing sweet potatoes onto the beach. The monkeys would do their best to shake off the sand before eating the potato. Then one day in 1953, an eighteen-month-old female named Imo got the idea to carry her sweet potato into the water and wash it off. This not only removed the gritty sand but also made the food saltier and tastier. Soon Imo's playmates picked up her trick. Their mothers slowly caught on, and then the males, except for a couple of the older ones—the monkeys weren't teaching one another, but they *were* watching and mimicking. Within a few years, virtually the entire community had developed the habit of washing their food. What's more, until that time, the macaques had avoided the water, but now they began to play in it. The behavior was passed down through the generations and continued for decades. Like beach communities of humans, these macaques had developed their own distinct culture. Over the years scientists have found evidence of culture in many species—animals as different as killer whales, crows, and, of course, other primates.

What sets us apart is that we humans seem to be the only animals able to *build* on the knowledge and innovations of the past. One day a human noted that round things roll and invented the wheel. Eventually we had carts, waterwheels, pulleys, and, of course, roulette. Imo, on the other hand, didn't build on prior chimp knowledge, nor did other chimps build on hers. We humans talk among ourselves, we teach each other, we seek to improve on old ideas and trade insights and inspiration. Chimps and other animals don't. Says archaeologist Christopher Henshilwood, "Chimps can show other chimps how to hunt termites, but they don't improve on it, they don't say, 'let's do it with a different kind of probe'—they just do the same thing over and over."

Anthropologists call the process by which culture builds upon prior culture (with relatively little loss) "cultural ratcheting." The cultural ratchet represents an essential difference between the cultures of humans and of other animals, and it is a tool that arose in the new settled societies, where the desire to be among like thinkers and to ponder with them the same issues became the nutrient upon which advanced knowledge would grow.

Archaeologists sometimes compare cultural innovations to viruses. Like viruses, ideas and knowledge require certain conditions—in this case, social conditions—to thrive. When those conditions are present, as in large, highly connected populations, the individuals of a society can infect one another, and culture can spread and evolve. Ideas that are useful, or perhaps simply provide comfort, survive and spawn a next generation of ideas.

Modern companies that depend on innovation for their success are well aware of this. Google, in fact, made a science of it, placing long, narrow tables in its cafeteria so that people would have to sit together and designing each food line to be three to four minutes long—not so long that its employees get annoyed and decide to dine on a cup of instant noodles, but long enough that they tend to bump into one another and talk. Or consider Bell Labs, which between the 1930s and the 1970s was the most innovative organization in the world, responsible for many of the key innovations that made the modern digital age possible—including the transistor and the laser. At Bell Labs, collaborative research was so highly valued that the buildings were designed to maximize the probability of chance encounters, and one employee's job description involved traveling to Europe each summer to act as an intermediary between scientific ideas there and in the United States. What Bell Labs was recognizing was that those who travel in wider intellectual groups have greater chances of dreaming up new innovations. As evolutionary geneticist Mark Thomas put it, when it comes to generating new ideas, "It's not how smart you are. It's how well-connected you are." Interconnectedness is a key mechanism in the cultural ratchet, and it is one of the gifts of the Neolithic revolution.

* * *

One evening, shortly after my father's seventy-sixth birthday, we took an after-dinner walk. The next day he was going to the hospital for surgery. He had been sick for several years, suffering from borderline diabetes, a stroke, a heart attack, and, worst of all—from his point of view—chronic heartburn and a diet that excluded virtually everything he cared to eat. As we slowly strolled that night, he leaned on his cane, raised his eyes from the street to the sky, and remarked that it was hard for him to grasp that this might be the last time he ever saw the stars. And then he began to unfold for me the thoughts that were on his mind as he faced the possibility that his death might be near.

Here on earth, he told me, we live in a troubled, chaotic universe. It had rewarded him at a young age with the cataclysms of the Holocaust and in his old age with an aorta that, against all design specs, was bulging dangerously. The heavens, he said, had always seemed to him like a universe that followed completely different laws, a realm of planets and suns that moved serenely through their age-old orbits and appeared perfect and indestructible. This was something we had talked about often over the years. It tended to come up whenever I was describing my latest adventures in physics, when he would ask me if I really believed that the atoms that make up human beings are subject to the same laws as the atoms of the rest of the universe—the inanimate and the dead. No matter how many times I said that, yes, I did really believe that, he remained unconvinced.

With the prospect of his own death in mind, I assumed that he would be less than ever inclined to believe in the impersonal laws of nature but would be turning, as people so often do at such times, to thoughts of a loving God. My father rarely spoke of God, because although he had grown up believing in the traditional God, and wanted to still, the horrors he had witnessed made that a difficult proposition. But, as he contemplated the stars that night, I thought that he might well be looking to God for solace. Instead he told me something that surprised me. He hoped I was right about the laws of physics, he said, because he now took comfort in the possibility that, despite the messiness of the human condition, he was made of the same stuff as the perfect and romantic stars.

We humans have been thinking about such issues since at least the time of the Neolithic revolution, and we still don't have the answers, but once we had awakened to such existential questions, the next milestone on the human path toward knowledge would be the development of tools—mental tools—to help us answer them.

The first tools don't sound grand. They're nothing like calculus, or the scientific method. They are the fundamental tools of the thinking trade, ones that have been with us for so long that we tend to forget that they weren't always part of our mental makeup. But for progress to occur, we had to wait for the introduction of professions that dealt with the pursuit of ideas rather than the procurement of food; for the invention of writing so that knowledge could be preserved and exchanged; for the creation of mathematics, which would become the language of science; and finally, for the invention of the concept of laws. Just as epic and transformative in their way as the so-called scientific revolution of the seventeenth century, these developments came about not so much as the product of heroic individuals thinking great thoughts but as the gradual by-product of life in the first true cities.

4

Civilization

One of the gems Isaac Newton is known for is his remark "If I have seen further it is by standing on the shoulders of giants." He wrote it in a 1676 letter to Robert Hooke, making the point that he had built upon Hooke's work and also that of René Descartes. (Hooke later became his bitter enemy.) Newton certainly did profit from the ideas of those who'd come before him; indeed, he seems to have profited from them even in composing the very sentence in which he said he profited from them—for in 1621, Vicar Robert Burton wrote, "A dwarf standing on the shoulders of a giant may see farther than a giant himself"; then, in 1651, poet George Herbert wrote, "A dwarf on a giant's shoulders sees farther of the two"; and in 1659, Puritan William Hicks wrote, "A pygmy upon a gyants shoulder may see farther than the [giant] himself." In the seventeenth century, it seems, dwarfs and pygmies mounted atop huge brutes were a staple of imagery about intellectual pursuits.

The antecedents Newton and the others were referring to were those of their more or less immediate past. On the other hand, the part played by the generations that preceded us by many thousands of years tends to be forgotten. But though we like to think of ourselves today as advanced, we got to this point only because of the profound innovations that occurred when the Neolithic villages evolved into the first true cities. The abstract knowledge and mental technologies that those ancient civilizations developed played a critical role in shaping our ideas about the universe—and in our ability to explore those ideas.

* * *

The first cities did not arise suddenly, as if nomads one day decided to band together and the next thing they knew they were hunting and gathering chicken thighs wrapped in Styrofoam and cellophane. Instead, the transformation of villages into cities was a gradual, natural development that occurred once the settled agricultural lifestyle had taken hold, over a period of hundreds or thousands of years. That slow evolution leaves room for interpretation regarding the precise time at which a village should be reclassified as a city. Despite that wiggle room, what are often *called* the first cities appeared in the Near East around 4000 B.C.

Perhaps the most prominent of those cities, and an important force in the trend toward urbanization, was the great walled city of Uruk, in what is today southeastern Iraq, near the city of Basra. Though the Near East was the earliest region to urbanize, it was not an easy land

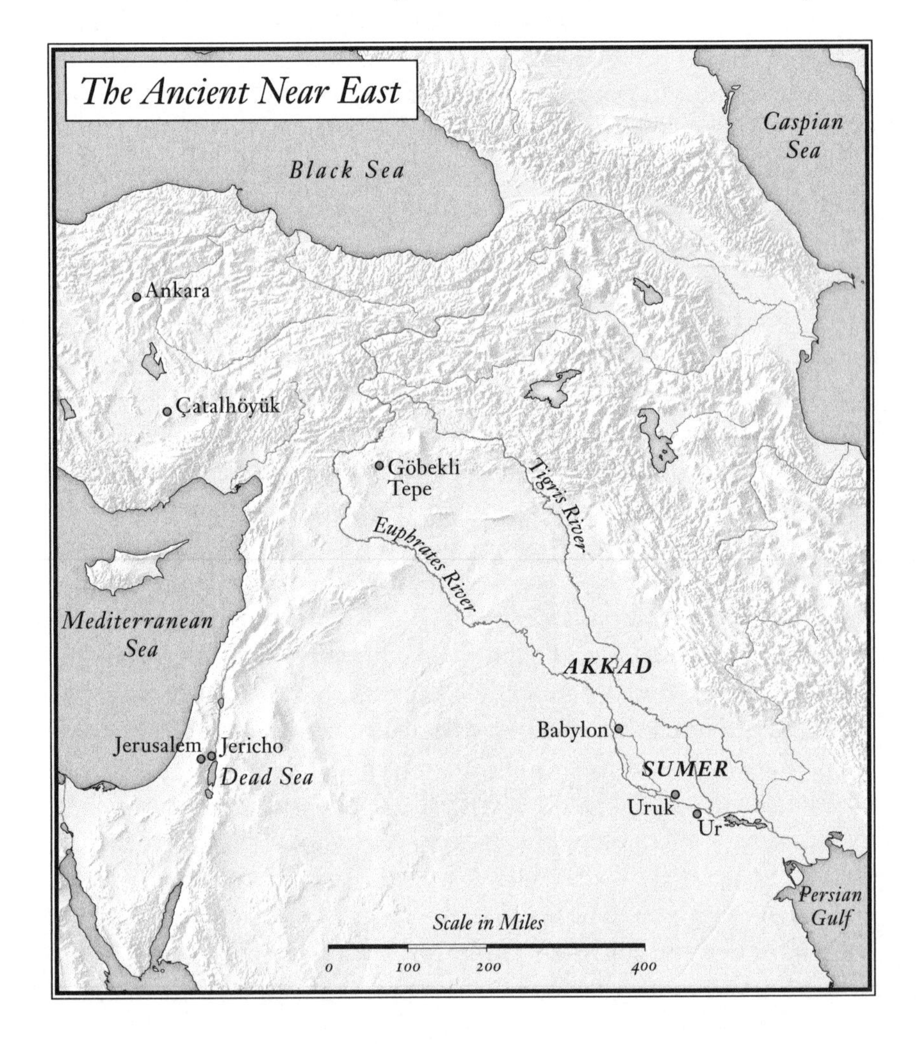

in which to carve out a living. The original settlers came for the water. That may seem misguided, given that much of the land is desert. But though the climate is unfavorable, the geography is inviting. For in the midst of the region is a long depression in the earth through which the Tigris and Euphrates Rivers and their tributaries flow, creating a rich and fertile plain. That plain is called Mesopotamia, ancient Greek for "between the rivers." The first settlements there were mere villages, limited in size by the boundaries of the rivers. Then, sometime after 7000 B.C., farming communities learned to dig canals and reservoirs to extend the rivers' reach, and with the expansion of the food supply, urbanization eventually became possible.

Irrigation wasn't easy. I don't know if you've ever tried to dig a ditch, but I did—in an attempt to lay some pipe for a lawn sprinkler. The first part went well—the part where I bought the shovel. Then the difficulties began. I raised the beautiful tool up high and brought it down with such authority it vibrated as it bounced off the hard ground. In the end, I was able to get the job done only by appealing to a higher authority: a guy with a gas-powered digger. Today's cities rely on excavations of all sorts, and few of us stop to admire them. But the irrigation canals of the ancient Near East, many miles long and up to seventy-five feet wide, and dug with crude tools and no help from machines, were a true wonder of the ancient world.

Bringing water to fields far from the river's natural boundaries required the exhausting labor of hundreds or thousands of workers, and planners and supervisors to direct them. There were several reasons farmers contributed their labor to this group effort. One was peer pressure. Another was that joining the effort was the only way to irrigate one's own land. Whatever their motivation, the farmers' efforts paid off. Food surpluses and the settled life meant families could support and care for more babies, and more of their offspring could survive. The birthrate accelerated, and infant mortality decreased. By 4000 B.C. the population was growing fast. Villages grew into towns, towns became cities, cities grew larger.

Uruk, built just inland of the marshes at the head of the Persian Gulf, was the most fruitful of these early cities. It came to dominate its region, far surpassing the size of any other settlement. Though the population of ancient cities is difficult to estimate, from the structures and remains archaeologists have found, it seems that Uruk had somewhere between fifty and a hundred thousand inhabitants, a tenfold increase in size since Çatalhöyük. That would make Uruk a small city in modern times, but Uruk was the New York, the London, the Tokyo, the São Paulo of its day.

The inhabitants of Uruk plowed their fields with a seed plow, a specialized and difficult-to-operate instrument that drops seeds into a furrow while it digs. They drained marshland and dug canals with hundreds of connecting channels. On the irrigated land, they grew an abundance of cereals and orchard fruits, primarily barley, wheat, and dates. They tended sheep, donkeys, cattle, and pigs and captured fish and fowl from the nearby marshes and turtles from the rivers. They herded goats and water buffalo for milk and drank a great deal of beer produced from barley. (Chemical tests of ancient pottery reveal evidence of beer as far back as 5000 B.C.)

What makes these developments important for us is that the rise of specialized professions required a new understanding of materials, chemicals, and the life cycles and requirements of plants and animals. Food production spawned fishermen, farmers, herdsmen, and hunters. The making of crafts morphed from being a part-time occupation in all households to a full-time occupation for a class of professionals devoted to particular skills. Bread became the product of bakers, and beer the province of brewers. Taverns arose, and with them tavern keepers, some of them women. From the remains of a workshop where molten metal seems to have been handled, we surmise the existence of smelters. Pottery, too, seems to have spawned a profession: thousands of simple beveled-rim bowls that appear to have been mass-produced in standard sizes suggest, if not a ninety-nine-cent store, then a centralized factory dedicated to making the ceramics.

Other specialized workers devoted their energies to clothing. Surviving artwork of the period depicts weavers, and anthropologists have discovered fragments of woven wool textiles. What's more, animal remains show that around that time, herdsmen began to keep many more sheep than goats. Since goats are better milk producers, the increase in sheep among the herds presumably reflects a heightened focus on wool. Also, the bones reveal that these herders butchered their sheep at an advanced age—not a great idea if you are interested in the meat, but wise if you are keeping them for their coats.

All these specialized professions were a boon to anyone who wanted beer or milk and the pottery to drink it from, but they also represented a glorious milestone in the history of the human intellect, for the combined strivings of all the new specialists generated an unprecedented explosion of knowledge. Yes, it was knowledge acquired for purely practical reasons and entangled with myth and ritual. And yes, the recipes for beer included instruction on how to curry favor from goddesses who governed the production and enjoyment of the brew. It was nothing you'd see published in *Nature*—but it *was* the embryonic material from which scientific knowledge, pursued for its own sake, would later grow.

* * *

In addition to the development of professions whose goal it was to produce things, there were also a handful of specialties that came into being around this time that were centered not around physical labor or the production of food and material goods, but around activities of the mind.

They say we feel more bound to those in our profession than to the members of almost any other group. I am as bad at most practical things as I am at digging ditches, and my main saving grace in the world of work is the ability to sit and think all day without tiring, a path I have been lucky to be able to follow. And so I feel a connection to these ancient tradespeople of the mind. Though multitheistic and superstitious, they are my kin, and the kin of all of us who are privileged to make a living from thinking and study.

The new "intellectual" professions developed because the urban lifestyle that took root in Mesopotamia during this era required some kind of centralized organization and that meant creating systems and rules, and gathering and recording data.

For example, urbanization demanded that systems of exchange be developed, and a body to oversee that exchange; increased but seasonal food production meant that communal systems of storage had to be created; and the fact that farmers and the people who depend on them, unlike nomadic tribes, cannot easily abandon their settlement when attacked meant that a militia or army became necessary. In fact, the Mesopotamian city-states were in a constant state of internecine warfare over land and water supplies.

There was also a great demand for the organization of laborers into a public-works force. For one, thick walls had to be constructed around the cities to deter would-be attackers. Also, roads had to be built to accommodate vehicles employing the newly invented wheel, while agriculture demanded ever larger irrigation projects. And of course the very existence of the new central authority required the construction of large buildings to accommodate the bureaucrats.

And then there is the need for police. When the population of settlements numbered only in the dozens or the hundreds, everyone might have known one another. But when the population expanded into the thousands, that was no longer possible, and so people frequently found themselves in situations where they interacted with strangers, which altered the nature of people's conflicts. Anthropologists, psychologists, and neuroscientists have studied how group dynamics change when groups grow larger, but on the most basic level it is easy to understand what happens. If I'm going to see someone all the time, even if I don't like him, I'd better pretend. And pretending to like someone generally precludes things like smashing a clay tablet on his head in order to steal his goat. But if I don't know the person and expect that we'll never again cross paths, the thought of all that delicious goat cheese may be too much to pass up. As a result, conflicts no longer occurred only among family, friends, or acquaintances but among strangers, and so formal methods of conflict resolution—and a police force—had to be created, producing another driving force for the formation of a centralized governing apparatus.

Who were the rulers of the world's first cities, the people who made all these centralized activities possible? The Mesopotamians looked to those who mediated between them and their gods, who officiated over their religious obligations and ceremonies, as their sources of authority.

Mesopotamians did not make the distinction we do between church and state—in Mesopotamia, they were inseparable. Each city was the home of a god or goddess, and each god or goddess was the patron deity of a city. The inhabitants in each city believed that their gods governed their existence and had built their city as their own dwelling. And if a city declined, they believed it was because their gods had abandoned them. And so religion became not just the belief system that held society together, but the executive power that enforced rules. What's more, due to the fear of the gods, religion was a useful tool in motivating obedience. "Goods were received by the god of the city, and redistributed to the people," wrote Near East scholar Marc Van De Mierop. "The temple, the house of the god, was the central institution that made the system work. . . . The temple, located in the city, was a focal point for all." As a result, at the very top of Urukian society, there arose a position of priest-king, whose authority derived from his role in the temple.

Authority means power, but to be effective, rulers must be able to gather data. For example, if the religious establishment was to oversee the exchange of goods and labor, to collect taxes, and to enforce contracts, it had to have people who could collect, process, and store information relevant to all those activities. Today we think of a government bureaucracy as having the intellectual heft of a Division I college football team, but it was from these first government bureaucracies that a specialized intellectual class arose. And it was for their bureaucratic needs that the most important mental technologies ever invented were developed: reading, writing, and arithmetic.

We look at the "three R's" as the most elementary of skills today, techniques we learn sometime after outgrowing diapers and before obtaining our first smartphone. But they seem elementary only because someone else invented them long ago, and they have been passed down ever since by teachers who took the pains to teach them to us. In ancient Mesopotamia, had anyone carried the title of professor, there would have been professors of reading, penmanship, counting, and addition, and those professors would have been teaching and studying ideas that were the most advanced of their day.

* * *

One striking difference between ourselves and the millions of other animal species on earth is that one human mind can influence the thoughts of another in a very complex and nuanced manner. The form of thought control I'm talking about occurs through language. Other animals can signal to one another fear or danger, hunger or affection, and in some cases they can signal these to us, but they have no ability to learn abstract concepts or to string more than a few words together in a meaningful way. A chimp, on command, can pick out a card that bears a picture of an orange, and a parrot can annoy you with an unending string of "Polly wants a cracker." But their ability to go beyond simple requests, commands, warnings, and identifications is virtually nil.

When in the 1970s scientists taught chimpanzees sign language to investigate whether they could master the underlying innate structure of grammar and syntax, linguist Noam Chomsky remarked, "It's about as likely that an ape will prove to have a language ability as [it is that] there is an island somewhere with a species of flightless birds waiting for human beings to teach them to fly." Decades later, it appears that Chomsky was right.

Just as no bird invented flying, and young birds do not have to go to flight school to learn it, language seems to be natural to humans—and solely to humans. Our species had to engage in complex cooperative behavior in order to survive in the wild, and—as I keep reminding my teenage children—pointing and grunting get you only so far. As a result, like the ability to stand upright or to see, language evolved as a biological adaptation, aided by a gene that has been present in human chromosomes so long that it has even been identified in ancient Neanderthal DNA.

Because the capacity for spoken language is innate, one would expect it to manifest widely, and indeed it appears to have been invented independently, over and over again, all around the globe, in every gathering of people who ever lived together as a group. Before the Neolithic revolution, in fact, there may have been as many languages as there were tribes. One reason we believe this is that, prior to the start of British colonization of Australia, in the late eighteenth century, five hundred tribes of indigenous peoples, with an average membership of five hundred each, roamed the continent of Australia, still living the pre-Neolithic lifestyle—and each tribe had its own language. In fact, as Steven Pinker has observed, "No mute tribe has ever been discovered, and there is no record that a region has served as a 'cradle' of language from which it spread to previously languageless groups."

Though language is an important defining trait of the human species, *written* language is a defining trait of human *civilization*, and one of its most important tools. Speaking enabled us to communicate with a small group in our immediate vicinity; writing enabled us to exchange ideas with people far away in both space and time. It made possible a vast accumulation of knowledge, a way for culture to build upon the past. In doing so it allowed us to outgrow the limits of our individual knowledge and memories. The telephone and the Internet have changed the world, but long before they came into being, writing was the first and most revolutionary communications technology.

Speech came naturally—it did not have to be invented. But writing did, and there are plenty of tribes that never took that step. Though we take it for granted, writing is one of the greatest inventions of all time, and one of the most difficult. The magnitude of the task is reflected in the fact that, though linguists have documented more than three thousand languages currently being spoken throughout the world, only about a hundred of those languages have been written down. What's more, over all of human history, writing was independently invented only a few times, and it made its way around the world mainly through cultural diffusion, being borrowed or adapted from existing systems rather than being repeatedly reinvented.

We believe that the first use of the written word, sometime before 3000 B.C., occurred in Sumer, in southern Mesopotamia. We are *certain* of only one other independent invention of a writing system, in Mexico sometime before 900 B.C. In addition, it is possible that Egyptian (3000 B.C.) and Chinese (1500 B.C.) writing systems also represent independent developments. All writing as we know it stems from one of these few inventions.

Unlike most people, I have personally had the experience of trying to "invent" the written word, because when I was eight or nine I was in the Cub Scouts and our pack leader gave us the assignment to take a stab at creating our own system of writing. When Mr. Peters was returning our work on the assignment, I could tell he was impressed with mine. What I had produced looked nothing like the work of the other kids. They had simply made small variations on the letters of the English alphabet. My writing system, on the other hand, looked completely novel.

Before handing my work back, Mr. Peters perused it a final time. He didn't like me, and I could tell he was trying to poke holes, to find a way to avoid praising the creative genius that seemed to underlie such an opus. "You did a . . . good job," he muttered. He hesitated before saying the word "good," as if using it on me meant that he had to pay its inventor a week's salary in royalties. And then, as he was extending the paper for me to take, he suddenly pulled it back. "You go to Sunday school, don't you?" he asked. I nodded. "Is this writing you invented based in any way on the Hebrew alphabet?" I couldn't lie. Yes, like the others, I had simply taken an alphabet I knew and made variations on the letters. Nothing to be ashamed of, but I was devastated. He had always defined me not as a kid, but as a Jewish kid, and now I had proved him right.

Our little Cub Scout enterprise may have seemed challenging to us, but we had a critical advantage over those who first invented writing, for we had already been taught how our spoken language can be parsed into elementary sounds and mapped onto individual letters. We had also learned that certain fundamental sounds, like *th* and *sh*, do not correspond to single letters, and we could distinguish between sounds like *p* and *b*, which might be difficult to do if we had had no prior experience with any kind of writing system.

You can get a taste for the difficulty involved if you try to identify the building blocks of sound you hear when people speak a foreign language. The more unfamiliar the language—for example, listening to Chinese if you speak an Indo-European language—the harder it is. You'll find it challenging to identify very many discrete sounds, much less to differentiate subtleties akin to the difference between *p* and *b*. But somehow that ancient Sumerian civilization overcame these challenges and created a written language.

When new technologies are invented, their initial application is often far different from the role they eventually come to play in society. In fact, for those who work in fields fueled by innovation and discovery, it is important to realize that the inventors of new technologies—like, as we'll see later, the inventors of scientific theories—often don't really understand the meaning of what they themselves have come up with.

If one thinks of writing as a technology—as the recording of spoken words on clay (and later other substrates, like paper)—it seems natural to compare its evolution to the development of the technology of sound recording. When Thomas Edison invented that technology, he had no idea people would eventually use it to record music. He thought it had little commercial value except perhaps to memorialize the utterances of people on their deathbeds or as an office dictating machine. Similarly, the initial role of writing was far different from the role it would eventually come to play in society. In the beginning, it was used simply to keep records and make lists, applications with no more literary content than an Excel spreadsheet.

* * *

The earliest written inscriptions we know of were made on clay tablets found at a temple complex at the site of Uruk. These list items such as sacks of grain and heads of cattle. Other tablets detail the division of labor. For instance, we know from them that the religious community at one temple employed eighteen bakers, thirty-one brewers, seven slaves, and a blacksmith. From partial translations, we also learn that laborers were given fixed rations of supplies such as barley, oil, and cloth and that one profession was titled "leader of the city," and another was called "great one of the cattle." Though one can imagine many reasons to write, 85 percent of the writing tablets that have been excavated pertain to accounting. Most of the remaining 15 percent are tablets aimed at educating future accountants. There was indeed much to learn, for the bookkeeping was complicated. For example, humans, animals, and dried fish were counted employing one system of numbers, and grain products, cheese, and fresh fish using another.

At the time of its origin, writing was limited to these purely utilitarian enterprises. There were no popcorn novels or written theories of the universe, just bureaucratic record-keeping documents such as invoices, lists of goods, and personal signs or "signatures" attesting to such things. That sounds mundane, but it had profound implications: without writing there could have been no urban civilization, because people would have lacked the ability to create and maintain the kind of complex symbiotic relationships that constitute the defining feature of city life.

In a city, we are all constantly giving and getting from one another—buying and selling, billing, shipping and receiving, borrowing and lending, paying for work and getting paid for it, and making and enforcing promises. Had there been no written language, all these reciprocal activities would have been mired in chaos and infused with conflict. Just imagine going through a week in your life in which no event, no transaction—not even your work production or hours of labor—could be recorded in any way. My guess is, we couldn't even get through a professional basketball game without the fans on both sides claiming victory.

The earliest writing systems were as primitive as their purpose. They employed generic slashes representing a number of some item, be it fruit, animals, or people. Eventually, to make it easier to distinguish which marks pertained to sheep, and which to sheep owners, it would have been natural to add more complexity by drawing small pictograms beside the numbers, and so the scripts began to use pictures for words. Scholars have identified the meanings of more than a thousand such early pictograms. For example, an outline of a cow's head was used to denote "cow," three semicircles arranged in a triangle meant "mountains," and a triangle with a mark for the vulva represented "woman." There were also compound signs, such as the sign for a female slave—literally, a woman from "over the mountains"—made from the sign for "woman" added to the sign for "mountain." Eventually, pictograms were also used to express verbs and make sentences. Pictograms for a hand and a mouth were thus placed alongside the sign for "bread," forming the pictogram "to eat."

The early scribes scratched their pictograms into flat clay tablets employing pointed tools. Later, the symbols were pressed into clay using reed styluses that made wedge-shaped marks. That kind of pictogram is called *cuneiform*, from the Latin words meaning "wedge-shaped." Thousands of these early clay tablets have been excavated from the ruins of Uruk, simple lists of things and numbers, devoid of grammar.

The downside of a written language based on pictograms is that, due to the large number of pictograms, it is exceedingly difficult to learn. This very complexity required the formation of a small literate class, members of the class of thinkers I spoke of earlier. These first professional scholars became a privileged caste that enjoyed high status, supported by a temple or palace. In Egypt, they even seem to have been exempt from taxation.

Archaeological remains indicate that around 2500 B.C., the need for scribes produced another great innovation: the world's first schools, known in Mesopotamia as "tablet houses." At first these were connected to temples, but later they were located in private buildings. The name came from the clay tablets that were the school's stock-in-trade—each schoolroom likely had shelves on which to lay the tablets out to dry, an oven to bake them in, and chests in which to store them. Because the writing systems were still very complex, aspiring scribes had to study for many years to memorize and learn to reproduce the thousands of intricate cuneiform characters. It is easy to underestimate the importance of that step in the march of human intellectual progress, but the idea that society should create a profession devoted to passing on knowledge, and that students should spend years acquiring it, was something entirely new—an epiphany for our species.

Over time, the Sumerians were able to simplify their written language while also using it to communicate ever more complex thoughts and ideas. They figured out that they could sometimes depict a word that is difficult to represent by adapting the symbol for a word that sounds the same but is easy to represent. For example, the pictogram for the word "to" could be made from the pictogram for the word "two," modified by a silent sign, called a determinative, to indicate the alternate meaning. Once they had invented this method, Sumerians began to create symbols denoting grammatical endings, for example by using a modified symbol for the word "shun" to represent the suffix *-tion*. They found they could use a similar trick to spell out longer words using shorter ones, as one might write the word "today" from the symbols for "two" and "day." By 2900 B.C. these innovations had enabled the number of distinct pictograms in the Sumerian language to be reduced from two thousand to about five hundred.

As the written language became a tool that was more flexible, easier to manipulate, and capable of more complex communication, the tablet houses were able to broaden their scope to include instruction in writing and arithmetic and, eventually, the specialized vocabulary of the emerging studies of astronomy, geology, mineralogy, biology, and medicine—not principles at first, just lists of words and their meanings. The schools also taught a kind of practical philosophy, "wise sayings," which were prescriptions for a successful life collected from town elders. These were decidedly blunt and pragmatic, such as "Don't marry a prostitute." Not exactly Aristotle, yet a step up from counting grains and goats, these were the beginnings of the pursuits and institutions that would later create the world of philosophy and the beginning of science.

By around 2000 B.C. the written culture of Mesopotamia had evolved yet again, this time with the development of a body of literature that spoke to the emotional components of the human condition. A stone tablet from this era that turned up at an archaeological site located about six hundred miles south of present-day Baghdad is inscribed with the oldest love poem ever found. In the voice of a priestess who professes her love for a king, its words describe feelings that are as fresh and recognizable today as they were four millennia ago:

> Bridegroom dear to my heart,
> Godly is your beauty, honeysweet,
> You have captivated me, let me stand trembling before you;
> Bridegroom, I would be taken to the bedchamber.
> Bridegroom, you have taken your pleasure of me,
> Tell my mother, she will give you delicacies; my father, he will give you gifts.

Sometime in the centuries following that poem came another innovation: the notion to represent the vocalizations that made up a word, rather than the thing the word represented. This radically altered the nature of writing, for the symbols now stood for syllables rather than ideas. It was a logical outgrowth of the old Sumerian trick of using, say, the word "shun" to stand for the syllable *-tion*. We don't know exactly how or when this advance occurred, but it's a good bet that the development of a more economical way of writing had to do with the flourishing of cosmopolitan trade, for writing commercial correspondence and business records in pictograms had to be cumbersome. So by 1200 B.C. the Phoenician script—the first great alphabet in human history—had appeared. What had once required the memorization of hundreds of intricate symbols could now be accomplished using only a couple dozen basic symbols in various combinations. The Phoenician alphabet would eventually be borrowed and adapted for use in Aramaic, Persian, Hebrew, and Arabic—and, around 800 B.C., Greek. And from Greece it spread, eventually, throughout Europe.

* * *

The first cities required, along with reading and writing, certain advances in mathematics. I've always thought that mathematics has a special place in the human heart. *Sure*, you may be thinking, *just like cholesterol*. It's true, mathematics has its detractors, and has had them throughout history. As early as A.D. 415, Saint Augustine wrote, "The danger . . . exists that the mathematicians have made a covenant with the devil to darken the spirit and confine man in the bounds of hell." Those who infuriated him were probably astrologers and numerologists—two of the main users of the dark art of mathematics in his day. But I think I've heard my own children say the same thing, if somewhat less eloquently, on several occasions. Still, love it or not, mathematics and logical thinking represent an important part of the human psyche.

Over the centuries mathematics has been put to many diverse uses, because, like science, mathematics as we define it today is less a specific endeavor than an approach to knowledge—a method of reasoning in which one carefully formulates concepts and assumptions and draws conclusions by applying rigorous logic. What is usually called the "first mathematics," though, is not mathematics in that sense, just as the writing in Sumerian record keeping is not writing in the sense of Shakespeare.

The earliest mathematics is like the mathematics my children and other students tire of studying in grade school: a set of rules that can be applied, more or less mindlessly, to solve specific kinds of problems. In the first cities of Mesopotamia, those problems pertained largely to the tracking of money, materials, and labor, to the arithmetic of weights and measures, and to the calculation of simple and compound interest—the same sort of mundane concerns that prompted the development of writing, and just as essential to the functioning of urban society.

Arithmetic is perhaps the most fundamental branch of mathematics. Even primitive peoples employ a system of counting, though they may not count beyond the five fingers on one hand. Young infants, too, seem to have been born with an ability to determine the number of objects in a collection, though only up to the number four. But to go beyond counting, which is a set of tools we possess shortly out of the womb, we must master addition, subtraction, multiplication, and division, skills we develop gradually throughout infancy and early childhood.

The first urban civilizations introduced formal and often elaborate rules and methods of arithmetical calculation, and they invented methods of solving equations that involved unknowns, which is what we do with algebra today. Compared with modern algebra, theirs was rudimentary at best, but they did develop recipes, if you will—perhaps hundreds of them—for performing complex calculations that involved solving quadratic and cubic equations. And they went beyond simple

The ruins of ancient Babylon, viewed from Saddam Hussein's former summer palace

business applications to apply their techniques to engineering. Before digging a canal, for example, an engineer in Babylon, a region of southern Mesopotamia, would calculate the labor required by computing the volume of dirt that had to be moved and dividing that by the amount of dirt a single digger could scoop out in a day. And before constructing a building, a Babylonian engineer would do analogous calculations to determine the amount of labor and the number of bricks that would be needed.

Despite its accomplishments, Mesopotamian mathematics fell short in an important practical respect. The practice of mathematics is an art, and the medium of that art is symbolic language. Unlike ordinary language, the symbols and equations of mathematics express not just ideas but the relations between ideas. And so, if there is an unsung hero of mathematics, it is notation. Good notation makes relationships precise and apparent and facilitates the human mind's ability to think about them; bad notation makes one's logical analysis inefficient and unwieldy. Babylonian mathematics falls into the latter category: all their recipes and calculations were phrased in the ordinary language of their day.

One Babylonian tablet, for example, contained the following calculation: "4 is the length and 5 is the diagonal. What is the breadth? Its size is not known. 4 times 4 is 16. 5 times 5 is 25. You take 16 from 25 and there remains 9. What times what shall I take in order to get 9? 3 times 3 is 9. 3 is the breadth." In modern notation, this would all be written: $x^2 + 4^2 = 5^2$; $x = \sqrt{(5^2 - 4^2)} = \sqrt{(25 - 16)} = \sqrt{9} = 3$. The great disadvantage of a mathematical statement like the one on the tablet is not just its lack of compactness but the fact that we cannot use algebraic rules to perform operations on equations written in prose.

Notational innovation didn't come until the classical age of Indian mathematics, beginning around A.D. 500. It is hard to overestimate the importance of what those Indian mathematicians achieved. They employed the base ten system and introduced zero as a number, with the properties that multiplication of any number by zero gives zero, while the addition of zero to any number leaves that number unchanged. They invented negative numbers—to represent debts, though, as one mathematician remarked, "people do not approve" of them. And most important, they employed symbols to designate unknowns. The first arithmetic abbreviations, however—*p* for "plus" and *m* for "minus"—weren't introduced to Europe until the fifteenth century, and the equal sign was not invented until 1557, when Robert Recorde of Oxford and

Cambridge chose the symbol we still use today because he felt that no two things could be more alike than parallel lines (and because parallel lines were already used as typographic ornaments, so printers did not have to cast a new form).

I've focused on number, but the thinkers of the world's first cities also made great progress in the mathematics of shapes—not just in Mesopotamia but also in Egypt. There, life centered on the Nile, which flooded its valley for four months each year, covering the land with rich silt but wreaking havoc with property lines. Each year, after the fields were flooded, officials would have to determine anew the boundaries of the farmers' holdings, and their surface areas, on which taxes were based. Due to the high stakes involved, the Egyptians developed reliable, though somewhat complicated, ways of calculating the area of squares, rectangles, trapezoids, and circles—as well as the volume of cubes, boxes, cylinders, and other figures related to their granaries. The term "geometry" comes from those land surveys—the word is Greek for "earth measurement."

So advanced was the practical geometry of the Egyptians that in the thirteenth century B.C., Egyptian engineers were able to level a fifty-foot beam in the pyramids with a margin of error of one-fiftieth of an inch. But like the arithmetic and rudimentary algebra of the Babylonians, the geometry of the ancient Egyptians had little in common with what we call mathematics today. It was created for practical use, not to satisfy their yearning for deeper truths about their world. So before geometry could achieve the heights that would later be demanded by the development of physical science, it had to be elevated from a practical to a theoretical enterprise. The Greeks, especially Euclid, accomplished this in the fourth and fifth centuries B.C.

The development of arithmetic, improved algebra, and geometry would enable the development of the theoretical laws of science centuries later, but as we try to picture that chain of discovery, there is one missing step that might not be apparent to those of us living today: before anyone could theorize about the particular laws of nature, there had to be invented the very concept of law.

* * *

Great technological advances that have enormous ramifications are easy to see as revolutionary. But new modes of thinking, new ways of approaching knowledge, can be less conspicuous. One mode of thinking whose origin we rarely consider is the idea of understanding nature in terms of laws.

We take the notion of scientific law for granted today, but like many great innovations, it has become obvious only since its development. To look at the workings of nature and intuit, as Newton did, that for every action there is an equal and opposite reaction—to think not in terms of individual instances but in terms of abstract patterns of behavior—was an enormous advance in human development. It is a way of thinking that evolved slowly, over time, and it had its roots not in science but in society.

The term "law" today has many and distinct meanings. Scientific laws provide descriptions of how physical objects behave but offer no explanation regarding why they follow those laws. No incentives for obedience, or penalties for the converse, apply to rocks or planets. In the social and religious realms, by contrast, laws described not how people *do* behave, but how they *should* behave, and they provide reasons to obey—in order to be good people, or to avoid punishment. The term "law" is used in both cases, but today those two concepts have little in common. When the idea first arose, however, no distinction was made between laws in the human and the inanimate realms. Inanimate objects were believed to be subject to laws in the same way that people were governed by religious and ethical codes.

The idea of law originated with religion. When the peoples of early Mesopotamia gazed around them, they saw a world on the brink of chaos, saved from it only by gods who favor order—albeit of a rather minimal and arbitrary kind. These were humanlike deities who acted, like we do, out of emotion and whim and constantly intervened in the lives of mortals. There were gods for everything—literally thousands of them—including a god for brewing, gods for farmers, scribes, merchants, and craftsmen. There was a god of livestock pens. There were demon gods—one caused epidemics; another was a female called Extinguisher, who killed young children. And each city-state had not only its own head god, but also a whole court of subordinate gods who played roles such as gatekeeper, gardener, ambassador, and hairdresser.

The worship of all these gods included the acceptance of a formal ethical code. It is difficult to imagine life without the protections of a legal system, but prior to the emergence of cities, nomadic humans had no formalized codes of law. Certainly, people knew what behaviors would be welcomed or condemned by others, but rules of conduct were not abstracted into edicts like "Thou shalt not kill." Behavior was governed not by a collection of general statutes but, in each particular instance, by concern about what others would think and the fear of reprisal from those who were more powerful.

The gods of urban Mesopotamia, though, made specific ethical demands, requiring that their flock follow formal rules ranging from "Help others" to "Don't vomit in streams." Here was the first instance of a higher power handing down what we might consider formalized laws. And violations were not to be taken lightly—they were said to bring upon the perpetrator problems such as illness or death, punishments administered by demonic gods with names like "Fever," "Jaundice," and "Cough."

The gods also worked through the city's earthly rulers, who derived their authority from their theological connection. By the time of the first Babylonian Empire, in the eighteenth century B.C., a more or less unified theological theory of nature had emerged, in which a transcendent god laid down laws that covered the actions of both people and what we would call the inanimate world. That set of human civil and criminal laws is called the Code of Hammurabi. It is named for the reigning Babylonian king, whom the great god Marduk commanded to "bring about the rule of righteousness in the land, to destroy the wicked and evildoers."

The Code of Hammurabi was issued about a year before Hammurabi's death, in 1750 B.C. It wasn't exactly a model for democratic rights: the upper classes and the royals were granted leniency and greater privileges, and slaves could be bought and sold, or killed. But the code did contain rules of justice, with the eye-for-an-eye harshness of the Torah, which would come perhaps a thousand years later. It decreed, for example, that anyone caught committing a robbery should be put to death; anyone who steals while helping to put out a fire shall be thrown into the fire; any "sister of god" who opens a tavern shall be burned to death; anyone causing a flood by being "too lazy" to keep his dam in working condition must replace any corn that is ruined; and anyone who swears to God that he was robbed while entrusted with someone else's money need not pay it back.

The laws of the Code of Hammurabi were carved onto an eight-foot-high block of black basalt, obviously intended for public viewing and reference. The block was discovered in 1901 and is now on display at the Louvre. It was not, like the pyramids, a great physical achievement, but it was a monumental intellectual achievement, an attempt to build a scaffold of order and rationality encompassing all the social interactions of Babylonian society—commercial, monetary, military, marital, medical, ethical, and so on—and, to date, it is the earliest known example of a ruler establishing an entire body of law for his people.

As I said, it was believed that the god Marduk did not just rule over people, but that he ruled over physical processes as well; he legislated to the stars just as he legislated to humans. And so in parallel to the Code of Hammurabi, Marduk was said to have created a kind of code for *nature* to follow. Those laws governing what we would call the inanimate world constituted the first scientific laws in the sense that they describe the workings of natural phenomena. They were not laws of nature in the modern sense, however, because they provide only a vague indication of *how* nature behaves; rather, like the Code of Hammurabi, the laws are commands and decrees that Marduk *orders* nature to follow.

The idea that nature "obeys" laws in the same sense that people do would persist for millennia. Anaximander, for example, one of the great natural philosophers of ancient Greece, said that all things arise from a primordial substance, and return to it, lest they have to "pay fine and penalty to each other for their iniquity according to time's order." Heraclitus, likewise, says that "the sun will not transgress his measures; otherwise [the goddess of justice] will find [and punish] him." The term "astronomy," in fact, has at its root the Greek word *nomos*, which means "law" in the sense of human law. It wasn't until Kepler, in the early seventeenth century, that the term "law" began to be used in the modern sense, to mean a generalization, based upon observation, that describes the behavior of some natural phenomenon but does not seek to assign to it a purpose or motive. Still, it was not a sudden transition, for though Kepler wrote occasionally of mathematical laws, even he believed that God ordered the universe to follow the principle of "geometric beauty," and he explains that the movement of planets probably results from the "mind" of the planet perceiving its angle and calculating its orbit.

* * *

Historian Edgar Zilsel, who studied the history of the idea of scientific law, wrote that "man seems to be inclined to interpret nature . . . after the pattern of society." Our attempts to formulate the laws of nature, in other words, seem to grow out of our natural inclination to understand our personal existence, and our experiences, and the culture in which we are raised, influence our approach to science.

Zilsel recognized that we all create mental stories to describe our lives, that we piece them together from what we are taught and what we experience, forming a vision of who we are and what our place in the universe is. We thus form a set of laws describing our personal world, and our life's meaning. Before the war, for example, the laws that seemed to govern my father's life led him to expect decency from his society, an approximation of justice from his courts, food from the market—and protection from his God. This was his view of the world, and about its validity he felt as blasé as a scientist who has seen his theory pass every test.

But though stars and planets pull one another reliably for billions of years, in the human world the laws can be turned upside down in mere hours. That's what happened to my father and to countless others in September 1939. In the prior months, my father had completed a course in fashion design in Warsaw, bought two new German sewing machines, and rented a small room in a neighbor's apartment where he opened a tailoring shop. Then the Germans invaded Poland and, on September 3, marched into his hometown, Częstochowa. The occupation government soon issued a series of anti-Semitic decrees resulting in the confiscation of anything of value—jewelry, cars, radios, furniture, money, apartments, even children's toys. Jewish schools were closed and outlawed. Adults were made to wear Star of David symbols. People were arbitrarily snatched from the street and made to perform compulsory labor. Others were shot and killed at any madman's whim.

That which destroyed the physical structure of my father's world also irrevocably altered the mental and emotional scaffolding surrounding it. And, sadly, the Holocaust is a story that has been duplicated, on various scales, many times before and since. So if our human experience informs our notion of scientific law, it is not surprising that for most of its history humankind has found it difficult to imagine that the world might be governed by neat, absolute regularities that are immune to whim, devoid of purpose, and not subject to divine intervention.

Even today, long after Newton produced his monumentally successful set of laws, many people continue not to believe that such laws are universally applicable. Yet centuries of progress have rewarded scientists, who recognized that physical law and human law follow distinctly different patterns.

Nine years before his death, at age seventy-six, Albert Einstein described his lifelong pursuit of an understanding of the physical laws of the universe this way: "Out yonder was this huge world, which exists independently of us human beings and which stands before us like a great, eternal riddle, at least partially accessible to our inspection and thinking. The contemplation of this world beckoned as a liberation . . . The road to this paradise . . . has shown itself reliable, and I have never regretted having chosen it." In a way, I think that my father, toward the end of his own life, found a similar feeling of "liberation" in that thought.

For our species, Uruk was the beginning of the long road toward deciphering that eternal riddle. The infant civilizations of the Near East established the rudiments of intellectual life—and then built on those to give us a class of thinkers who created mathematics, written language, and the concept of laws. The next step in the blossoming and maturing of the human mind was taken in Greece, more than a thousand miles away. The great Greek miracle gave birth to the idea of the mathematical proof, the disciplines of science and philosophy, and the concept of what we today call "reason"—some two thousand years before Newton.

5

Reason

In 334 B.C., Alexander, the twenty-two-year-old king of the Greek state of Macedon, led an army of seasoned citizen warriors across the Hellespont at the beginning of a long campaign to conquer the vast Persian Empire. By chance, as I write this, I myself have a twenty-two-year-old son whose name—Alexei—has the same Greek root. They say that kids grow up faster today than ever, but one thing I cannot imagine my Alexei doing is leading an army of seasoned Greek citizen warriors into Mesopotamia to confront the Persian Empire. There are several ancient accounts of how the young Macedonian king achieved his victory, most of which involve drinking large quantities of wine. However he did it, his long course of conquest took him all the way to the Khyber Pass and beyond. By the time he died at age thirty-three, he had accomplished enough in his brief existence that he has ever since been referred to as Alexander the Great.

At the time of Alexander's invasion, the Near East was peppered with cities like Uruk that had existed for thousands of years. To put this in perspective: if the United States had been in existence as long as Uruk, we would now be on roughly our six hundredth president.

To walk the streets of those ancient cities Alexander conquered must have inspired awe, for one would have found oneself roaming among immense palaces, vast gardens irrigated by special channels, and grand stone buildings graced by columns topped with carvings of griffins and bulls. These were vibrant and complex societies, not at all in decline. Yet their cultures had been surpassed intellectually by the Greek-speaking world that conquered them, epitomized by its young leader—a man who had been schooled by Aristotle himself.

With the conquest of Mesopotamia by Alexander, the feeling that all things Greek were superior quickly spread throughout the Near East. Children, always at the vanguard of a cultural shift, learned the Greek language, memorized Greek poetry, and took up the sport of wrestling. Greek art grew popular in Persia. Berosus, a priest of Babylon, Sanchuniathon, a Phoenician, and Flavius Josephus, a Jew, all wrote histories of their people aimed at showing their compatibility with Greek ideas. Even taxes were Hellenized—they began to be recorded in the relatively new Greek alphabet, and on papyrus rather than in cuneiform on tablets. But the greatest aspect of the Greek culture that Alexander brought with him had nothing to do with arts or administration. It was what he had learned firsthand from Aristotle: a new, rational approach to the struggle to know our world, a magnificent turning point in the history of human ideas. And Aristotle himself was building on the ideas of several generations' worth of scientists and philosophers who had begun to challenge the old verities about the universe.

* * *

In the early years of ancient Greece, the Greek understanding of nature was not very different from that of the Mesopotamians. Inclement weather might have been explained by saying that Zeus had indigestion, and if farmers had a bad crop, people would have thought it was because the gods were angry. There might not have been a creation myth stating that the earth is a droplet in the sneeze of the hay fever god, but there might as well have been, for, in the millennia after writing was invented, the body of recorded human words reveals a wild profusion of stories about how the world came into being and what forces governed it. What all had in common was the description of a turbulent universe created by an inscrutable god out of some type of formless void. The word "chaos" itself comes from the Greek term for the nothingness that was said to have preceded the creation of the universe.

If before creation all was chaos, after creating the world, the gods of Greek mythology didn't seem to put a lot of energy into their effort to bring order to it. Lightning, windstorms, droughts, floods, earthquakes, volcanoes, infestations, accidents, disease—all these and many other irregular plagues of nature took their toll on human health and life. Egoistic, treacherous, and capricious, the gods were thought to be constantly causing calamities through their anger or just through carelessness, as if they were bulls in a china shop and we were the china. This is the primitive theory of the cosmos that passed orally from generation to generation in Greece, until finally being written down by Homer and Hesiod around 700 B.C., a century or so after writing finally spread to Greek culture. From then on, it was a staple of Greek education, forming the accepted wisdom of generations of thinkers.

For those of us living in modern society, the beneficiaries of a long history of scientific thought, it is difficult to understand how nature could have appeared this way to those ancient peoples. The idea of structure and order in nature seems as obvious to us as the idea that the gods controlled everything seemed to them. Today, our daily activities are quantitatively mapped, assigned certain hours and minutes. Our lands are delineated by latitude and longitude, our addresses marked by street names and numbers. Today, if the stock market goes down three points, a pundit will give us an explanation, such as that the decline was due to new worries over inflation. True, another expert might say it was due to developments in China, and a third may pin it on unusual sunspot activity, but, right or wrong, our explanations are expected to be based on cause and effect.

We demand from our world causality and order because these concepts are ingrained in our culture, in our very consciousness. Unlike us, however, the ancients lacked a mathematical and scientific tradition, and so the conceptual framework of modern science—the idea of precise numerical predictions, the notion that repeated experiments should give identical results, the use of time as a parameter to follow the unfolding of events—would have been difficult to grasp or accept. To the ancients nature appeared to be ruled by tumult, and to believe in orderly physical laws would have seemed as outlandish to them as the tales of their wild and capricious gods seem to us (or, perhaps, as our own dear theories will seem to historians who study them a thousand years from now).

Why should nature be predictable, explainable in terms of concepts that can be discovered by the human intellect? Albert Einstein, a man who wouldn't have been surprised to find that the space-time continuum could be warped into the shape of a salted pretzel, was astonished by the much simpler fact that nature has order. He wrote that "one should expect a chaotic world, which cannot be grasped by the mind in any way." But he went on to write that, contrary to his expectation, "the most incomprehensible thing about the universe is that it is comprehensible."

Cattle don't understand the forces that hold them to the earth, nor do crows know anything about the aerodynamics that give them flight. With his statement, Einstein was expressing a momentous and uniquely *human* observation: that order rules the world, and that the rules governing nature's order don't have to be explained by myths. They are knowable, and humans have the ability, unique among all the creatures on earth, to decipher nature's blueprint. That lesson has profound implications, for if we can decipher the design of the universe, we can use that knowledge to understand our place in it, and we can seek to manipulate nature to create products and technologies that make our lives better.

The new rational approach to nature originated during the sixth century B.C., with a group of revolutionary thinkers who lived in greater Greece, on the shores of the Aegean, that large Mediterranean bay that separates present-day Greece and Turkey. Several hundred years before Aristotle, at the same time that Buddha was bringing a new philosophical tradition to India, and Confucius to China, these earliest of the Greek philosophers made the paradigmatic shift to viewing the universe as ordered, not random—as Cosmos, not Chaos. It is hard to overstate what a profound shift that was, or the degree to which it has shaped human consciousness ever since.

The area that gave rise to these radical thinkers was a magical land of grapevines, fig orchards, and olive trees and of prosperous, cosmopolitan cities. Those cities lay at the mouths of rivers and gulfs that emptied into the sea, and at the ends of roads that ran inland. According to Herodotus, it was a paradise where "the air and climate are the most beautiful in the whole world." It was called Ionia.

The Greeks had founded many city-states on what is now the Greek mainland and in southern Italy, but they were merely provinces—the center of Greek civilization was in Turkish Ionia, just hundreds of miles west of Göbekli Tepe and Çatalhöyük. And the vanguard of the Greek enlightenment was to be found in the city of Miletus, located on the shores of a gulf, the Gulf of Latmus, which gave it access to the Aegean and hence the Mediterranean.

According to Herodotus, at the turn of the first millennium B.C. Miletus had been a modest settlement populated by Carians, a people of Minoan descent. Then, around 1000 B.C., soldiers from Athens and its vicinity overran the area. By 600 B.C. the new Miletus had become a kind of ancient New York City, attracting, from all over Greece, poor, hardworking refugees seeking a better life.

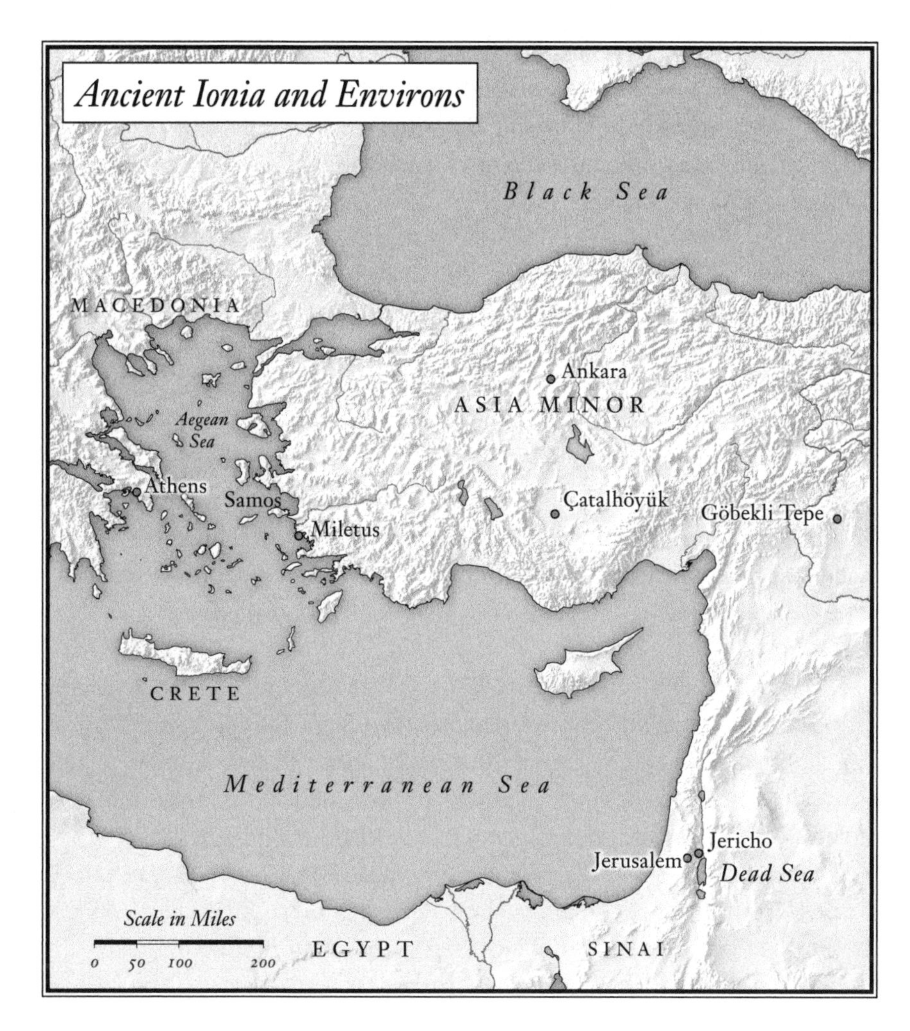

Over the centuries, the population of Miletus ballooned to 100,000, and the city developed into a center of great wealth and luxury, becoming the richest of the Ionian cities, indeed the richest city in the entire Greek world. From the Aegean, the fishermen of Miletus harvested bass, red mullet, and mussels. From the rich soil, farmers harvested corn and figs—the only fruit known to the Greeks that they could keep for any length of time—while orchards provided olives, for food as well as for pressing into oil, the ancient Greeks' version of butter, soap, and fuel. What's more, access to the sea made Miletus an important center of trade. Commodities like flax, timber, iron, and silver were brought in from the dozens of colonies the citizens of Miletus had established as far away as Egypt, while its skilled artisans created pottery, furniture, and fine woolens to ship abroad.

But Miletus was not just a crossroads for the exchange of goods; it was also a place for the sharing of ideas. Within the city, people from dozens of scattered cultures met and spoke, and Milesians also traveled widely, exposing them to many disparate languages and cultures. And so, as its inhabitants argued over the price of salted fish, tradition met tradition and superstition confronted superstition, creating an openness to new ways of thinking and fostering a culture of innovation—in particular, the all-important willingness to question conventional wisdom. What's more, the wealth of Miletus created leisure, and with leisure came the freedom to devote time to pondering the issues of our existence. Thus, through the confluence of so many favorable circumstances, Miletus became a sophisticated, cosmopolitan paradise and a center of scholarship, creating a perfect storm of all the factors necessary for a revolution in thought.

It was in this environment, in Miletus and eventually in wider Ionia, that there emerged a group of thinkers who began to question the religious and mythological explanations of nature that had been passed down for thousands of years. They were the Copernicuses and Galileos of their day, the formative pioneers of both philosophy and science.

The first of these scholars, according to Aristotle, was a man named Thales, born around 624 B.C. Many Greek philosophers were said to live in poverty. Indeed, if ancient times were anything like today, even a *famous* philosopher could have achieved a more prosperous existence by finding a better job, like selling olives at the side of the road. Tradition has it, though, that Thales was an exception, a cunning and wealthy merchant who had no trouble financing his time to think and ponder. It is said that, in one instance, he made a fortune by cornering the market on olive presses and then charging exorbitant prices for the oil, like a one-man OPEC. He is also said to have been very involved in his city's politics and to have known its dictator, Thrasybulus, intimately.

Thales used his wealth to travel. In Egypt, he found that although the Egyptians had the expertise to build the pyramids, they lacked the insight to measure their height. As we've seen, however, they had developed a novel set of mathematical rules that they used to determine the area of plots of land for purposes of taxation. Thales adapted those Egyptian techniques of geometry to calculate the heights of the pyramids—

and also showed how, using them, one could determine the distance of ships at sea. This made him quite a celebrity in ancient Egypt.

When Thales returned to Greece, he brought Egyptian mathematics with him, translating its name to his native tongue. But in Thales's hands, geometry was not just a tool for measuring and calculating; it was a body of theorems connected by logical deduction. He was the first to prove geometric truths, rather than simply stating as facts conclusions that seemed to work, and the great geometer Euclid would later include some of Thales's theorems in his *Elements*. Still, as impressive as Thales's mathematical insight was, his real claim to fame was his approach to explaining the phenomena of the physical world.

Nature, in Thales's view, wasn't the stuff of mythology; it operated according to principles of science that could be used to explain and predict all the phenomena hitherto attributable to the intervention of the gods. He was said to be the first person to understand the cause of eclipses, and he was the first Greek to propose that the moon shone by reflected sunlight.

Even when he was off base, Thales was remarkable in the originality of his thinking and his ideas. Consider his explanation of earthquakes. In Thales's day, these were thought to occur when the god Poseidon became irritated and struck the earth with his trident. But Thales held what must have seemed like an oddball view: that earthquakes had nothing to do with the gods. His explanation wasn't one I'd hear from any of my Caltech seismologist friends—he believed that the world was a hemisphere floating on an endless expanse of water, and that earthquakes occurred when the water sloshed around. Thales's analysis is nonetheless groundbreaking in its implications, because he attempted to account for earthquakes as a consequence of a natural process, and he employed empirical and logical arguments to back up his idea. Perhaps most important of all is his having focused on the question of why earthquakes occur in the first place.

In 1903, the poet Rainer Maria Rilke gave advice to a student that holds as true for science as it does for poetry: "Be patient toward all that is unsolved in your heart and try to love the questions," he wrote, and "live the questions." The greatest skill in science (and often in business as well) is the ability to ask the right questions—and Thales practically invented the idea of asking scientific questions. Everywhere he looked, including in the heavens, he saw phenomena that begged to be explained, and his intuition led him to ponder phenomena that would eventually shed light on the fundamental workings of nature. He asked questions not just about earthquakes but about the size and shape of the earth, the dates of the solstices, and the relation of the earth to the sun and moon—the very same questions that two thousand years later led Isaac Newton to his great discovery of gravity and the laws of motion.

In acknowledgment of what a radical break with the past Thales had made, Aristotle referred to Thales and later Ionian thinkers as the first of the *physikoi*, or physicists—the group to which I am proud to belong, and to which Aristotle felt he himself belonged. The term comes from the Greek *physis*, meaning "nature," a term Aristotle chose to describe those who sought natural explanations for phenomena, in contrast to the *theologoi*, or theologians, who sought supernatural explanations.

Aristotle had less admiration, however, for members of another radical group: those who used mathematics to model nature. Credit for that innovation goes to a thinker of the generation following Thales, who lived not far away from him on the Aegean island of Samos.

* * *

Some of us spend our work hours trying to understand how the universe functions. Others haven't mastered algebra. In Thales's day, members of the former group were also members of the latter, for, as we've seen, algebra as we know it—and most of the rest of mathematics—hadn't yet been invented.

To today's scientist, understanding nature without equations would be like trying to understand your partner's feelings when all he ever says is "I'm fine." For mathematics is the vocabulary of science—it is how a theoretical idea communicates. We scientists may not always be good at using language to reveal intimate personal thoughts, but we've gotten very adept at communicating our theories through mathematics. The language of mathematics enables science to delve deeper into theories, and with more insight and precision than ordinary language, for it is a language with built-in rules of reasoning and logic that keep extending the meaning, allowing it to unfold and reverberate in sometimes quite unexpected ways.

Poets describe their observations through language; physicists describe theirs with math. When a poet completes a poem, the poet's job is done. But when the physicist sets down a mathematical "poem," that is just the beginning of the job. By applying the rules and theorems of mathematics, the physicist must then coax that poem into revealing new lessons of nature that its own author might never have imagined. For equations not only embody ideas; they offer the consequences of those ideas to anyone with sufficient skill and persistence to extract them. That is what the language of mathematics achieves: it facilitates the *expression* of physical principles, it illuminates the *relationships* between them, and it guides human *reasoning* about them.

At the beginning of the sixth century B.C., however, no one knew this. The human species hadn't yet come up with the idea that mathematics could help us understand how nature operates. It was Pythagoras (c. 570–c. 490 B.C.)—founder of Greek mathematics, inventor of the term "philosophy," and curse of middle school students the world round who must stop texting long enough to learn the meaning of $a^2 + b^2 = c^2$—who is said to have first helped us use mathematics as the language of scientific ideas.

The name Pythagoras, in ancient times, not only was associated with genius but also carried a magical and religious aura. He was looked upon as Einstein might have been had he been not just a physicist but also the pope. We have, from many later writers, a lot of information on the life of Pythagoras, and several biographies. But by the first centuries after Christ, the tales had become unreliable, tainted by ulterior religious and political motives that caused writers to distort his ideas and magnify his place in history.

One thing that does seem to be true is that Pythagoras grew up on Samos, across the bay from Miletus. Also, all his ancient biographers agree that sometime between the ages of eighteen and twenty Pythagoras visited Thales, who was by then very old and near death. Aware that his earlier brilliance had faded considerably, Thales is said to have apologized for his diminished mental state. Whatever lessons Thales imparted, Pythagoras went away impressed. Many years later, he could sometimes be spotted sitting at home, singing songs of praise to his late teacher.

Like Thales, Pythagoras traveled a great deal, probably to Egypt, Babylon, and Phoenicia. He left Samos at age forty, finding life under the island's tyrant, Polycrates, unbearable, and landed in Croton, in what is now southern Italy. There, he attracted a large number of followers. It was also there that he was said to have had his epiphany about the mathematical ordering of the physical world.

Nobody knows how language was first developed, though I have always imagined some caveman stubbing his toe and spontaneously blurting, *Ow!*, at which time someone thought, *What a novel way to express one's feelings*, and soon everyone was talking. The origin of mathematics as the language of science is also shrouded in mystery; but in that case we do at least have a legend that describes it.

According to the legend, while walking past a blacksmith's shop one day, Pythagoras heard the sound of the blacksmith's hammers ringing out, and he noticed a pattern in the tones produced by different hammers pounding on the iron. Pythagoras ran into the forge and experimented with the hammers, noting that the differences in tone did not depend on the force employed by the man delivering the blow, nor on the precise shape of the hammer, but rather on the hammer's size or, equivalently, its weight.

Pythagoras returned home and continued his experimentation, not on hammers but on strings of different lengths and tensions. He had, like other Greek youths, been schooled in music, especially the flute and the lyre. Greek musical instruments at the time were the product of guesswork, experience, and intuition. But in his experiments Pythagoras is said to have discovered a mathematical law governing stringed instruments that could be used to define a precise relationship between the length of musical strings and the tones they produce.

Today we would describe the Pythagorean relation by saying that the frequency of the tone is inversely proportional to the length of the string. Suppose, for example, that a string produces a certain note when plucked. Hold the string down at the halfway point and it produces a note one octave higher—that is, of twice the frequency. Hold it down at one-fourth the length and the tone goes up another octave, to four times the original frequency.

Did Pythagoras actually discover this relationship? No one knows to what extent the legends about Pythagoras are true. For example, he probably did *not* prove the "Pythagorean theorem" that plagues middle school students—it is believed that it was one of his followers who first *proved* it, but the formula had already been *known* for centuries. Regardless, the real contribution of Pythagoras was not in deriving any specific laws but in promoting the idea of a cosmos that was structured according to numerical relationships, and his influence came not from discovering the mathematical connections in nature but from celebrating them. As classicist Carl Huffman put it, Pythagoras was important "for the honor he gives to number and for removing it from the practical realm of trade and instead pointing to correspondences between the behavior of number and the behavior of things."

Where Thales said that nature follows orderly rules, Pythagoras went even further, asserting that nature follows *mathematical* rules. He preached that it is mathematical law that is the fundamental truth about the universe. Number, the Pythagoreans believed, is the essence of reality.

Pythagoras's ideas had a great influence on later Greek thinkers—most notably on Plato—and on scientists and philosophers throughout Europe. But of all the Greek champions of reason, of all the great Greek scholars who believed that the universe could be understood through rational analysis, by far the most influential as far as the future development of science is concerned was not Thales, who invented the approach, nor Pythagoras, who brought mathematics to it, nor even Plato, but rather that student of Plato's who later became the tutor of Alexander the Great: Aristotle.

* * *

Born in Stagira, a town in northeastern Greece, Aristotle (384–322 B.C.) was the son of a man who had been the personal physician to Alexander's grandfather, King Amyntas. He was orphaned young and sent to Athens to study at Plato's Academy when he was seventeen. After Plato, the word "academy" came to mean a place of learning, but in his day it was simply the name of a public garden on the outskirts of Athens that harbored the grove of trees where Plato and his students liked to assemble. Aristotle remained there for twenty years.

At Plato's death, in 347 B.C., Aristotle left the Academy, and a few years later he became Alexander's tutor. It's unclear why King Philip II chose him to tutor his son, since Aristotle had not yet made his reputation. To Aristotle, though, becoming the tutor of the heir apparent to the king of Macedon must have seemed like a good idea. He was paid handsomely, and reaped other benefits when Alexander went on to conquer Persia and much of the rest of the world. But after Alexander succeeded to the throne, Aristotle, then nearly fifty, returned to Athens, where, over the course of thirteen years, he produced most of the works for which he is known. He never again saw Alexander.

The kind of science that Aristotle taught probably wouldn't have been identical to what he himself had learned from Plato. Aristotle was a prize pupil at the Academy, but he was never comfortable with Plato's emphasis on mathematics. His own bent was toward detailed, natural observation, not abstract laws—very different both from Plato's kind of science and from science as it is practiced today.

When I was in high school, I loved my courses in chemistry and physics. Seeing how passionate I was about them, my father sometimes asked me to explain those sciences to him. Coming from a poor Jewish family that could afford to send him only to the local religious school, he had gotten an education focused more on theories of the Sabbath than on theories of science, and since he had never progressed beyond the seventh grade, I had my work cut out for me.

I began our exploration by saying that physics is largely the study of one thing: change. My father pondered for a moment and then grunted. "You know nothing of change," he told me. "You're too young, and you've never experienced it." I protested that of course I'd experienced change, but he answered with one of those old Yiddish expressions that sounds either deep or idiotic, depending on your tolerance level for old Yiddish expressions. "There is change," he said, "and there is *change*."

Aristotle, pictured with Plato (left), from a fresco by Raphael

I dismissed his aphorism in the way only a teenager can. In physics, I said, there is not change and *change*—there is only CHANGE. In fact, one might say that Isaac Newton's central contribution in creating physics as we know it today was his invention of a unified mathematical approach that could be used to describe *all* change, whatever its nature. Aristotle's physics—which originated in Athens two thousand years before Newton—has its roots in a far more intuitive and less mathematical approach to understanding the world, which I thought might be more accessible to my father. And so, hoping I would find something there that would make it easier to explain things to him, I began to read about Aristotle's concept of change. After much effort, I learned that although Aristotle spoke Greek and had never uttered a word of Yiddish, what he essentially believed was this: "There is change, and there is *change*."

In my father's version, the second invocation of the word "change" sounded ominous, and he meant it to convey the kind of violent change he had experienced when the Nazis invaded. That distinction between ordinary or natural change, on the one hand, and violent *change*, on the other, is the same distinction that Aristotle made: he believed that all transformations one observes in nature could be categorized as either natural or violent.

In Aristotle's theory of the world, natural change was that which originated *within the object* itself. In other words, the cause of natural change is intrinsic to the nature or composition of the object. For example, consider the type of change we call motion—the change of position. Aristotle believed that everything is made of various combinations of four fundamental elements—earth, air, fire, and water—each of which has a built-in tendency to move. Rocks fall toward the earth and rain falls toward the oceans, according to Aristotle, because the earth and sea are the natural resting places of those substances. To cause a rock to fly upward requires external intervention, but when a rock falls, it is following its built-in tendency and executing "natural" motion.

In modern physics, no cause is required to explain why an object remains at rest or in uniform motion with a constant speed and direction. Similarly, in Aristotle's physics, it was not necessary to explain why objects execute natural motion—why things made of earth and water fall, or why air and fire rise. This analysis reflects what we see in the world around us—in which bubbles rise out of water, flames seem to rise into the air, massive objects fall from the sky, oceans and seas rest upon the land, and the atmosphere lies above all.

To Aristotle, motion was but one of many natural processes, like growth, decay, and fermentation, all of which are governed by the same principles. He viewed natural change, in all its various forms—the burning of a log, the aging of a person, the flight of a bird, the falling of an acorn—as the fulfillment of inherent potential. Natural change, in Aristotle's system of beliefs, is what propels us through our daily lives. It is the kind of change that wouldn't raise eyebrows, that we tend to take for granted.

But sometimes the natural course of events is disrupted, and motion, or change, is imposed by something external. This is what happens when a rock is tossed into the air, when grapevines are ripped out of the earth or chickens slaughtered for food, or when you lose your job, or fascists take over a continent. These are the kinds of change that Aristotle termed "violent."

In violent change, according to Aristotle, an object changes or moves in a direction that violates its nature. Aristotle sought to understand the cause of that kind of change, and he chose a term for it: he called it "force."

Like his concept of natural change, Aristotle's doctrine of violent change corresponds well with what we observe in nature—solid matter, for example, plummets downward on its own, but to get it going in any other direction, such as upward or sideways, requires force, or effort.

Aristotle's analysis of change was remarkable because although he saw the same environmental phenomena as the other great thinkers of his time, unlike the others, he rolled up his sleeves and made observations about change in unprecedented and encyclopedic detail—the changes both in people's lives and in nature. Trying to discover what all the different kinds of change have in common, he studied the causes of accidents, the dynamics of politics, the motion of oxen hauling heavy burdens, the growth of chicken embryos, the eruption of volcanoes, the alterations in the Nile River delta, the nature of sunlight, the rising of heat, the motion of the planets, the evaporation of water, the digestion of food in animals with multiple stomachs, the way things melt and burn. He dissected animals of all sorts, sometimes far past their sell-by dates, but if others objected to the foul smell, he simply scoffed.

Aristotle called his attempt to create a systematic account of change *Physics*—thus associating himself with the heritage of Thales. His physics was vast in scope, encompassing both the living and the inanimate, and the phenomena of both the heavens and the earth. Today the different categories of change he studied are the subjects of entire branches of science: physics, astronomy, climatology, biology, embryology, sociology, and so on. In fact, Aristotle was a prolific writer—a veritable one-man Wikipedia. His contributions include some of the most comprehensive studies ever undertaken by a person never diagnosed with OCD. All told, he produced—according to records from antiquity—170 scholarly works, about one-third of which have been preserved until today. There was *Meteorology; Metaphysics; Ethics; Politics; Rhetoric; Poetics; On the Heavens; On Generation and Corruption; On the Soul; On Memory; On Sleep and Sleeplessness; On Dreams; On Prophesying, Longevity, Youth and Age; On the History and Parts of Animals;* and on and on.

While his former pupil Alexander went on to conquer Asia, Aristotle returned to Athens and established a school called the Lyceum. There, while strolling along a public walk or pacing in a garden, he would teach his students what he had learned over the years.* But though he was a great teacher and a brilliant and prolific observer of nature, Aristotle's approach to knowledge was far different from the approach of what we call science today.

* * *

According to philosopher Bertrand Russell, Aristotle was "the first to write like a professor . . . a professional teacher, not an inspired prophet." Russell said that Aristotle is Plato "diluted by common sense." Indeed, Aristotle placed a great value on that trait. Most of us do. It's what keeps us from responding to those kind fellows in Nigeria whose emails promise that if we wire them one thousand dollars today, they will wire us one hundred billion dollars tomorrow. However, looking back on Aristotle's ideas, and knowing what we know now, one might argue that it is precisely in Aristotle's devotion to conventional ideas that we find one of the greatest differences between today's approach to science and Aristotle's—and one of the greatest shortcomings of Aristotelian physics. For though common sense is not to be ignored, sometimes what is needed is uncommon sense.

*Afterward, the students would be rubbed with oil. I've always thought that offering an option like that would be an easy way to increase my popularity with my own students, but unfortunately it would probably have the opposite effect on the university administration.

In science, in order to make progress, you often have to defy what historian Daniel Boorstin referred to as "the tyranny of common sense." It is common sense, for example, that if you push an object it will slide, then slow down and stop. But to perceive the underlying laws of motion, you must look beyond the obvious, as Newton did, and envision how an object in a theoretically frictionless world would move. Similarly, to understand the ultimate mechanism of friction, you must be able to look past the facade of the material world, to "see" how objects might be made of invisible atoms, a concept that had been proposed by Leucippus and Democritus about a century earlier, but which Aristotle did not accept.

Aristotle also showed great deference to common opinion, to the institutions and ideas of his time. He wrote, "What everyone believes is true." And to the doubters, he said, "Whoever destroys this faith will hardly find a more credible one." A vivid example of Aristotle's reliance on conventional wisdom—and the way it distorted his vision—is his somewhat tortured argument that slavery, which he and most of his fellow citizens accepted, is inherent in the nature of the physical world. Employing the kind of argument that is strangely reminiscent of his writings in physics, Aristotle asserted that "in all things which form a composite whole and which are made up of parts . . . a distinction between the ruling and the subject element comes to light. Such a duality exists in living creatures, but not in them only; it originates in the constitution of the universe." Because of that duality, Aristotle argued, there are men who are, by their nature, free, and men who are by nature slaves.

Today scientists and other innovators are often portrayed as odd and unconventional. I suppose there is some truth to that stereotype. One physics professor I knew selected his lunch each day from the free offerings at the cafeteria's condiments table. Mayonnaise provided fat, ketchup was his vegetable, and saltines his carbs. Another friend loved cold cuts but hated bread, and at a restaurant had no qualms ordering for lunch a lonely pile of salami, which he would eat with a knife and fork as if it were a steak.

Conventional thinking is not a good attitude for a scientist—or anyone who wants to innovate—and sometimes that has its costs in the way people view you. But as we will see repeatedly, science is the natural enemy of preconceived notions, and of authority, even the authority of the scientific establishment itself. For revolutionary breakthroughs necessarily require a willingness to fly in the face of what everyone else believes to be true, to replace old ideas with credible new ones. In fact, if there is one barrier to progress that stands out in the history of science, and of human thought in general, it is an undue allegiance to the ideas of the past—and present. And so if I were hiring for a creative position, I'd beware of too much common sense, but I'd count oddball traits in the plus column, and keep that condiments table well supplied.

* * *

Another important clash between Aristotle's approach and that of later science is that it was qualitative, not quantitative. Today physics is, even in its simple high school form, a science of quantity. Students taking even the most elementary versions of physics learn that a car moving sixty miles per hour is going eighty-eight feet each second. They learn that if you drop an apple, its speed will increase by twenty-two miles per hour each second that it falls. They do mathematical calculations such as computing that when you plop down into a chair, the force exerted on your spine as the chair stops you can be—for a split second—more than a thousand pounds. Aristotle's physics was nothing like that. On the contrary, he complained loudly about philosophers who sought to turn philosophy "into mathematics."

Any attempt to turn natural philosophy into a quantitative pursuit in Aristotle's day was, of course, impeded by the state of knowledge in ancient Greece. Aristotle had no stopwatch, no clock with a second hand, nor was he ever exposed to thinking of events in terms of their precise durations. Also, the fields of algebra and arithmetic that would be needed to manipulate such data were no more advanced than they had been in Thales's time. As we've seen, the plus, minus, and equal signs had not yet been invented, nor had our number system or concepts like "miles per hour." But scholars in the thirteenth century and later made progress in quantitative physics with instruments and mathematics that were not terribly more advanced, so these were not the only barriers to a science of equations, measurement, and numerical prediction. More important was the fact that Aristotle was, like everyone else, simply not interested in quantitative description.

Even when studying motion, Aristotle's analysis was only qualitative. For example, he had just a vague understanding of speed—as in "some things go farther than others in a similar amount of time." That sounds to us like a message we might find inside a fortune cookie, but in Aristotle's time, people considered it precise enough. And with only a qualitative notion of speed, there could be only the foggiest notion of acceleration, which is change in speed or direction—and which we teach as early as middle school. Given these profound differences, if someone with a time machine had gone back and given Aristotle a text on Newton's physics, it would have meant no more to him than a book of microwave pasta recipes. Not only would he have been unable to understand what Newton meant by "force" or "acceleration"—he would not have cared.

What did interest Aristotle, as he conducted his thorough observations, was that motion and other kinds of change seemed to happen toward some *end*. He understood movement, for example, not as something to be *measured* but as a phenomenon whose *purpose* could be discerned. A horse pulls on a cart to move it down the road; a goat walks in order to find food; a mouse runs to avoid being eaten; boy rabbits defile girl rabbits to make more rabbits.

Aristotle believed that the universe was one large ecosystem designed to function harmoniously. He saw purpose everywhere he looked. Rain falls because plants need water to grow. Plants grow so that animals can eat them. Grape seeds grow into grapevines, and eggs turn into chickens, to actualize the potential that exists within those seeds and eggs. From time immemorial, people had always arrived at their understanding of the world through projections of their own experience. And so it was that in ancient Greece it was far more natural to analyze the purpose of events in the physical world than it would have been to try to explain them through the mathematical laws being developed by Pythagoras and his followers.

Here again we see the importance in science of the particular questions you choose to ask. For even if Aristotle had embraced Pythagoras's notion that nature obeys quantitative laws, he would have missed the point, because he was simply less interested in the quantitative specifics of the laws than in the question of why objects follow them. What *compels* the string in a musical instrument, or a falling rock, to behave with numerical regularity? These were the issues that would have excited Aristotle, and it is here that we see the greatest disconnect between his philosophy and the way science is conducted now—for while Aristotle perceived what he interpreted as *purpose* in nature, today's science does not.

That characteristic of Aristotle's analysis—his search for *purpose*—had a huge influence on later human thought. It would endear him to many Christian philosophers through the ages, but it impeded scientific progress for nearly two thousand years, for it was completely incompatible with the powerful principles of science that guide our research today. When two billiard balls collide, the laws that were first set forth by Newton—not a grand underlying purpose—determine what happens next.

Science first arose from the fundamental human desire to know our world and to find meaning in it, so it's not surprising that the yearning for purpose that motivated Aristotle still resonates with many people today. The idea that "everything happens for a reason" may give comfort to those seeking to understand a natural disaster or other tragedy. And for such people the scientist's insistence that the universe is not guided by any sense of purpose can make the discipline of science seem cold and soulless.

Yet there is another way of looking at this, and it's one that I am familiar with from my father. When the issue of purpose would arise, my father often referred not to anything that had befallen him but to a particular incident my mother had experienced before they met, when she was just seventeen. The Nazis had occupied her city, and one of them, for a reason my mother never knew, ordered a few dozen Jews, my mother among them, to kneel in a neat row in the snow. He then walked from one end to the other, stopping every few steps to shoot one of his captives through the head. If this was part of God's or nature's grand plan, my father wanted nothing to do with such a God. For people like my father, there can be relief in the thought that our lives, however tragic or triumphant they may be, are all the result of the same indifferent laws that create exploding stars and that, good or bad, they are ultimately a gift, a miracle that somehow springs from those sterile equations that rule our world.

* * *

Though Aristotle's ideas dominated thinking about the natural world until Newton's day, over the years there were plenty of observers who cast doubt on his theories. For example, consider the idea that objects not executing their natural motion will move only when a force is acting upon them. Aristotle himself realized that this raised the question of what propels an arrow, a javelin, or any other projectile after the initial impetus. His explanation was that, due to the fact that nature "abhors" a vacuum, particles of air rush in behind the projectile after the initial impetus and push it along. The Japanese seem to have successfully adapted that idea to pack passengers into Tokyo subway cars, but even Aristotle himself was only lukewarm about his theory. Its weakness became more obvious than ever during the fourteenth century, when the proliferation of cannons made the idea of air particles rushing behind heavy cannonballs to push them along seem absurd.

Just as important, the soldiers firing those cannons cared little about whether it was particles of air or tiny, invisible nymphs that pushed their cannonballs along. What they did desire to know was the trajectory their projectiles would follow, and in particular whether that trajectory would end on the heads of their enemies. This disconnect illustrates the real chasm that stood between Aristotle and those who would later call themselves scientists: issues like the trajectory of a projectile—its speed and position at various instants—were, for Aristotle, beside the point. But if one wants to apply the laws of physics to make predictions, then those issues are critical. And so the sciences that would eventually replace Aristotle's physics, the ones that would make it possible, among other things, to calculate the trajectory of a cannonball, were concerned with the quantitative details of the processes at work in the world—measurable forces, speeds, and rates of acceleration—not with the purpose or philosophical reasons for those processes.

Aristotle knew that his physics was not perfect. He wrote, "Mine is the first step and therefore a small one, though worked out with much thought and hard labor. It must be looked at as a first step, and judged with indulgence. You, my readers or hearers of my lectures, if you think I have done as much as can fairly be expected of an initial start . . . will acknowledge what I have achieved and will pardon what I have left for others to accomplish." Here Aristotle is voicing a feeling he shared with most of the later geniuses of physics. We think of them, the Newtons and Einsteins, as all-knowing, and confident, even arrogant, about their knowledge. But as we'll see, like Aristotle, they were confused about a lot of things, and, also like Aristotle, they knew it.

* * *

Aristotle died in 322 B.C., at age sixty-two, apparently of a stomach ailment. A year earlier, he had fled Athens when its pro-Macedonian government was overthrown after the death of his former student Alexander. Though Aristotle spent twenty years in Plato's Academy, he had always felt like an outsider in Athens. Of that city, he wrote, "the same things are not proper for a stranger as for a citizen; it is difficult to stay." With Alexander gone, however, the issue of whether to stay became a critical one, for there was a dangerous backlash directed at anyone associated with Macedon, and Aristotle was aware that the politically motivated execution of Socrates had established the precedent that a cup of hemlock is a potent rebuttal to any philosophical argument. Always the deep thinker, Aristotle had the idea to run rather than risk becoming a martyr. He supplied a lofty reason for his decision—to keep the Athenians from once again sinning "against philosophy"—but the decision, like Aristotle's approach to life in general, was a very practical one.

After Aristotle's death, his ideas were passed along by generations of students at the Lyceum and by others who wrote commentaries on his works. His theories faded, along with all learning, during the early Middle Ages but became prominent again during the High Middle Ages among Arab philosophers, from whom later Western scholars learned of it. With some modifications, his thinking eventually became the official philosophy of the Roman Catholic Church. And so, for the next nineteen centuries, to study nature meant to study Aristotle.

We've seen how our species developed both a brain for asking questions and a propensity to ask them, as well as the tools—writing, mathematics, and the idea of laws—with which to begin to answer them. With the Greeks, by learning to use reason to analyze the cosmos, we reached the shores of a glorious new world of science. But that was only the beginning of a greater adventure of exploration that lay ahead.

Part II

The Sciences

The dogmas of the quiet past are inadequate . . . so we must think and act anew.

—Abraham Lincoln, Second Annual Message, December 1, 1862

6

A New Way to Reason

I've had the enriching experience of collaborating on two books with friends—physicist Stephen Hawking and spiritual leader Deepak Chopra. Their worldviews could not have been further apart if they existed in different universes. My vision of life is pretty much the same as Stephen's—that of the scientist. But it is far different from Deepak's, which is probably why we chose to title our book *War of the Worldviews* and not *Isn't It Wonderful How We Agree on Everything?*

Deepak is passionate about his beliefs, and in the time we spent traveling together, he was always trying to convert me and questioning my approach to understanding the world. He called it *reductionist*, because I believe that the mathematical laws of physics can, ultimately, explain everything in nature—including human beings. In particular, as I've said, I believe, as do most scientists today, that everything—again including us—is made of atoms and the elementary particles of matter, which interact through the four fundamental forces of nature, and that if one understands how all that works, one can, in principle, at least, explain everything that happens in the world. In practice, of course, we have neither enough information about our surroundings nor a powerful enough computer to use our fundamental theories to analyze phenomena like human behavior, so the question of whether Deepak's mind is ruled by the laws of physics must remain an open question.

I didn't object in principle to Deepak's characterizing me as a reductionist, but I would bristle when he said it, because the way he said it made me feel embarrassed and defensive, as if a soulful person couldn't believe as I did. In fact, at gatherings of Deepak's supporters, I sometimes felt like an Orthodox rabbi at a convention of pork producers. I was always being asked leading questions like "Do your equations tell you what I am experiencing when I look at a painting by Vermeer or listen to a symphony by Beethoven?" Or: "If my wife's mind is really both particles and waves, how do you explain her love for me?" I had to admit I couldn't explain her love for him. But then again, I couldn't explain any love—using equations. To me, that is beside the point. For, as a tool for understanding the physical world, if not our mental experience (at least not yet), the application of mathematical equations has had unprecedented success.

We may not be able to calculate next week's weather by tracing the motion of each atom and applying the fundamental principles of atomic and nuclear physics, but we do have a science of meteorology that uses higher-level mathematical models and isn't too bad at predicting the weather tomorrow. Similarly, we have applied sciences that study the ocean, light and electromagnetism, the properties of materials, diseases, and dozens of other aspects of our everyday world in ways that allow us to put our knowledge to extraordinary practical uses, undreamt of until the past few hundred years. Today, at least among scientists, there is virtually universal agreement about the validity of the mathematical approach to understanding the physical world. Yet it took a very long time for that view to prevail.

The acceptance of modern science as a metaphysical system based on the idea that nature behaves according to certain regularities originated with the Greeks; but science didn't achieve the first of its convincing successes at making use of these laws until the seventeenth century. That jump, from the ideas of philosophers like Thales, Pythagoras, and Aristotle to those of Galileo and Newton, was a great leap. Still, it needn't have taken two thousand years.

* * *

The first great stumbling block in the path to accepting and building on the Greek heritage was the Roman conquest of Greece in 146 B.C. and Mesopotamia in 64 B.C. The rise of Rome was the beginning of centuries of declining interest in philosophy, mathematics, and science—even among the Greek-speaking intellectual elite—because the practical-minded Romans placed little stock in those fields of study. A remark by Cicero expresses nicely the Roman disdain for theoretical pursuits: "The Greeks," he said, "held the geometer in the highest honor; accordingly, nothing made more brilliant progress among them than mathematics. But we have established as the limit of this art its usefulness in measuring and counting." And indeed, during the roughly thousand-year duration of the Roman Republic and its successor, the Roman Empire, the Romans undertook vast and impressive engineering projects that were dependent, no doubt, upon much measuring and counting; but, as far as we know, they produced not one Roman mathematician of note. That is an astounding fact, and it stands as evidence of the enormous effect of culture on developments in mathematics and science.

Though Rome did not provide an environment conducive to science, after the dissolution of the Western Roman Empire, in A.D. 476, things got even worse. Cities shrank, the feudal system arose, Christianity dominated Europe, and rural monasteries and later the cathedral schools became the centers of intellectual life, which meant that scholarship was focused on religious issues, and inquiries into nature were considered frivolous or unworthy. Eventually, the intellectual heritage of the Greeks was lost to the Western world.

Fortunately for science, in the Arab world the Muslim ruling class *did* find value in Greek learning. That's not to say that they pursued knowledge for its own sake—that stance was no more endorsed by Islamic ideology than by Christianity. But wealthy Arab patrons *were* willing to fund translations of Greek science into Arabic, in the belief that Greek science was useful. And indeed, there was a period of hundreds of years during which medieval Islamic scientists made great progress in practical optics, astronomy, mathematics, and medicine, overtaking the Europeans, whose own intellectual tradition lay dormant.*

But by the thirteenth and fourteenth centuries, a time when the Europeans were awakening from their long slumber, science in the Islamic world had gone into serious decline. There seem to have been several factors at play. For one, conservative religious forces came to impose an ever narrower understanding of practical utility, which they considered to be the only acceptable justification for scientific investigation. Also, for science to flourish, society must be prosperous and offer

*The medieval period runs from A.D. 500 to 1500 (or, in some definitions, 1600). In either case it spans, with some overlap, the era between the cultural achievements of the Roman Empire and the flourishing of science and art in the Renaissance. It was a time some in the nineteenth century dismissed as "a thousand years without a bath."

possibilities for private or governmental patronage, because most scientists do not have the resources to support their work in an open market. In late medieval times, however, the Arab world came under attack by external forces ranging from Genghis Khan to the Crusaders and was torn asunder by internal factional warfare as well. Resources that might once have been devoted to the arts and sciences were now diverted to warfare—and the struggle for survival.

Another reason that the study of science began to stagnate was that the schools that came to dominate a significant segment of intellectual life in the Arab world didn't value it. Known as madrassas, these were charitable trusts sustained by religious endowments, whose founders and benefactors were suspicious of the sciences. As a result, all instruction had to center on religion, to the exclusion of philosophy and science. Any teaching of those subjects thus had to occur outside the colleges. With no institution to support them or bring them together, scientists became isolated from one another, which created a huge barrier to specialized scientific training and research.

Scientists cannot exist in a vacuum. Even the greatest profit immensely from their interaction with others in their field. The lack of peer-to-peer contact in the Islamic world prevented the cross-fertilization of ideas necessary for progress. Also, without the salutary benefits of mutual criticism, it became hard to control the proliferation of theories that lacked empirical grounding, and difficult to find a critical mass of support for those scientists and philosophers with views that challenged the conventional wisdom.

Something comparable to this kind of intellectual suffocation occurred in China, another grand civilization that could conceivably have developed modern science before the Europeans. In fact, China had a population of more than 100 million people during the High Middle Ages (1200–1500), roughly double the population of Europe at the time. But the educational system in China proved, like that of the Islamic world, far inferior to the one that was developing in Europe, at least with respect to science. It was rigidly controlled and focused on literary and moral learning, with little attention paid to scientific innovation and creativity. That situation remained virtually unchanged from the early Ming dynasty (around 1368) until the twentieth century. As in the Arab world, only modest advances in science (as opposed to technology) were achieved, and they came despite, and not because of, the educational system. Thinkers who were critical of the intellectual status quo and who attempted to develop and systematize the intellectual tools necessary to push the life of the mind forward were strongly discouraged, as was the use of data as a means of advancing knowledge. In India, too, a Hindu establishment focused on caste structure insisted on stability at the expense of intellectual advance. As a result, though the Arab world, China, and India did produce great thinkers in other realms, they produced no scientists equivalent to those who, in the West, would create modern science.

* * *

The revival of science in Europe began toward the end of the eleventh century, when the Benedictine monk Constantinus Africanus began to translate ancient Greek medical treatises from Arabic to Latin. As had been the case in the Arab world, the motivation to study Greek wisdom lay in its utility, and these early translations whetted the appetite for the translation of other practical works in medicine and astronomy. Then, in 1085, during the Christian reconquest of Spain, whole libraries of Arabic books fell into Christian hands, and over the next decades, large numbers were translated, thanks in part to generous funding from interested local bishops.

It is hard to imagine the impact of the newly available workers. It was as if a contemporary archaeologist had stumbled on and translated tablets of ancient Babylonian text and found that they presented advanced scientific theories far more sophisticated than our own. Over the next few centuries, sponsorship of the translations became a status symbol among the social and commercial elite of the Renaissance. As a result, the recovered knowledge spread beyond the Church and became a kind of currency, collected as the wealthy today might collect art—and indeed, the wealthy would display their books and maps as one might today display a sculpture or painting. Eventually, the new value placed on knowledge independent of its utilitarian value led to an appreciation of scientific inquiry. In time, this undermined the Church's "ownership" of truth. In competition with truth as revealed in the Scriptures and in Church tradition, there was now a rival truth: the truth as revealed by nature.

But merely translating and reading ancient Greek works does not a "scientific revolution" make. It was the development of a new institution—the university—that would really transform Europe. It would become the driver of the development of science as we know it today, and it would keep Europe in the forefront of science for many centuries, producing the greatest strides in science the world had ever known.

The revolution in education was fueled by growing affluence and a multitude of career opportunities for the well educated. Cities like Bologna, Paris, Padua, and Oxford acquired reputations as centers of learning, and both students and teachers gravitated to them in large numbers. Teachers would set up shop, either independently or under the auspices of an existing school. Eventually, they organized into voluntary associations modeled after the trade guilds. But though the associations called themselves "universities," these were at first simply alliances, owning no real estate and having no fixed location. Universities in the sense that we know them came decades later—Bologna in 1088, Paris around 1200, Padua around 1222, and Oxford by 1250. There, natural science, not religion, would become the focus, and scholars would come together to interact and stimulate one another.

That's not to say that the university in medieval Europe was the Garden of Eden. As late as 1495, for example, German authorities saw the need for a statute explicitly forbidding anyone associated with the university from drenching freshmen with urine, a statute that no longer exists but that I still require my own students to adhere to. Professors, for their part, often had no dedicated classroom and were thus forced to lecture in rooming houses, churches, even brothels. Worse, professors were commonly paid directly by the students, who could also hire and fire them. At the University of Bologna, there was another bizarre twist on what is the norm today: students fined their professors for unexcused absence or tardiness, or for not answering difficult questions. And if a lecture was not interesting or proceeded too slowly or too quickly, they would jeer and become rowdy. Aggressive tendencies got so out of hand in Leipzig that the university had to pass a rule against throwing stones at professors.

Despite these practical hardships, the European universities were great enablers of scientific progress, in part because of the way they brought people together to share and debate ideas. Scientists can withstand distractions like jeering students and perhaps even occasional flying urine, but to go without endless academic seminars—that's just unthinkable. Today, most scientific advances stem from university research, as they must, because that is where the lion's share of funding for basic research goes. But just as important, historically, has been the role of the university as a gathering place of minds.

The scientific revolution that would distance us from Aristotelianism, transform our views of nature and even society, and lay the groundwork for who we are today is often said to have begun with Copernicus's heliocentric theory and to have culminated in Newton's physics. But that picture is oversimplified—though I use the term "scientific revolution" as a convenient shorthand, the scientists involved had a wide variety of goals and beliefs, rather than being a united bunch deliberately striving together to create a new system of thought. Even more important, the changes the "scientific revolution" refers to were actually gradual: the great scholars of 1550 to 1700 who built the grand cathedral of knowledge whose pinnacle was Newton did not emerge from nowhere. It was the medieval thinkers at the early universities of Europe who did the backbreaking work of digging the foundation for them.

The greatest of that work was accomplished by a group of mathematicians at Merton College, Oxford, between 1325 and 1359. Most people know, at least vaguely, that the Greeks invented the idea of science and that it was in the time of Galileo that modern science came into being. Medieval science, though, gets little respect. That's a shame, because medieval scholars made surprising progress, despite living in an age in which people routinely judged the truth of statements not according to empirical evidence but by how well they fit into their preexisting system of religion-based beliefs—a culture that is inimical to science as we know it today.

Philosopher John Searle wrote about an incident that illustrates the fundamentally different terms in which we and medieval thinkers see the world. He told of a Gothic church in Venice called the Madonna del Oro (Madonna of the Orchard). The original plan was to call it the church of San Christoforo, but while it was being built a statue of the Madonna mysteriously turned up in an adjoining orchard. The name change came because the statue was assumed to have fallen from heaven, an event that was considered a miracle. There was no more doubt, then, about that supernatural explanation than there would be, now, about the earthly interpretation that we would assign such an incident. "Even if the statue were found in the gardens of the Vatican," Searle wrote, "the church authorities would not claim it had fallen out of heaven."

I brought up the accomplishments of medieval scientists at a party

The library at Merton College, Oxford

once. I said I was quite impressed by their work, given their culture and the hardships they faced. We scientists today complain about the time "wasted" applying for grants, but at least we have heated offices and don't have to hunt cats for dinner when our town's agricultural production takes a dip. Not to mention having to escape the Black Death, which came in 1347 and killed half the population.

The party I was at was heavy on academics, so the person I was talking to didn't react to my musings the way most people would have—by suddenly realizing she needed to go refill her chardonnay. Instead she said, incredulously, "Medieval scientists? Come on. They operated on patients without anesthetic. They made healing potions out of lettuce juice, hemlock, and gall from a castrated boar. And didn't even the great Thomas Aquinas himself believe in witches?" She had me there. I had no idea. But I looked it all up later, and she was right. Yet despite her apparently encyclopedic knowledge of certain aspects of medieval medical scholarship, she hadn't heard of their more lasting ideas in the realm of physical science, which seemed to me all the more miraculous, given the state of medieval knowledge in other fields. And so, though I had to concede that one would not want to be treated by a medieval doctor who traveled to the present-day world in a time machine, I stood my ground with regard to the progress those medieval scholars had made in the physical sciences.

What did they do, these forgotten heroes of physics? To start with, among all the types of change considered by Aristotle, they singled out change of position—that is, motion—as being most fundamental. That was a deep and prescient observation, because most types of change we observe are specific to the substances involved—the rotting of meat, the evaporation of water, the falling of leaves from a tree. As a result, they wouldn't yield much to a scientist looking for the universal. The laws of motion, on the other hand, are fundamental laws that apply to *all* matter. But there's another reason the laws of motion are special: on the submicroscopic level, they are the cause of all the macroscopic changes we experience in our lives. That's because, as we now know—and as some of the ancient Greek atomists had speculated—the many kinds of change we experience in the everyday world can ultimately be understood by analyzing the laws of motion that operate on the fundamental building blocks of materials: atoms and molecules.

Though the Merton scholars did not discover those comprehensive laws of motion, they did intuit that such laws existed, and they set the stage for others to discover them centuries later. In particular, they created a rudimentary theory of motion that had nothing to do with the science of other types of change—and nothing to do with the idea of purpose.

* * *

The task the Merton scholars assumed wasn't an easy one, given that the mathematics required for even the simplest analysis of motion was still primitive at best. But there was another serious handicap, and to overcome it was an even greater triumph than to succeed with the limited mathematics of the era, for it was not a technical barrier but a limitation imposed by the way people thought about the world: the Merton scholars were, like Aristotle, hampered by a worldview in which time played a mostly qualitative and subjective role.

We who are steeped in the culture of the developed world experience the passage of time in a way that people of earlier eras would not recognize. For most of humankind's existence, time was a highly elastic framework that stretched and contracted in a completely private manner. Learning to think of time as anything but inherently subjective was a difficult and far-reaching step forward, as great an advance for science as the development of language or the realization that the world could be understood through reason.

For example, to search for regularities in the timing of events—to imagine that for a rock to fall sixteen feet will always require one second—would have been a revolutionary concept in the era of the Merton scholars. For one, no one had any idea how to measure time with any precision, and the concept of minutes and seconds was virtually unheard of. In fact, the first clock to record hours of equal length wasn't invented until the 1330s. Before that, daylight, however long, had been divided into twelve equal intervals, which meant that an "hour" might be more than twice as long in June as in December (in London, for example, it varied from 38 to 82 of today's minutes). That this didn't bother anyone reflects the fact that people had very little use for anything but a vague and qualitative notion of time's passage. In light of that, the idea of speed—distance traveled per unit of time—must have seemed an oddball one indeed.

Given all the obstacles, it seems miraculous that the Merton scholars managed to create a conceptual foundation for the study of motion. And yet they went so far as to state the first-ever quantitative rule of motion, the "Merton rule": *The distance traversed by an object that accelerates from rest at a constant rate is equal to the distance traversed by an object that moves for the same amount of time at half the accelerating object's top speed.*

Admittedly, that's a mouthful. Although I've long been familiar with it, looking at it now, I had to read it twice just to follow what it was saying. But the opacity of the rule's phrasing serves a purpose, for it illustrates how much easier science would become once scientists learned to use—and invent, if necessary—the appropriate mathematics.

In today's mathematical language, *the distance traversed by an object that accelerates from rest at a constant rate* can be written as $\frac{1}{2} a \times t^2$. The second quantity, *the distance traversed by an object that moves for the same amount of time at half the accelerating object's top speed*, is simply $\frac{1}{2} (a \times t) \times t$. Thus, the above statement of the Merton rule, translated mathematically, becomes: $\frac{1}{2} a \times t^2 = \frac{1}{2} (a \times t) \times t$. That's not just more compact, it also makes the truth of the statement instantly obvious, at least to anyone who has had pre-algebra.

If those days are far behind you, just ask a sixth grader—he or she will understand it. In fact, the average sixth grader today knows far more mathematics than even the most advanced fourteenth-century scientist. Whether an analogous statement will be said of twenty-eighth-century children and twenty-first-century scientists is an interesting question. Certainly human mathematics prowess has, for centuries, been steadily progressing.

An everyday example of what the Merton rule is saying is this: If you accelerate your car steadily from zero to one hundred miles per hour, you'll go the same distance as if you had driven at fifty the entire time. It sounds like my mother nagging me about driving too fast, but though the Merton rule is today common sense, the Merton scholars were unable to prove it. Still, the rule made quite a splash in the intellectual world, and it quickly diffused to France, Italy, and other parts of Europe. The proof came soon after, from across the Channel, where the French counterparts of the Merton scholars worked at the University of Paris. Its author was Nicole Oresme (1320–1382), a philosopher and theologian who would eventually rise to the post of bishop of Lisieux. To achieve his proof, Oresme had to do what physicists throughout history have done repeatedly: invent new mathematics.

If mathematics is the language of physics, a lack of appropriate mathematics makes a physicist unable to speak or even reason about a topic. The complex and unfamiliar math Einstein needed to use to formulate general relativity may be why he once advised a young schoolgirl, "Do not worry about your difficulties in mathematics: I can assure you that mine are still greater." Or, as Galileo put it, "the book [of nature] cannot be understood unless one first learns to comprehend the language and read the letters in which it is composed. It is written in the language of mathematics, and its characters are triangles, circles, and other geometrical figures, without which it is humanly impossible to understand a single word of it; without these, one is wandering around in a dark labyrinth."

To shine a light on that dark labyrinth, Oresme invented a type of diagram intended to represent the physics of the Merton rule. Though he didn't understand his diagrams in the same way we do today, one might consider this to be the first geometric representation of the physics of motion—and, thus, the first graph.

I've always found it strange that many people know who invented calculus, though few people ever use it, while few people know who invented the graph, though everyone uses them. I suppose that's because today the idea of graphs seems obvious. But in medieval times, the idea of representing quantities with lines and forms in space was strikingly original and revolutionary, maybe even a little nutty.

To get an idea of the difficulty in achieving even a simple change in the way people think, I like to remember the story of another nutty invention, a decidedly nonmathematical one: Post-it notes, those small pieces of paper with a strip of reusable adhesive on one side that allows them to be easily attached to things. The Post-it note was invented in 1974 by Art Fry, a chemical engineer at the 3M Company. Suppose, though, that they had not been invented back then, and that today I came to you, an investor, with the idea and a prototype pad of notes. Surely you'd recognize the invention as a gold mine and jump at the opportunity to invest, right?

Outlandish as it may seem, most people probably wouldn't, as evidenced by the fact that when Fry presented his idea to the marketing people at 3M—a company known for both their adhesives and their innovations—they were unenthusiastic and believed they'd have a hard time selling something that would have to command a premium price compared with the scratch paper it was intended to replace. Why didn't they rush to embrace the treasure Fry was offering them? Because in the pre-Post-it era, the notion that you might want to stick scraps of paper striped with weak adhesive on things was beyond people's imagination. And so Albert Fry's challenge was not just to invent the product but to change the way people thought. If that was an uphill battle with regard to the Post-it note, one can only imagine the degree of difficulty one faces when trying to do the same in a context that really matters.

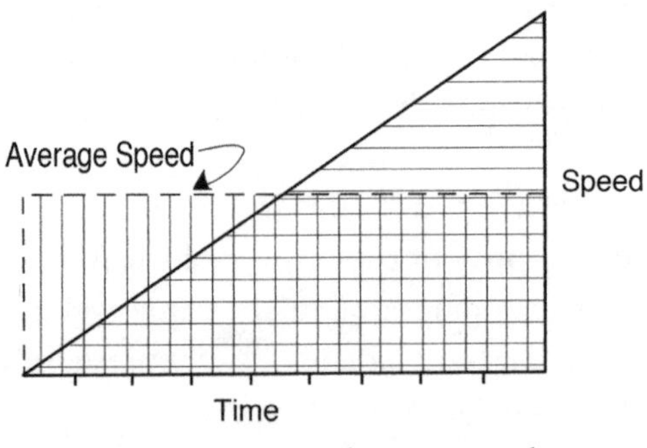

A graph showing the Merton rule

Luckily, Oresme didn't need Post-its for his proof. Here is how we would interpret his arguments. To begin, place time along the horizontal axis and velocity along the vertical axis. Now suppose the object you are considering starts at time zero and moves for a while at constant speed. That motion is represented by a horizontal line. If you shade in the area below that line, you get a rectangle. Constant acceleration, on the other hand, is represented by a line that rises at some angle because as time increases, so does velocity. If you shade in the region below that line, you have a triangle.

The area under these curves—the shaded areas—represents speed multiplied by time, which is the distance the object has traveled. Employing this analysis, and knowing how to calculate the areas of rectangles and triangles, it is then easy to demonstrate the validity of the Merton rule.

One reason Oresme doesn't get the credit he deserves is that he did not publish many of his works. In addition, though I've explained how we would interpret his work today, the conceptual framework he actually used was not nearly as detailed and quantitative as I described, and was utterly different from our modern understanding of the relation between mathematics and physical quantities. That new understanding would spring from a series of innovations regarding the concepts of space, time, velocity, and acceleration that would be one of the most significant contributions of the great Galileo Galilei (1564–1642).

* * *

Though medieval scholars working in the universities of the thirteenth and fourteenth centuries had progressed in furthering the tradition of a rational and empirical scientific method, the grand explosion of European science did not immediately follow. Instead it was the inventors and engineers who transformed European society and culture in late medieval Europe, a period concurrent with the first stirrings of the Renaissance, which spanned roughly from the fourteenth to the seventeenth centuries.

These early Renaissance innovators created the first great civilization not powered primarily by the exertion of human muscle. Waterwheels, windmills, new kinds of mechanical linkages, and other devices were developed or improved and incorporated into village life. They powered sawmills, flour mills, and a variety of clever machines. Their technological innovations had little to do with theoretical science, but they did lay the groundwork for later advances by generating new wealth that helped foster a rise in learning and literacy, and by promoting the realization that an understanding of nature can aid the human condition.

The entrepreneurial spirit of the early Renaissance also saw the invention of one technology that had a direct and major impact on later science, as well as on society in general: the printing press. Though the Chinese had developed movable type centuries earlier—around 1040—it was relatively impractical, due to the use of pictograms in Chinese writing, which created the need for thousands of different characters. In Europe, however, the appearance of mechanical movable-type printing around 1450 changed everything. In 1483, for example, to set up a book for printing the Ripoli Press charged three times what a scribe would charge to copy the book. But with their setup, Ripoli would produce a thousand or more copies, while a scribe would produce only one. As a result, within just a few decades, more books had been printed than the scribes of Europe had produced in all the preceding centuries combined.

The printing press strengthened the emerging middle class and revolutionized the circulation of thoughts and information throughout Europe. Knowledge and information suddenly became available to a far wider group of citizens. Within a few years the first math texts were printed, and by 1600 almost a thousand had been published. In addition, there came a new wave in the recovery of ancient texts. Just as important, those with new ideas suddenly had a far larger forum for their views, and those who, like scientists, thrived on scrutinizing and furthering the ideas of others soon had far greater access to the work of their cohorts.

As a result of these changes in European society, its establishment was less fixed and uniform than that of the Islamic world, China, or India. Those societies had become rigid and focused on a narrow orthodoxy. The European elite, meanwhile, found themselves tugged and bent by the competing interests of town and country, church and state, pope and emperors, as well as the demands of a new lay intelligentsia and growing consumerism. And so, as European society evolved, its arts and its science had more freedom to change, and they did, resulting in a new and more practical interest in nature.

In both the arts and science, that new emphasis on natural reality became the soul of the Renaissance. The term itself is French for "rebirth," and indeed the Renaissance represented a new beginning in both physical existence and culture: it began in Italy just after the Black Death killed between a third and half of Europe's population, then spread slowly, not reaching northern Europe until the sixteenth century.

In art, Renaissance sculptors learned anatomy, and painters learned geometry, both focusing on creating more faithful representations of reality based on keen observation. Human figures were now rendered in natural surroundings, and with anatomical accuracy, and three-dimensionality was suggested through the use of light and shadow and linear perspective. Painters' subjects also now showed realistic emotion, their faces no longer displaying the flat, otherworldly quality of earlier medieval art. Renaissance musicians, meanwhile, studied acoustics, while architects scrutinized the harmonic proportions of buildings. And scholars interested in natural philosophy—what we today call science—placed a new emphasis on gathering data and drawing conclusions from it, rather than employing pure logical analysis biased by a desire to confirm their religious worldview.

Leonardo da Vinci (1452–1519) perhaps best epitomizes the scientific and humanist ideal of that period, which did not recognize a stark separation between science and the arts. A scientist, engineer, and inventor, he was also a painter, sculptor, architect, and musician. In all these pursuits, Leonardo tried to understand the human and natural worlds through detailed observation. His notes and studies in science and engineering consume more than ten thousand pages, while as a painter he wasn't content to merely observe posing subjects but also studied anatomy and dissected human corpses. Where prior scholars had viewed nature in terms of general qualitative features, Leonardo and his contemporaries invested enormous effort in perceiving the fine points of nature's design—and placed less emphasis on the authority of both Aristotle and the Church.

It was into this intellectual climate, toward the end of the Renaissance, that Galileo was born, in Pisa in 1564, just two months before the birth of another titan, William Shakespeare. Galileo was the first of seven children of Vincenzo Galilei, a well-known lute player and music theorist.

Vincenzo came from a noble family—not the kind of noble family that we think of today, people who go on fox hunts and sip tea every afternoon, but the kind that has to use its name to get a job. Vincenzo probably wished he were a nobleman of the first kind, as he loved the lute and played it whenever possible—walking in town, riding a horse, standing at the window, lying in bed—a practice that brought in little in cold cash.

Hoping to guide his son toward a lucrative way of life, Vincenzo sent young Galileo to study medicine at the University of Pisa. But Galileo was more interested in math than in medicine, and he began to take private lessons in the works of Euclid and Archimedes, and even Aristotle. He told friends many years later that he would have preferred to forgo university training and instead take up drawing and painting. Vincenzo, however, had pushed him toward a more practical pursuit on the age-old fatherly theory that it is worth certain compromises to avoid a life where dinner means hemp-seed soup and beef entrails.

When Vincenzo heard that Galileo had turned toward mathematics rather than medicine, it must have seemed as if he'd chosen to major in living off his inheritance, as inadequate as it would be. But it hardly mattered. In the end Galileo didn't complete a degree in medicine, math, or anything else. He dropped out and began a life journey that would indeed find him chronically short of money and often in debt.

After quitting school, Galileo at first supported himself by giving private mathematics lessons. He eventually got wind of an opening for a junior position at the University of Bologna. Though he was twenty-three, he applied, and in a novel twist on rounding he reported his age as "around twenty-six." The university apparently wanted someone "around" a little older and hired a thirty-two-year-old who had also actually finished his degree. Still, even centuries later, it has to be comforting to anyone who has ever been turned down for an academic job that it's an experience you share with the great Galileo.

Two years later, Galileo did become a professor, in Pisa. There, he taught his beloved Euclid and also a course on astrology aimed at helping medical students determine when to bleed patients. Yes, the man who did so much to further the scientific revolution also advised aspiring doctors on what the position of Aquarius means for the placement of leeches. Today astrology has been discredited, but in the age before we knew very much about the laws of nature, the idea that heavenly bodies affect our lives here on earth seemed to be a reasonable one. After all,

Galileo Galilei, as painted by
Flemish artist Justus Sustermans in 1636

it is true of the sun and also the moon, which had long been known to have a mysterious correlation with the tides.

Galileo made astrological forecasts both out of personal interest and for profit, charging his students twelve scudi for a reading. At five per year, he could double his sixty-scudi teaching salary, a sum he could get by on, but just barely. He also liked to gamble, and in an era before anyone knew much about the mathematics of probability, Galileo was not only a pioneer in calculating odds but also a good bluffer.

In his late twenties, tall and stocky, with a fair complexion and reddish hair, Galileo was generally well liked. But his tenure at Pisa didn't last long. Though generally respectful of authority, he was prone to sarcasm and could be scathing to both his intellectual adversaries and administrators who rubbed him the wrong way. The rubbing that inflamed him at Pisa came when the university stubbornly insisted that professors don their academic gowns around town as well as when they were teaching.

Galileo, who liked to write poetry, retaliated by writing a poem for the university authorities. Its subject was clothing—Galileo came out against it. It is, he argued, a source of deceit. For example, without clothes, his verses proclaimed, a bride could look at her prospective mate and "See if he is too small, or has French diseases [and] Thus informed, take or leave him as she pleases." Not a poem likely to endear you to Parisians. It didn't go over well in Pisa, either, and young Galileo was again on the job market.

As it turned out, that was all for the good. Galileo promptly received an appointment near Venice, at the university in Padua, starting at 180 scudi per year, triple his former salary—and he would later describe his sojourn there as the best eighteen years of his life.

By the time Galileo got to Padua, he was disenchanted with Aristotelian physics. For Aristotle, science consisted of observation and theorizing. For Galileo, that was missing a crucial step, the use of experiments, and in Galileo's hands, experimental physics advanced as much as its theoretical side. Scholars had been performing experiments for centuries, but they were generally done to illustrate ideas that they already accepted. Today, on the other hand, scientists perform experiments to rigorously test ideas. Galileo's experiments fell somewhere in between. They were explorations—more than illustrations, but not quite rigorous tests.

Two aspects of Galileo's approach to experiments are especially important. First, when he got a result that surprised him, he didn't reject it—he questioned his own thinking. Second, his experiments were quantitative, a rather revolutionary idea at the time.

Galileo's experiments were much like those you might see performed in a high school science class today, though of course his lab differed from what you'd find in a high school in that it lacked electricity, gas, water, and fancy equipment—and by "fancy equipment" I mean, for example, a clock. As a result, Galileo had to be a sixteenth-century MacGyver, making complex devices from the Renaissance equivalent of duct tape and a toilet bowl plunger. For instance, to create a stopwatch, Galileo poked a small hole in the bottom of a large bucket. When he needed to time an event, he would fill the vessel with water, collect what leaked out, and weigh it—the weight of the water was proportional to the duration of the event.

Galileo employed this "water clock" to attack the controversial issue of free fall—the process by which an object falls to the earth. To Aristotle, free fall was a type of natural motion governed by certain rules of thumb, such as: "If half the weight moves the distance in a given time, its double [the whole weight] will take half the time." In other words, objects fall at a constant speed, which is proportional to their weight.

If you think about it, that's common sense: a rock falls faster than a leaf. And so, given the lack of measuring or recording instruments and the little that was known about the concept of acceleration, Aristotle's description of free fall must have seemed reasonable. But if you think again, it also violates common sense. As Jesuit astronomer Giovanni Riccioli would point out, even the mythological eagle that killed Aeschylus by dropping a turtle on his head knew instinctively that an object dropped on your head will do more damage if dropped from higher up—and that implies that objects speed up as they fall. As a result of such considerations, there was a long tradition of back-and-forth about this issue, with various scholars over the centuries having expressed skepticism about Aristotle's theory.

Galileo was familiar with the criticisms and wanted to do his own investigation into the matter. However, he knew his water clock was not precise enough to allow him to experiment with falling objects, so he had to find a process that was slower yet demonstrated the same physical principles. He settled upon measuring the time it took for highly polished bronze balls to roll down smooth planes that were inclined at various angles.

Studying free fall by taking measurements of balls rolling down ramps is something like buying an outfit according to how it looks on the Internet—there's always the chance that the clothes will look different on the actual you than on the gorgeous model. Despite the dangers, such reasoning lies at the core of the way modern physicists think. The art of designing a good experiment lies largely in knowing which aspects of a problem are important to preserve and which you can safely ignore—and in how to interpret your experimental results.

In the case of free fall, Galileo's genius was to design the rolling-ball experiment with two criteria in mind. First, he had to slow things down enough so that he could measure them; equally important, he sought to minimize the effects of air resistance and friction. Although friction and air resistance are part of our everyday experience, he felt that they obscure the simplicity of the fundamental laws that govern nature. Rocks might fall faster than feathers in the real world, but the underlying laws, Galileo suspected, dictated that in a vacuum they would fall at the same rate. We must "cut loose from these difficulties," he wrote, "and having discovered and demonstrated the theorems, in the case of no resistance . . . use them and apply them [to the real world] . . . with such limitations as experience will teach."

For small tilts, the balls in Galileo's experiment rolled rather slowly, and the data was relatively easy to measure. He noted that with these small angles, the distance covered by the ball was always proportional to the square of the time interval. One can show mathematically that this means the ball had gained speed at a constant rate—that is, the ball was undergoing constant acceleration. Moreover, Galileo noted that the ball's rate of fall did not depend on how heavy it was.

What was striking was that this remained true even as the plane was tilted at ever steeper angles; no matter what angle the tilt, the distance the ball covered was independent of the ball's weight and proportional to the square of the time it took to roll. But if that is true for a tilt of forty, fifty, sixty, even seventy or eighty degrees, why not ninety? And so now comes Galileo's very modern-sounding reasoning: he said that his observations of the ball rolling down the plane must also hold true for free fall, which one can consider as equivalent to the "limiting case" in which the plane is tilted at ninety degrees. In other words, he hypothesized that if he tilted the plane all the way—so that it was vertical and the ball was actually falling rather than rolling—it would still gain speed at a constant rate, which would mean that the law he observed for inclined planes also holds for free fall.

In this way, Galileo replaced Aristotle's law of free fall with his own. Aristotle had said that things fall at a speed that is proportional to their weight, but Galileo, postulating an idealized world in which the fundamental laws reveal themselves, came to a different conclusion: in the absence of the resistance provided by a medium such as air, all objects fall with the same constant acceleration.

* * *

If Galileo had a taste for mathematics, he also had a penchant for abstraction. It was so well developed that he sometimes liked to watch scenes play out entirely in his imagination. Nonscientists call these fantasies; scientists call them thought experiments, at least when they pertain to physics. The good thing about imagining experiments conducted purely in your mind is that you avoid the pesky issue of setting up apparatus that actually works but are nevertheless able to examine the logical consequences of certain ideas. And so, in addition to sinking Aristotle's theory about free fall through his practical experiments with inclined planes, Galileo also employed thought experiments to join the debate regarding one of the other central criticisms of Aristotle's physics, the criticism that concerned the motion of projectiles.

What is it that, after the initial force applied when a projectile is fired, continues to propel it forward? Aristotle had guessed it might be particles of air that rush in behind the projectile and continually push it, but even he was skeptical of that explanation, as we've seen.

Galileo attacked the issue by imagining a ship at sea, with men playing catch in a cabin, butterflies fluttering about, fish swimming in a bowl at rest upon a table, and water dripping from a bottle. He "noted" that these all proceed in exactly the same manner when the ship is in steady motion as when the ship is at rest. He concluded that because everything on a ship moves along with the ship, the ship's motion must be "impressed" upon the objects, so that once the ship is moving, its motion becomes a sort of baseline for everything on it. Couldn't, in the same way, the motion of a projectile be impressed upon the projectile? Could that be what keeps the cannonball going?

Galileo's ruminations led him to his most profound conclusion, another radical break with Aristotelian physics. Denying Aristotle's assertion that projectiles require a reason for their motion—a force—Galileo proclaimed that all objects that are in uniform motion tend to maintain that motion, just as objects at rest tend to stay at rest.

By "uniform motion" Galileo meant motion *in a straight line* and *at a constant speed.* The state of "rest" then, is simply an example of uniform motion in which the velocity happens to be zero. Galileo's observation came to be called the law of inertia. Newton later adapted it to become his first law of motion. A few pages after stating the law, Newton adds that it was Galileo who discovered it—a rare instance of Newton's giving credit to someone else.

The law of inertia explains the problem of the projectile that had plagued the Aristotelians. According to Galileo, once fired, a projectile will remain in motion unless some force acts to stop it. Like Galileo's law of free fall, this law is a profound break from Aristotle: Galileo was asserting that a projectile does not require the continual application of force to keep it in motion; in Aristotle's physics, continued motion in the absence of a force, or "cause," was inconceivable.

On the basis of what I had told him about Galileo, my father, who liked to compare any important person you were talking about to some figure from Jewish history, called Galileo the Moses of science. He said it because Galileo led science out of the Aristotelian desert and toward a promised land. The comparison is especially apt because, like Moses, Galileo did not make it to the promised land himself: he never went so far as to identify gravity as a force or to decipher its mathematical form—that would have to wait for Newton—and he still clung to some of Aristotle's beliefs. For example, Galileo believed in a kind of "natural motion" that isn't uniform and yet needn't be caused by a force: motion in circles around the center of the earth. Galileo apparently believed that it is that type of natural motion that allows objects to keep up with the earth as it rotates.

These last vestiges of Aristotle's system would have to be abandoned before a true science of motion could arise. For reasons such as this, one historian described Galileo's concept of nature as "an impossible amalgam of incompatible elements, born of the mutually contradictory world views between which he was poised."

* * *

Galileo's contributions to physics were truly revolutionary. What he is most famous for today, however, is his conflict with the Catholic Church, based on his assertion that, contrary to the view of Aristotle (and Ptolemy), the earth is not the center of the universe but just an ordinary planet, which, like the others, orbits the sun. The idea of a heliocentric universe had existed as far back as Aristarchus, in the third century B.C., but the modern version can be credited to Copernicus (1473–1543).

Copernicus was an ambivalent revolutionary whose goal was not to challenge the metaphysics of his day, but simply to fix ancient Greek astronomy: what bothered him was that in order to make the earth-centered model work, one had to introduce a great many complicated ad hoc geometric constructions. His model, on the other hand, was much more refined and simpler, even artful. In the spirit of the Renaissance, he appreciated not just its scientific relevance but its aesthetic form. "I think it easier to believe this," he wrote, "than to confuse the issue by assuming a vast number of Spheres, which those who keep Earth at the center must do."

Copernicus first wrote about his model privately in 1514 and then spent decades making astronomical observations that supported it. But like Darwin, centuries later, he circulated his ideas discreetly, only among his most trusted friends, for fear of being scorned by the populace and the Church. And yet if Copernicus felt the danger, he also knew that with the proper politicking, the Church's reaction could be tempered, and when Copernicus finally did publish, he dedicated the book to the pope, with a long explanation of why his ideas were not heresy.

In the end, the point was a moot one, for Copernicus didn't publish his book until 1543, and by then he lay stricken on his deathbed—some say he didn't even see a final printed version of his book until the very day he died. Ironically, even after his book was published, it had little immediate impact until later scientists like Galileo adopted it and began to spread the word.

Though Galileo didn't invent the idea that the earth is not the center of the universe, he contributed something just as important—he used a telescope (which he improvised, based on a much more rudimentary version that had been invented not long before) to find startling and convincing evidence for that view.

It all started by accident. In 1597, Galileo was writing and lecturing in Padua about the Ptolemaic system, giving little indication that he had any doubts about its validity.* Meanwhile, at around the same time, an incident occurred in Holland that reminds us of the importance in science of being in the right place (Europe) at the right time (in this case, just decades after Copernicus). The incident, which would eventually cause Galileo to change his thinking, took place when two children who were playing in the shop of an obscure spectacle maker named Hans Lippershey put two lenses together and looked through them at a distant weathervane atop the town's church. It was magnified. According to what Galileo later wrote about this event, Lippershey looked through the lenses, "one convex and the other concave . . . and noted the unexpected result; and thus [invented] the instrument." He had created a spyglass.

We tend to think of the development of science as a series of discoveries, each leading to the next through the efforts of some solitary intellectual giant with a clear and extraordinary vision. But the vision of the great discoverers of intellectual history is more often muddled than clear, and their accomplishments more indebted to their friends and

*He did, however, have some sympathy for a version of Copernicus's ideas that had been developed by German astronomer (and astrologer) Johannes Kepler, mainly because it supported his own pet theory of the tides (which he ascribed, incorrectly, to the action of the sun). Still, when Kepler urged Galileo to speak out in support, Galileo refused.

colleagues—and luck—than the legends show and than the discoverers themselves often wish to admit. In this instance, Lippershey's spyglass had a magnifying power of only two or three, and when Galileo first heard about it some years later, in 1609, he was unimpressed. He became interested only because his friend Paolo Sarpi, described by historian J. L. Heilbron as a "redoubtable anti-Jesuit polymathic monk," saw potential in the device—he thought that if the invention could be improved, it could have important military applications for Venice, an unwalled city dependent for its survival on early detection of any impending enemy attack.

Sarpi turned for help to Galileo, who, among his many and various ventures to supplement his income, had a sideline making scientific instruments. Neither Sarpi nor Galileo had any expertise in the theory of optics, but, through trial and error, in a few months Galileo developed a nine-power instrument. He gifted it to an awestruck Venetian Senate in exchange for a lifetime extension of his appointment and a doubling of his then salary, to one thousand scudi. Galileo would eventually improve his telescope to a magnifying power of thirty, the practical limit for a telescope of that design (a plano-concave eyepiece and a plano-convex objective).

Around December 1609, by which time Galileo had already developed a telescope with a magnifying power of twenty, he turned it skyward and aimed it at the largest body in the night sky, the moon. That observation, and others he would make, provided the best evidence to date that Copernicus was correct about the earth's place in the cosmos.

Aristotle had claimed that the heavens form a separate realm, made of a different substance and following different laws, which cause all heavenly bodies to move in circles around the earth. What Galileo saw, though, was a moon that was "uneven, rough, and full of cavities and prominences, being not unlike the face of the earth, relieved by chains of mountains and deep valleys." The moon, in other words, did not seem to be of a different "realm." Galileo saw, too, that Jupiter had its own moons. The fact that these moons orbited Jupiter and not the earth violated Aristotle's cosmology, while supporting the idea that the earth was not the center of the universe but merely one planet among many.

I should note here that when I say Galileo "saw" something, I don't mean that he simply put the scope to his eyes, pointed it somewhere, and feasted on a revolutionary new set of images, as if he were watching a show at the planetarium. Quite the contrary, his observations required long periods of difficult and tedious effort, for he had to squint for hours through his imperfect, poorly mounted (by today's standards) glass, and struggle to make sense of what he saw. When he gazed at the moon, for example, he could "see" mountains only by painstakingly noting and interpreting the movement, over a period of weeks, of the shadows they cast. What's more, he could see only one one-hundredth of the surface at a time, so to create a composite map of the whole, he had to make numerous, scrupulously coordinated observations.

Such difficulties illustrate that, with regard to the telescope, Galileo's genius lay not so much in perfecting the instrument as in the way he applied it. For example, when he perceived what seemed to be, say, a lunar mountain, he wouldn't simply trust appearances; he would study the light and shadows and apply the Pythagorean theorem to estimate the mountain's height. When he observed Jupiter's moons, he at first thought they were stars. Only after multiple careful and meticulous observations, and a calculation involving the known motion of the planet, did he realize that the positions of the "stars" relative to Jupiter were changing in a manner that suggested they were circling.

Having made these discoveries, Galileo, though reluctant to enter the theological arena, became eager to be recognized for them. And so he began to devote much energy to publicizing his observations and crusading to replace the accepted cosmology of Aristotle with the sun-centered system of Copernicus. Toward that end, in March 1610, he published *The Starry Messenger*, a pamphlet describing the wonders he had seen. The book was an instant best seller, and though it was (in modern format) only about sixty pages long, it astounded the world of scholars, for it described marvelous, never-seen-before details of the moon and planets. Soon Galileo's fame spread throughout Europe, and everyone wanted to peer through a telescope.

That September, Galileo moved to Florence to take the prestigious position of "chief mathematician of the University of Pisa and philosopher of the grand duke." He retained his prior salary but had no obligation to teach or even reside in the city of Pisa. The grand duke in question was the Grand Duke Cosimo II de' Medici of Tuscany, and Galileo's appointment was as much the result of a campaign to curry favor with the Medicis as it was due to his grand accomplishments. He had even named the newly discovered moons of Jupiter the "Medicean planets."

Soon after the appointment, Galileo fell terribly ill and remained bedridden for months. Ironically, he probably had the "French disease," syphilis, a product of his attraction to Venetian prostitutes. But even while ill, Galileo continued striving to persuade influential thinkers of the validity of his findings. And by the following year, when he had regained his health, his star had risen so high that he was invited to Rome, where he gave lectures about his work.

In Rome, Galileo met Cardinal Maffeo Barberini and was granted an audience at the Vatican with Pope Paul V. It was a triumphant trip in every way, and Galileo seems somehow to have finessed his differences with official Church doctrine so as to cause no offense—perhaps because most of his lectures focused on the observations he had made with his telescope, without much discussion of their implications.

It was inevitable, though, that in his subsequent politicking Galileo would eventually come into conflict with the Vatican, for the Church had endorsed a version of Aristotelianism created by Saint Thomas Aquinas that was incompatible with Galileo's observations and explanations; in addition, unlike his politic predecessor Copernicus, Galileo could be insufferably arrogant, even when consulting theologians regarding Church doctrine. And so, in 1616, Galileo was summoned to Rome to defend himself before various high-ranking officials of the Church.

The visit seemed to end in a draw—Galileo was not condemned, nor were his books banned, and he even had another audience with Pope Paul; but the authorities forbade him to teach that the sun, not the earth, is the center of the universe, and that the earth moves around it rather than vice versa. In the end, the episode would prove to have caused him a huge problem, for much of the evidence used against Galileo in his Inquisition trial seventeen years later would be drawn from the meetings during which the Church officials had explicitly forbidden him to teach Copernicanism.

For a while, however, the tensions eased, especially after Galileo's friend Cardinal Barberini became Pope Urban VIII in 1623. For unlike Pope Paul, Urban held a generally positive view of science, and in the early years of his reign he welcomed audiences with Galileo.

Encouraged by the friendlier atmosphere, with Urban's rise, Galileo began work on a new book, which he finished when he was sixty-eight, in 1632. The fruit of that labor was entitled *Dialogo Sopra i due Massimi Sistemi del Mondo* (in English, *Dialogue Concerning the Two Chief World*

Systems: Ptolemaic and Copernican). But the "dialogue" was extremely one-sided, and the Church reacted—with good reason—as if the book's title were *Why Church Doctrine Is Wrong and Pope Urban Is a Moron.*

Galileo's *Dialogue* took the form of a conversation between friends: Simplicio, a dedicated follower of Aristotle; Sagredo, an intelligent neutral party; and Salviati, who made persuasive arguments for the Copernican view. Galileo had felt comfortable writing the book because he had told Urban about it, and Urban had seemed to approve. But Galileo had assured the pope that his purpose in writing it was to defend the Church and Italian science from the charge that the Vatican had banned heliocentrism out of ignorance—and Urban's approval was based on the proviso that Galileo would present the intellectual arguments on both sides without judgment. If Galileo had indeed striven for that, he failed miserably. In the words of his biographer J. L. Heilbron, Galileo's *Dialogue* "dismissed the fixed-earth philosophers as less than human, ridiculous, small-minded, half-witted, idiotic, and praised Copernicans as superior intellects."

There was another insult. Urban had wanted Galileo to include in the book a disclaimer, a passage affirming the validity of Church doctrine; but instead of stating the disclaimer in his own voice, as Urban had asked, Galileo had the affirmation of religion voiced by his character Simplicio, described by Heilbron as a "nincompoop." Pope Urban, being no nincompoop himself, was deeply offended.

When the stardust had settled, Galileo was convicted of violating the Church's 1616 edict against teaching Copernicanism and forced to renounce his beliefs. His offense was as much about power and the control, or "ownership," of truth as it was about the specifics of his worldview.* For most of those who formed the intellectual elite of the Church recognized that the Copernican view was probably correct; what they objected to was a renegade who would spread that word and challenge the doctrine of the Church.

On June 22, 1633, dressed in the white shirt of penitence, Galileo knelt before the tribunal that had tried him, and bowed to the demand that he affirm the authority of the Scriptures, declaring: "I, Galileo, son of the late Vincenzo Galilei, Florentine, aged seventy years . . . swear

*Indeed, while Galileo was prohibited from teaching Copernicanism, he was allowed to continue his work and use a telescope during his years of house arrest.

that I have always believed, do believe, and by God's help will in the future believe, all that is held, preached, and taught by the Holy Catholic and Apostolic Roman Church."

Despite proclaiming that he had always accepted Church doctrine, however, Galileo went on to confess that he had advocated the condemned Copernican theory even "after an injunction had been judicially intimated" to him by the Church, to the effect that he must, in the Church's words, "abandon the false opinion that the sun is the center of the world and immovable, and that the earth is not the center of the world, and moves . . ."

What's really interesting is the wording of Galileo's confession: "I wrote and printed a book," he said, "in which I discuss this new doctrine already condemned, and adduce arguments of great cogency in its favor." So even while pledging allegiance to the Church's version of the truth, he still defends the content of his book.

In the end, Galileo capitulates by saying that "desiring to remove from the minds of your Eminences, and all faithful Christians, this strong suspicion, justly conceived against me, with sincere heart and unfeigned faith I abjure, curse, and detest the aforesaid errors and heresies . . . and I swear that in the future I will never again say or assert, verbally or in writing, anything that might furnish occasion for a similar suspicion regarding me."

Galileo would not receive as harsh a punishment as the Inquisition had leveled on Giordano Bruno, who had also declared that the earth revolved around the sun and for his heresy was burned at the stake in Rome in 1600. But the trial made the position of the Church quite clear.

Two days later, Galileo was released to the custody of the Florentine ambassador. He spent his last years under a kind of house arrest in his villa at Arcetri, near Florence. While living in Padua, Galileo had fathered three children out of wedlock. Of them, the daughter to whom he had been extremely close had died of the plague in Germany, and his other daughter was estranged from him; but his son, Vincenzo, lived nearby and took loving care of him. And although he was a prisoner, Galileo was allowed visitors, even heretics—provided they were not mathematicians. One of them was the young English poet John Milton (who would later refer to Galileo and his telescope in *Paradise Lost*).

Ironically, it was during his time in Arcetri that Galileo recorded his most fully worked-out ideas on the physics of motion, in the book he considered to be his greatest work: *Discourses and Mathematical Demonstrations Relating to Two New Sciences.* The book could not be published in Italy, due to the pope's ban on his writings, so it was smuggled to Leiden and published in 1638.

By then Galileo's health was failing. He had become blind in 1637, and the next year he began to experience debilitating digestive problems. "I find everything disgusting," he wrote, "wine absolutely bad for my head and eyes, water for the pain in my side . . . my appetite is gone, nothing appeals to me and if anything should appeal [the doctors] will prohibit it." Still, his mind remained active, and a visitor who saw him shortly before his death commented that—despite the prohibition on visitors of that profession—he had recently enjoyed listening to two mathematicians argue. He died at age seventy-seven, in 1642, the year Newton was born, in the presence of his son, Vincenzo—and, yes, a few mathematicians.

Galileo had wanted to be buried next to his father in the main Basilica of Santa Croce, in Florence. Grand Duke Cosimo's successor, Ferdinando, had even planned to build a grand tomb for him there, across from that of Michelangelo. Pope Urban let it be known, however, that "it is not good to build mausoleums to [such men] . . . because the good people might be scandalized and prejudiced with regard to Holy authority." And so Galileo's relatives deposited his remains instead in a closet-size chamber under the church's bell tower and held a small funeral attended only by a few friends, relatives, and followers. Still, many, even within the Church, felt the loss. Galileo's death, the librarian at the court of Cardinal Barberini, in Rome, courageously wrote, "touches not just Florence, but the whole world and our whole century that from this divine man has received more splendor than from almost all the other ordinary philosophers."

7

The Mechanical Universe

When Galileo published his *Discourses and Mathematical Demonstrations Relating to Two New Sciences*, he had brought human culture only to the brink of a new world. Isaac Newton took the last giant steps and in the process completed the blueprint for an entirely new way of thinking. After Newton, science abandoned the Aristotelian view of nature driven by purpose and embraced instead a Pythagorean universe driven by numbers. After Newton, the Ionian assertion that the world could be understood through observation and reason was transformed into a grand metaphor: the world is like a clock, its mechanisms governed by numerical laws that make every aspect of nature precisely predictable, including, many believed, human interactions.

In faraway America, the founding fathers embraced Newtonian thinking in addition to theology to assert, in the Declaration of Independence, that "the Laws of Nature and of Nature's God entitle" people to political self-determination. In France, after the Revolution and its antagonism toward science, Pierre-Simon de Laplace took Newtonian physics to a new level of mathematical sophistication and then proclaimed that, employing Newtonian theory, a superior intellect could "embrace in the same formula the motions of the greatest bodies in the universe and those of the slightest atoms; nothing would be uncertain for it, and the future, like the past, would be present to its eyes."

Today we all reason like Newtonians. We speak of the force of a person's character, and the acceleration of the spread of a disease. We talk of physical and even mental inertia, and the momentum of a sports team. To think in such terms would have been unheard of before Newton; not to think in such terms is unheard of today. Even those who know nothing of Newton's laws have had their psyches steeped in his ideas. And so to study the work of Newton is to study our own roots.

Because Newton's vision of the world is now second nature to us, it takes effort to appreciate the astonishing brilliance of his creation. In fact, in high school, when I was first introduced to "Newton's laws," they appeared so simple that I wondered what all the fuss was about. I found it odd that it had taken one of the smartest people in the history of science many years to create what I, a boy of fifteen, could learn in just a few lectures. How could concepts so easily accessible to me have been so difficult to grasp a few hundred years ago?

My father seemed to understand. While I tell my kids stories of inventions like Post-it notes, my father usually turned to tales of the old country. When people looked at the world hundreds of years ago, he told me, they saw a reality very different from what we perceive today. He told me of the time when, as a teenager in Poland, he and some friends had put sheets over a goat, which then raced through his family's home. The elders all thought they had seen a ghost. Okay, it *was* the night of the Jewish holiday Purim, and the elders were all rather drunk, but my father didn't use their inebriation to explain away their reaction—he said they were merely interpreting what they saw in terms of the context of their beliefs, and the ghost concept was one they were used to and comfortable with. I might consider that ignorant, he said, but what Newton would tell the world about the mathematical laws of the universe probably seemed just as strange to people of that day as his elders' ghosts do to me. It's true: today, even if you've never taken a course in physics, a bit of Isaac Newton's spirit resides within you. But had we not been raised in a Newtonian culture, those laws that are now so self-evident to us all would have been, for most of us, incomprehensible.

* * *

Describing his life, shortly before his death, Newton put his contributions this way: "I don't know what I may seem to the world, but, as to myself, I seem to have been only like a boy playing on the sea shore, and diverting myself in now and then finding a smoother pebble or a prettier shell than ordinary, whilst the great ocean of truth lay undiscovered before me."

Each of Newton's pebbles could have constituted a monumental career for scholars less brilliant and productive than him. In addition to his work on gravity and motion, he devoted many years to uncovering the secrets of optics and light, and he invented physics as we know it today, as well as calculus. When I told this to my father—who hadn't heard of Newton until I began studying his work—he frowned and said, "Don't be like him. Stick to one field!" At first I reacted to that with the kind of condescension teenagers specialize in. But actually my father might have had a good point. Newton did come perilously close to being a genius who started much and finished nothing. Luckily, as we'll see, fate intervened, and today Newton is credited with ushering in an entire revolution of thought.

One thing Newton never did do, actually, was play at the seashore. In fact, though he profited greatly from occasional interaction with scientists elsewhere in Britain and on the Continent—often by mail—he never left the vicinity of the small triangle connecting his birthplace, Woolsthorpe, his university, Cambridge, and his capital city, London. Nor did he seem to "play" in any sense of the word that most of us mean by it. Newton's life did not include many friends or family he felt close to, or even a single lover, for, at least until his later years, getting Newton to socialize was something like convincing cats to gather for a game of Scrabble. Perhaps most telling was a remark by a distant relative, Humphrey Newton, who served as his assistant for five years and said that he saw Newton laugh only once—when someone asked him why anyone would want to study Euclid.

Newton had a purely disinterested passion for understanding the world, not a drive to improve it for the benefit of humankind. He achieved much fame in his lifetime but had no one to share it with. He achieved intellectual triumph but never love. He received the highest of accolades and honors but spent much of his time in intellectual quarrel. It would be nice to be able to say that this giant of intellect was an empathetic, agreeable man, but if he had any such tendencies, he did a good job suppressing them and coming off as an arrogant misanthrope. He was the kind of man who, if you said it was a gray day, would say, "No, actually the sky is blue." Even more annoying, he was the kind who could prove it. Physicist Richard Feynman (1918–1988) voiced the feelings of many a self-absorbed scientist when he wrote a book titled *What Do You Care What Other People Think?* Newton never wrote a memoir, but if he had, he probably would have called it *I Hope I Really Pissed You Off*, or maybe *Don't Bother Me, You Ass.*

Stephen Hawking once told me that there was a sense in which he was glad to be paralyzed, because it allowed him to focus much more intensely on his work. I suppose that Newton could have said, for the same reason, that there were marvelous advantages to living entirely in his own world rather than wasting time by sharing it with anyone else. In fact, a recent study reports that students who are brilliant in math have a significantly greater tendency to go into a scientific career if they don't have great verbal skills. I have long suspected that having poor social skills is also correlated to success in science. I've certainly known quite a few successful scientists who'd have been considered too odd to employ anywhere except at a major research university. One fellow graduate student wore the same pants and white T-shirt every day, though it was rumored he actually had two sets of them, so his clothes did occasionally get washed. Another fellow, a famous professor, was so shy that when you spoke to him, he'd generally avert his eyes, speak very softly, and step back if he noticed that you were standing closer than four feet away. The latter two behaviors caused issues during post-seminar chitchat, because they made him difficult to hear. At our first encounter, during my graduate school days, I made the mistake of approaching too closely, and then naively following him as he backtracked, with the result that he almost fell over a chair.

Science is a subject of magnificent beauty. But although progress in science requires the cross-fertilization of ideas that can result only from interaction with other creative minds, it also demands long hours of isolation, which may confer a distinct advantage on those who would rather not socialize, or even prefer to live apart. As Albert Einstein wrote, "One of the strongest motives that leads men to art and science is escape from everyday life with its painful crudity and hopeless dreariness. . . . Each makes this cosmos and its construction the pivot of his emotional life in order to find in this way the peace and security which he cannot find in the narrow whirlpool of personal experience."

Newton's disdain for the everyday pursuits of the world allowed him to pursue his interests with few distractions, but it also led him to withhold much of his scientific work, choosing not to publish the vast bulk of his writings. Fortunately, he didn't throw them out, either—he was a pack rat worthy of his own reality show, only, instead of hoarding pet carcasses, old magazines, and shoes he'd outgrown at age seven, Newton's "stuff" consisted of scribblings on everything from mathematics, physics, alchemy, religion, and philosophy to accounts of every penny he ever spent and descriptions of his feelings about his parents.

Newton saved virtually everything he wrote, even sheets of throwaway calculations and old school notebooks, making it possible, for those who wished to dig, to understand to an unprecedented degree the evolution of Newton's scientific ideas. Most of his scientific papers were eventually donated to the library at Cambridge, his intellectual home. But other papers, totaling millions of words, were eventually sold at Sotheby's, where economist John Maynard Keynes was one of the bidders, buying most of Newton's writings on alchemy.

Newton's biographer Richard Westfall spent twenty years studying his life, concluding in the end that Newton was "not finally reducible to the criteria by which we comprehend our fellow human beings." But if Newton was an alien, at least he was an alien who left behind diaries.

* * *

Newton's struggle to understand the world stemmed from an extraordinary curiosity, an intense drive toward discovery that seemed to come completely from within, like the impulse that drove my father to trade his piece of bread for the solution to that math puzzle. But in Newton's case there was something else that fueled that drive. Though he is revered as a model of scientific rationality, his inquiries into the nature of the universe were, like those of others going all the way back to Göbekli Tepe, intricately tied to his spirituality and religion. For Newton believed that God is revealed to us through both his word and his works, so that to study the laws of the universe is to study God, and the zeal for science just a form of religious zeal.

Newton's penchant for solitude and his long hours of work were, at least from the point of view of his intellectual achievements, great strengths. If his retreat into the realm of the mind was a boon for science, however, it came at a great cost to the man and seems to have been connected to the loneliness and pain of his childhood.

When I was in school, I felt bad for the kids who weren't popular, especially since I was one of them. Newton, though, had it worse. He was unpopular with his own *mother.* He had come into the world on December 25, 1642, like one of those Christmas gifts you hadn't put on your list. His father had died a few months earlier, and his mother,

Hannah, must have thought that Isaac's existence would prove a short-lived inconvenience, for he was apparently premature and not expected to survive. More than eighty years later, Newton told his niece's husband that he was so tiny at birth that he could have fit into a quart pot, and so weak he had to have a bolster around his neck to keep it on his shoulders. So dire was the little bobblehead's situation that two women who were sent for supplies a couple of miles away dawdled, certain that the child would be dead before they returned. But they were wrong. The neck bolster was all the technology needed to keep the infant alive.

If Newton never saw the use of having people in his life, perhaps that was because his mother never seemed to have much use for *him*. When he was three, she married a wealthy rector, the Reverend Barnabas Smith. More than twice Hannah's age, Smith wanted a young wife, but not a young stepson.

One can't be sure what kind of family atmosphere this led to, but it's probably safe to assume there were some tensions, since, years later, in notes he wrote about his childhood, Isaac recalled "threatening my father and mother Smith to burne them and the house over them." Isaac did not say how his parents had reacted to his threat, but the record shows that he was soon banished to the care of his grandmother. Isaac and she got along better, but the bar had been set pretty low. They certainly weren't close—in all the writings and scribbles Isaac left behind, there is not a single affectionate recollection of her. On the bright side, there are also no recollections of his wanting to set her on fire and burn the house down.

When Isaac was ten, the Reverend Smith died and he returned home briefly, joining a household that now included the three young children from his mother's second marriage. A couple of years after Smith's death, Hannah shipped him off to a Puritan school in Grantham, eight miles from Woolsthorpe. While studying there, he boarded in the home of an apothecary and chemist named William Clark, who admired and encouraged Newton's inventiveness and curiosity. Young Isaac learned to grind chemicals with a mortar and pestle; he measured the strength of storms by jumping into and against the wind and comparing the distance of his leaps; he built a small windmill adapted to be powered by a mouse running on a treadmill, and a four-wheeled cart he would sit in and power by turning a crank. He also created a kite that carried a lit lantern on its tail, and flew it at night, frightening the neighbors.

Though he got along well with Clark, his classmates were a different story. At school, being different and clearly intellectually superior brought Newton the same reaction as it would today: the other kids hated him. The lonely but intensely creative life he led as a boy was preparation for the creative but tortured and isolated life he would lead for most—though happily not all—of his adult life.

As Newton approached the age of seventeen, his mother pulled him out of school, determined that he should return home to manage the family estate. But Newton did not do well as a farmer, proving that you can be a genius at calculating the orbits of the planets and a total klutz when it comes to growing alfalfa. What's more, he didn't care. As his fences fell into disrepair and his swine trespassed in cornfields, Newton built waterwheels in a brook or just read. As Westfall writes, he rebelled against a life spent "herding sheep and shoveling dung." Most physicists I know would.

Fortunately, Newton's uncle and his old schoolmaster from Grantham intervened. Recognizing Isaac's genius, they had him sent off to Trinity College, Cambridge, in June 1661. There he would be exposed to the scientific thinking of his time—only to one day rebel and overturn it. The servants celebrated his parting—not because they were happy for him but because he had always treated them harshly.

* * *

Cambridge would remain Newton's home for more than three and a half decades, ground zero for the revolution in thought that he launched during that time. Though that revolution is often portrayed as consisting of a series of epiphanies, his struggle to master the secrets of the universe was really more like trench warfare—one grueling intellectual battle after another, in which ground was won gradually and at great costs in energy and time. No one of lesser genius, or less fanatical dedication, could possibly have prevailed in that struggle.

At first, even Newton's living conditions were a source of struggle. When Isaac went off to Cambridge, his mother granted him a stipend of just ten pounds—though she herself had a comfortable yearly income of more than seven hundred pounds. The stipend landed him at the very bottom of the Cambridge University social structure.

A sizar, in the rigid Cambridge hierarchy, was a poor student who received free food and tuition and earned a small amount of money by attending to wealthier students: dressing their hair, cleaning their boots, bringing them bread and beer, and emptying their chamber pots. To be a sizar would have been a promotion for Newton: he was what was called a subsizar, someone who had the same menial duties as a sizar but had to pay for his own food as well as for the lectures he attended. It must have been hard for Newton to swallow becoming the servant of the same breed of boys who'd always tormented him at the Grantham school. And so at Cambridge he got a taste of what life was like "downstairs."

In 1661, Galileo's *Discourses and Mathematical Demonstrations Relating to Two New Sciences* was just over two decades old, and like his other works, it hadn't yet had much effect on the Cambridge curriculum. Which meant that in exchange for his service and his fees, Newton was treated to lessons that covered everything scholars knew about the world, as long as those scholars were Aristotle: Aristotelian cosmology, Aristotelian ethics, Aristotelian logic, Aristotelian philosophy, Aristotelian physics, Aristotelian rhetoric . . . He read Aristotle in the original, he read textbooks on Aristotle, he read all the books in the established curriculum. He finished none of them, for, like Galileo, he did not find Aristotle's arguments convincing.

Still, Aristotle's writings constituted the first sophisticated approach to knowledge that Newton was exposed to, and even as he refuted them, he learned from the exercise how to approach the diverse issues of nature and to think about them in an organized and coherent manner—and with astounding dedication. In fact, Newton, who was celibate and rarely engaged in recreational activities, worked harder than anyone I've ever heard of—eighteen hours per day, seven days a week. It was a habit he would adhere to for many decades.

Dismissive of all the Aristotle studies that made up the Cambridge curriculum, Newton began his long journey toward a new way of thinking in 1664, when his notes indicate that he had initiated his own program of study, reading and assimilating the works of the great modern European thinkers, among them Kepler, Galileo, and Descartes. Not a terribly distinguished student, Newton nonetheless managed to graduate in 1665 and to be awarded the title of scholar, along with four years of financial support for additional studies.

Then, in the summer of 1665, a terrible outbreak of plague afflicted Cambridge and the school closed down, not to reopen again until spring

1667. While the school was closed, Newton retreated to his mother's home in Woolsthorpe and continued his work in solitude. In some histories, the year 1666 is called Newton's annus mirabilis. According to that lore, Newton sat at the family farm, invented calculus, figured out the laws of motion, and, after seeing a falling apple, discovered his universal law of gravitation.

True, that wouldn't have been a bad year. But it didn't happen that way. The theory of universal gravitation wasn't as simple as a single bright idea that could be had through an epiphany—it was an entire body of work that formed the basis of a whole new scientific tradition. What's more, that storybook image of Newton and the apple is destructive, because it makes it seem as if physicists make progress through huge and sudden insights, like someone who's been hit on the head and can now predict the weather. In reality, even for Newton, progress required many hits on the head, and many years in which to process his ideas and come to a true understanding of their potential. We scientists endure the headaches from those hits because, like football players, we love our sport more than we hate the pain.

One reason most historians doubt the story of the miraculous epiphanies is that Newton's insights into physics during the plague period came not all at once, but over a period of three years—1664 to 1666. Moreover, there was no Newtonian revolution at the end of that period: in 1666, even Newton was not yet a Newtonian. He still thought of uniform motion as arising from something internal to the moving body, and by the term "gravity," he meant some inherent property arising from the material an object is made of, rather than an external force exerted by the earth. The ideas he developed then were only a beginning, a beginning that left him baffled and floundering about many things, including force, gravity, and motion—all the basics that would eventually constitute the subject of his great work, *Principia Mathematica.*

We have a pretty good idea what Newton was thinking on the farm in Woolsthorpe because, as was his habit, he wrote it all down in a huge, mostly blank notebook he had inherited from Reverend Smith. Newton was lucky to have that notebook and, in his later years, to have sufficient paper for the millions of words and mathematical notations in which he recorded his work.

I've mentioned innovations such as the university and the use of mathematical equations, but there are other unsung enablers of the scientific revolution that we take for granted, prominent among them the growing availability of paper. Fortunately for Newton, the first commercially successful paper mill in England had been established in 1588. Just as important, the Royal Mail service was opened to the public in 1635, making it possible for the antisocial Newton to correspond on paper with other scientists, even those in far-flung places. But paper was still dear in Newton's day, and he treasured the notebook, which he called the "Waste Book." In it we find the details of Newton's approach to the physics of motion, a rare glimpse into the still developing ideas of a brilliant mind.

We know, for example, that on January 20, 1665, Newton began entering into his Waste Book an extensive mathematical—rather than philosophical—investigation of motion. Crucial to the analysis was his development of calculus, a new kind of mathematics designed for the analysis of change.

In the tradition of Oresme, Newton conceived of change as the slope of a curve. For example, if you graph the distance an object has traveled on the vertical axis, against time on the horizontal axis, then the slope of the graph represents its speed. A flat line thus represents an unchanging position, while a steep line or curve indicates that the object's position is changing drastically—that it is moving at high speed.

But Oresme and others interpreted graphs in a more qualitative man-

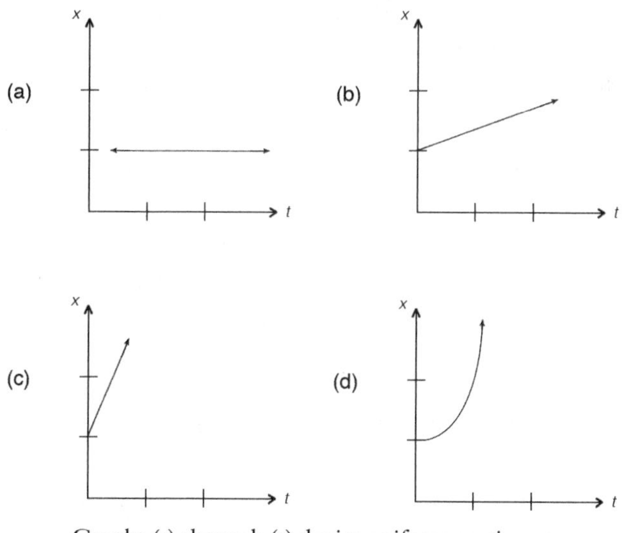

Graphs (a) through (c) depict uniform motion at (a) zero speed (stationary), (b) low speed, and (c) high speed. Graph (d) indicates accelerated motion.

ner than we do today. A graph of distance versus time, for example, was not understood to represent, *at each point*, the distance traveled at the time indicated by the coordinate on the horizontal axis. Nor was the slope of that graph understood to represent the object's speed *at each moment*. Instead, speed to physicists before Newton meant *average* speed—the total distance traveled, divided by the length of time the journey took. Those were pretty crude calculations, because the times they considered in those calculations were typically hours, days, or even weeks. In fact, it wasn't possible to measure short times with any precision until 1670, when English clockmaker William Clement invented the pendulum-based grandfather's clock, which finally made it possible to measure time to the nearest second.

To move beyond averages to the values of graphs and their slopes at each individual point was the revelation in Newton's analysis. He addressed an issue no one had ever addressed before: How do you define the *instantaneous* speed of the object, its speed at *each instant*? How do you divide the distance traveled by the time it took when the time interval involved is actually a single point? Does this even make sense? In his Waste Book, Newton attacked this problem.

If Galileo liked to envision "limiting cases," such as a plane that is tilted more and more until it approaches vertical, Newton took the idea to a whole new extreme. To define the instantaneous speed at a given point of time, he imagined first calculating the average speed in the traditional way, over some interval of time that included the point in question. Then he imagined something new and abstract: shrinking that interval more and more, until, in the limiting case, its size approaches zero.

In other words, Newton imagined that you could make a time interval so small that it would be tinier than any finite number—and yet greater than zero. Today we call the length of such an interval "infinitesimal." If you calculate the average speed over a time interval and shrink that interval down to the infinitesimal, you get the speed of the object at an instant, its instantaneous speed.

The mathematical rules for finding the instantaneous speed at a given time—or, more generally, the slope of a curve at a given point—form the basis of calculus.* If atoms are the indivisibles from which chemical

*Differential calculus, to be precise. There is also the reverse of the process: integral calculus. The term "calculus" used alone is usually meant to encompass both.

compounds are made, infinitesimals are like indivisibles that build space and time.

With his calculus, Newton had invented the mathematics of change. With regard to motion in particular, he introduced a sophisticated understanding of instantaneous velocity to a culture that had only recently created its first-ever way of measuring speed—by tossing a knotted rope anchored by a log off the stern of a ship and counting the number of *knots* that go past in a given time. Now, for the first time, it made sense to speak of an object's speed—or the change of anything—at a particular moment.

Today calculus is used to describe changes of all sort—the way air flows over airplane wings; the way populations grow and weather systems evolve; the ups and downs of the stock market; the evolution of chemical reactions. In any enterprise in which you might graph a quantity, in all areas of modern science, calculus is a key tool.*

Calculus would eventually allow Newton to relate the amount of force applied to an object at any given time to its change in velocity at that moment. Moreover, it would show how one could add up all those infinitesimal changes in velocity to derive an object's path as a function of time. But those laws and methods would not come for decades.

In physics, as in mathematics, Newton's Waste Book went far beyond anything yet envisioned. Prior to Newton, for example, the collision of objects was looked at as a kind of contest between the internal constitutions of the two, like two muscular gladiators vying to toss each other out of the arena. In Newton's way of thinking, though, each body is analyzed only in terms of the external cause applied to it—that is, force.

Despite that advance in thinking, in the more than one hundred axioms of his Waste Book related to this problem, Newton provides only a flawed and convoluted picture of what he meant by "force." In particular, he gives no clue as to how to quantify force, such as that exercised by the pull of the earth or that causing the "change of motion" of an object. The picture Newton began to paint during the years at Woolsthorpe is a picture he wouldn't perfect for almost twenty years—and a far cry from the spark needed for the Newtonian revolution.

*Technically, population growth and stock market prices, being discrete and not continuous quantities, are not governed by calculus, but these systems are often approximated as continuous.

* * *

Physicist Jeremy Bernstein tells a story about a visit to the United States in 1958 by Austrian physicist Wolfgang Pauli. Pauli presented a theory to an audience at Columbia University that included Niels Bohr, who seemed skeptical. Pauli conceded that at first sight his theory might look somewhat crazy, but Bohr replied that, no, the problem was that the theory was not crazy *enough*. At that, Pauli turned to the audience and argued, "Yes, my theory *is* crazy enough!" Then Bohr insisted, "No, your theory is *not* crazy enough!" Soon the two famous physicists were both stalking around the front of the room, yelling and sounding like fifth graders.

I bring this up to make the point that all physicists—and all innovators—have more wrong ideas than right ones, and if they are very good at what they do, they also have crazy ideas, which are the best kind—though only if they are right, of course. Sorting out the right from the wrong is not an easy process; it can take a lot of time and effort. We should therefore have some sympathy for people with outlandish ideas. In fact, Newton was one of them: after having made such an auspicious start during the plague, he spent much of the next phase of his life pursuing wrong ideas that many later scholars who studied his work have thought of as crazy.

It all started off well enough. In spring 1667, shortly after Cambridge reopened, Newton returned to Trinity College. That fall, Trinity held an election. We all, at times, face discrete situations that will have an outsize effect on our future—personal challenges, job interviews that can change our lives, college or professional school entrance exams that may have a great influence on our later opportunities. The Trinity College election was, for Newton, all this rolled into one: its result would determine whether the twenty-four-year-old would be able to remain at the university in a higher position called "fellow" or have to return to a life of herding sheep and shoveling dung. His prospects didn't look good, for there hadn't been elections at Trinity College in three years, and there were only nine openings, but far more candidates, many of whom had political connections. Some even had letters signed by the king *commanding* their election. But Newton did get elected.

With his farming career now behind him for good, one might think Newton would have buckled down and proceeded with the business of turning his Waste Book thoughts on calculus and motion into Newton's Laws. But he didn't. Instead, over the next few years, Newton did notable work in two very different fields—optics and mathematics, in particular algebra. The latter paid off handsomely in that he was soon looked upon as a genius by the small community of Cambridge mathematicians. As a result, when the influential Isaac Barrow quit his prestigious post as Lucasian professor of mathematics—the position Stephen Hawking would hold a few centuries later—Barrow effectively arranged for Newton to take his place. The salary was magnificent by the standards of the era: Newton's university was now willing to grant him ten times what his mother had been willing to provide—one hundred pounds per year.

Newton's efforts in optics didn't work out as well for him. While still a student, he had read recent works on optics and light by Oxford scientists Robert Boyle (1627–1691), who was also a pioneer in chemistry, and Robert Hooke (1635–1703), a "crooked and pale faced" man who was a good theorist but a brilliant experimenter, as he had shown in his work as Boyle's assistant. The work of Boyle and Hooke inspired Newton, though he never admitted it. But soon he was not only calculating; he was experimenting, and he was grinding glass and making improvements to the telescope.

Newton attacked the study of light from all angles. He stuck a needle-like bodkin in his eye and pressed with it until he saw white and colored circles. Did light come from pressure? He stared at the sun for as long as he could stand it—so long it took him days to recover—and he noted that when he looked away from the sun, colors were distorted. Was light real or a product of the imagination?

To study color in the laboratory, Newton made a hole in the shutter of the single window in his study and let in a sunbeam. Its white light, philosophers thought, was the purest kind, light that is completely colorless. Hooke had sent such light through prisms and observed that from the prisms came colored light. He concluded that transparent substances like the prism produce color. But Newton sent the light through prisms, too, and he reached a different conclusion. He noted that although prisms split white light into colors, when fed with colored light, they left the colors unchanged. In the end, Newton concluded that the glass was not *producing* color, but—by bending the beam differently for different colors—it was *separating* white light into the colors that compose it. White light was not pure, Newton proclaimed, but a mix.

Such observations led Newton to a theory of color and light, which he worked out between 1666 and 1670. The result was the conclusion—it infuriated him when Hooke called it a hypothesis—that light is made up of rays of tiny "corpuscles," like atoms. We know now that the specifics of Newton's theory are wrong. True, the idea of light corpuscles would be resurrected by Einstein a few hundred years later—today they are called photons. But Einstein's light corpuscles are quantum particles and don't obey Newton's theory.

Though Newton's work on the telescope brought him fame, the idea of light corpuscles was met in Newton's day, as it would be in Einstein's, by great skepticism. And in the case of Robert Hooke, whose theory had described light as consisting of waves, it was met with hostility. What's more, Hooke complained about Newton creating mere variations on the experiments Hooke had previously performed and passing them off as his own.

Years of skipped meals and sleepless nights investigating optics had for Newton led to an intellectual battle that quickly became bitter and vicious. To make matters worse, Hooke was a brash man who shot from the hip and composed his responses to Newton in just a few hours, while Newton, meticulous and careful in all things, felt the need to put a great deal of work into his replies. In one instance he spent months.

Personal animosity aside, here was Newton's introduction to the social side of the new scientific method—the public discussion and disputation of ideas. Newton had no taste for it. Already one who tended toward isolation, he withdrew.

Bored by mathematics and furious with the criticism of his optics, by the mid-1670s Newton—in his early thirties but with hair that was already gray and usually uncombed—virtually cut himself off from the entire scientific community. He would remain cut off for the next decade.

Conflict averseness was not the only cause for his new, almost total isolation: over the prior several years, even as he'd worked on mathematics and optics, Newton had begun to turn the focus of his hundred-hour workweeks to two new interests, interests he wasn't anxious to discuss with anyone. These are the "crazy" research programs for which he has since been so often criticized. And indeed, they were decidedly outside the mainstream: the mathematical and textual analysis of the Bible, and alchemy.

To later scholars, Newton's decision to devote himself to work on theology and alchemy has often seemed incomprehensible, as if he had given up submitting articles to *Nature* in favor of writing brochures for Scientologists. Those judgments, though, don't take into account the true scope of the enterprise, for what united his efforts in physics, theology, and alchemy was a common aim: the struggle to learn the truth about the world. It is interesting to consider those efforts briefly—not because they proved correct, nor because they prove that Newton had bouts of insanity, but because they highlight the often thin line between scientific inquiry that proves fruitful and that which does not.

Newton believed the Bible promised that the truth would be revealed to men of piety, though certain elements of it might not be apparent from a simple reading of the text. He also believed that pious men of the past, including great alchemists like the Swiss physician Paracelsus, had divined important insights and included them in their works in a coded form to hide them from the unfaithful. After Newton derived his law of gravity, he even became convinced that Moses, Pythagoras, and Plato had all known it before him.

That Newton would turn his ideas into a mathematical analysis of the Bible is understandable, given his talents. His work led him to what he considered to be precise dates for the creation, Noah's Ark, and other biblical events. He also calculated, and repeatedly revised, a Bible-based prediction for the end of the world. One of his final predictions was that the world would end sometime between 2060 and 2344. (Don't know if that will prove true, but, strangely, it does fit neatly into some scenarios of global climate change.)

In addition, Newton came to doubt the authenticity of a number of passages and was convinced that a massive fraud had corrupted the legacy of the early Church to support the idea of Christ as God—an idea he considered idolatrous. In short, he did not believe in the Trinity, which is ironic, given that he was a professor at Trinity College. It was also dangerous, for he almost certainly would have lost his post, and perhaps much more, had word of his views gotten out to the wrong people. But while Newton was committed to reinterpreting Christianity, he was very circumspect about allowing his work to be exposed to the public—despite the fact that it was this work on religion, and not his revolutionary work in science, that Newton regarded as his most important.

Newton's other passion in those years—alchemy—also consumed an enormous amount of time and energy, and those studies would continue for thirty years, far more time than he ever devoted to his work on physics. They consumed money as well, for Newton assembled both an alchemy laboratory and a library. Here, too, we would be mistaken to simply dismiss his efforts as nonscientific, for, as in his other pursuits, his investigations were carefully undertaken and, given Newton's underlying beliefs, well reasoned. Again, Newton came to conclusions that are difficult for us to understand, because his reasoning was embedded in a larger context that we find completely unfamiliar.

Today we think of alchemists as robed, bearded men who chant incantations as they try to turn nutmeg into gold. Indeed, alchemy's earliest known practitioner was an Egyptian named Bolos of Mendes, who lived around 200 B.C. and ended every "experiment" with the incantation "One nature rejoices in another. One nature destroys another. One nature masters another." Sounds like he's listing the different things that can happen when two people get married. But the natures Bolos was talking about were chemicals, and he did indeed have some understanding of chemical reactions. Newton believed that in the distant past, scholars like Bolos had discovered profound truths that had since been lost but could be recovered by analyzing Greek myths, which he was convinced were alchemical recipes written in code.

In his alchemical investigations, Newton maintained his meticulous scientific approach, conducting myriad careful experiments and taking copious notes. And so the future author of the *Principia*—often called the greatest book in the history of science—also spent years scribbling notebooks full of laboratory observations such as these: "Dissolve volatile green lion in the central salt of Venus and distill. This spirit is the green lion the blood of the green lion Venus, the Babylonian Dragon that kills everything with its poison, but conquered by being assuaged by the Doves of Diana, it is the Bond of Mercury."

When I started my career in science, I idolized all the usual heroes—the Newtons and Einsteins of history and the contemporary geniuses like Feynman. To enter a field that has produced all those greats can put a lot of pressure on a young scientist. I felt that pressure when I first received my faculty appointment at Caltech. It felt like the night before my first day at junior high, when I worried about attending gym class, and especially about showering in front of all the other boys. For in theoretical physics you bare yourself—not physically, but intellectually, and others do look on, and they judge.

Such insecurities are rarely spoken of or shared, yet they are common. Every physicist has to find his or her own way to deal with that pressure, but the one consequence everyone has to avoid, if they are to be successful, is the tendency to fear being wrong. Thomas Edison is often said to have advised, "To have a great idea, have a lot of them." And indeed, any innovator goes down more dead ends than glorious boulevards, so to be afraid to take a wrong turn is to guarantee never going anywhere interesting. And so I would have loved to have known, at that stage in my career, of all of Newton's wrong ideas and wasted years.

For those who, like me, take comfort in learning that people who are sometimes brilliantly right are also sometimes wrong, it is reassuring to know that even a genius such as Newton could be led astray. He might have figured out that heat is the result of the motion of the minute particles from which he believed all matter was constructed, but when he thought he had tuberculosis, he drank a "cure" of turpentine, rosewater, beeswax, and olive oil. (The cure was also supposed to be good against sore breasts and the bite of a mad dog.) Yes, he invented calculus, but he also thought that the floor plan of the lost Temple of King Solomon, in Jerusalem, contained mathematical hints regarding the end of the world.

Why did Newton drift so far off course? When one examines the circumstances, one factor jumps out above all others: Newton's isolation. Just as intellectual isolation led to the proliferation of bad science in the medieval Arab world, the same thing seemed to be hampering Newton, though in his case the isolation was self-imposed, for he kept his beliefs regarding religion and alchemy private, not willing to chance ridicule or even censure by opening the discussion to intellectual debate. There was not a "good Newton" and a "bad Newton," a rational and an irrational Newton, wrote Oxford philosopher W. H. Newton-Smith. Rather, Newton went astray by failing to subject his ideas to discussion and challenge "in the public forum," which is one of the most important "norms of the institution of science."

Allergic to criticism, Newton was equally hesitant to share the revolutionary research he had done on the physics of motion during the plague years. Fifteen years into his term as Lucasian professor, those ideas remained an unpublished, unfinished work. As a result, in 1684, at the age of forty-one, this maniacally hardworking former prodigy had produced merely a pile of disorganized notes and essays on alchemy and religion, a study littered with unfinished mathematical treatises, and a theory of motion that was still confused and incomplete. Newton had performed detailed investigations in a number of fields but arrived at no sound conclusions, leaving ideas on math and physics that were like a supersaturated solution of salt: thick with content but not yet crystallized.

Such was the state of Newton's career at that time. Says historian Westfall, "Had Newton died in 1684 and his papers survived, we would know from them that a genius had lived. Instead of hailing him as a figure who shaped the modern intellect, however, we would at most mention him in brief paragraphs lamenting his failure to reach fulfillment."

That this wasn't Newton's fate was not due to any conscious decision by the man to finish and publicize his work. On the contrary, in 1684 the course of scientific history was altered by an almost chance encounter, an interaction with a colleague that provided just the ideas and stimulus Newton needed. Were it not for that encounter, the history of science, and the world today, would be far different, and not for the better.

* * *

The seed that would grow into the greatest advancement in science the world had ever seen sprouted after Newton met with a colleague who happened to be passing through Cambridge in the heat of late summer.

In January of that fateful year, astronomer Edmond Halley—of comet fame—had sat at a meeting of the Royal Society of London, an influential learned society dedicated to science, discussing a hot issue of the day with two of his colleagues. Decades earlier, employing planetary data of unprecedented accuracy collected by the Danish nobleman Tycho Brahe (1546–1601), Johannes Kepler had discovered three laws that seemed to describe the orbits of the planets. He declared that the planets' orbits were ellipses with the sun at one of the foci, and he identified certain rules those orbits obey—for example, that the square of the time it takes for a planet to complete one orbit is proportional to the cube of its average distance from the sun. In a sense, his laws were beautiful and concise descriptions of how the planets move through space, but in another sense they were empty observations, ad hoc statements that provided no insight about why such orbits should be followed.

Halley and his two colleagues suspected that Kepler's laws reflected some deeper truth. In particular, they conjectured that Kepler's laws would all follow if one assumed that the sun pulled each planet toward it with a force that grew weaker in proportion to the square of the planet's distance, a mathematical form called an "inverse square law."

That a force that emanates in all directions from a distant body like the sun should diminish in proportion to the square of your distance from that body can be argued from geometry. Imagine a gigantic sphere so large that the sun appears as a mere dot at its center. All points on the surface of that sphere will be equidistant from the sun, so, in the absence of any reason to believe otherwise, one would guess that the sun's physical influence—essentially, its "force field"—should be spread equally over the sphere's surface.

Now imagine a sphere that is, say, twice as large. The laws of geometry tell us that doubling the sphere's radius yields a surface area that is four times as large, so the sun's attractive force will now be spread over four times the square footage. It would make sense, then, that at any given point on that larger sphere, the sun's attraction would be one-fourth as strong as before. That's how an inverse square law works: when you go farther out, the force decreases in proportion to the square of your distance.

Halley and his colleagues suspected that an inverse square law stood behind Kepler's laws, but could they prove it? One of them—Robert Hooke—said he could. The other, Christopher Wren, who is best known today for his work as an architect but was also a well-known astronomer, offered Hooke a prize in exchange for the proof. Hooke refused it. He was known to have a contrary personality, but the grounds he gave were suspicious: he said he would hold off revealing his proof so that others, by failing to solve the problem, might appreciate its difficulty. Perhaps Hooke really had solved the problem. Perhaps he also designed a dirigible that could fly to Venus. In any case, he never did provide the proof.

Seven months after that encounter, Halley, finding himself in Cambridge, decided to look up the solitary Professor Newton. Like Hooke, Newton said he had done work that could prove Halley's conjecture.

Like Hooke, he didn't come up with it. He rummaged through some papers but, not finding his proof, promised to look for it and send it to Halley later. Months passed and Halley received nothing. One can't help but wonder what Halley was thinking. He asks two sophisticated grown men if they can solve a problem, and one says, "I know the answer but I'm not telling!" while the other says, effectively, "The dog ate my homework." Wren held on to the reward.

Newton did find the proof he was looking for, but when he examined it again, he discovered that it was in error. But Newton did not give up—he reworked his ideas, and eventually he succeeded. That November, he sent Halley a treatise of nine pages showing that all three of Kepler's laws were indeed mathematical consequences of an inverse square law of attraction. He called the short tract *De Motu Corporum in Gyrum* (*On the Motion of Bodies in Orbit*).

Halley was thrilled. He recognized Newton's treatment as revolutionary, and he wanted the Royal Society to publish it. But Newton demurred. "Now I am upon this subject," he said, "I would gladly know the bottom of it before I publish my papers." Newton would "gladly know"? Since what would follow was a herculean effort resulting in perhaps the most significant intellectual discovery that had ever been made, those words constituted one of the great understatements in history. Newton would get to the "bottom of it" by demonstrating that underlying the issue of planetary orbits is a universal theory of motion and force that applies to all objects, both in the heavens and on earth.

For the next eighteen months, Newton did nothing but work on expanding the treatise that would become the *Principia*. He was a physics machine. It had always been his habit, when engaged by a topic, to skip meals and even sleep. His cat, it was once said, had grown fat on the food he left sitting on his tray, and his old college roommate reported that he would often find Newton in the morning just where he had left him the night before, still working on the same problem. But this time Newton was even more extreme. He cut himself off from practically all human contact. He seldom left his room, and on the rare occasions he ventured out to the college dining hall, he'd often have just a nibble or two while still standing, and then quickly return to his quarters.

At last Newton had shut the door to his alchemical lab and shelved his theological investigations. He did continue to lecture, as was required of him, but those lectures seemed strangely obscure and impossible to follow. It was later discovered why: Newton had simply shown up at each class session and read from rough drafts of the *Principia.*

* * *

Newton might not have pushed his work on force and motion forward very much in the decades since he was voted a fellow at Trinity; but he was, in the 1680s, a far greater intellect than he had been in the plague years of the 1660s. He possessed far more mathematical maturity and, from his studies in alchemy, more scientific experience. Some historians even believe that it was his years studying alchemy that enabled the final breakthrough in the science of motion that led him to write the *Principia.*

Ironically, one of the catalysts for Newton's breakthrough was a letter he recalled receiving five years earlier from Robert Hooke. The idea Hooke proposed was that orbital motion could be looked at as the sum of two different tendencies. Consider an object (such as a planet) that is orbiting in a circular path around some other object that is attracting it (such as the sun). Suppose the orbiting body had a tendency to continue in a straight line—that is, to fly off its curved orbit and shoot straight ahead, like a car whose driver has missed a curve in the rain. This is what mathematicians call going off in the tangential direction.

Now suppose, also, that the body had a second tendency, an attraction toward the orbit's center. Mathematicians call motion in that direction radial motion. A tendency toward radial motion, said Hooke, can complement a tendency toward tangential motion, so that, together, these two tendencies produce orbital motion.

It is easy to see how that idea would have resonated with Newton. Recall that, in improving on Galileo's law of inertia, Newton had proposed in his Waste Book that all bodies tend to continue moving in a straight line unless acted upon by an external cause, or force. For an orbiting body, the first tendency—to move off the orbit in a straight line—arises naturally from that law. Newton realized that if you add to that picture a force that attracts a body toward the center of the orbit, then you've also provided the cause of the radial motion that was Hooke's second necessary ingredient.

But how do you describe this mathematically, and in particular, how do you make the connection between the particular mathematical form

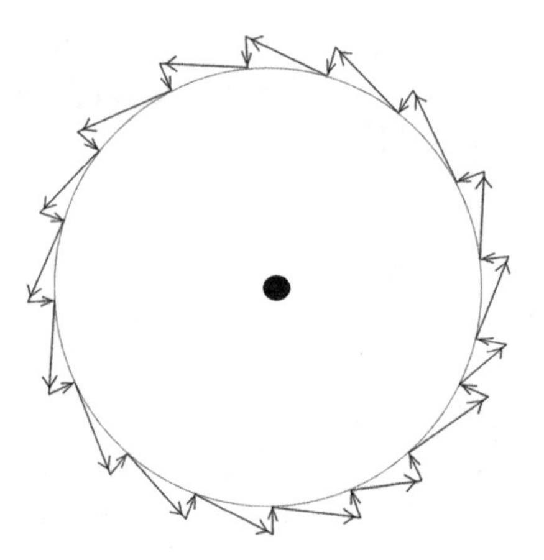

Circular motion seen as arising
from tangential and radial motion

of the inverse square law and the particular mathematical properties of orbits that Kepler discovered?

Imagine dividing time into tiny intervals. In each interval, the orbiting object can be thought of as moving tangentially by a small amount and, at the same time, radially by a small amount. The net of these two motions is to return it to its orbit, but a bit farther along the circle than where it had started. Repeating this many times results in a jagged circle-like orbit, as shown above.

If, in an orbit like this, the time intervals are tiny enough, the path can be made as close to a circle as one might wish. Here is where Newton's work on calculus applied: if the intervals are *infinitesimal*, the path, for all practical purposes, *is* a circle.

That is the description of orbits that Newton's new mathematics allowed him to create. He put together a picture in which an orbiting body moves tangentially, and "falls" radially, creating a jagged path—and then he took the limiting case, in which the straight-line segments of the jags become vanishingly small. That effectively smooths the jagged sawtooth path into a circle.

Orbital motion, in this view, is just the motion of some body that is continually deflected from its tangential path by the action of a force pulling it toward some center. The proof was in the pudding: employing the inverse square law to describe centripetal force in his mathematics of orbits, Newton produced Kepler's three laws, as Halley had asked.

Showing that free fall and orbital motion are both instances of the same laws of force and motion was one of Newton's greatest triumphs, because it once and for all disproved Aristotle's claim that the heavens and earth form different realms. If Galileo's astronomical observations revealed that the features of other planets are much like those of the earth, Newton's work showed that the *laws* of nature also apply to other planets and are not unique to planet Earth.

Even in 1684, however, Newton's insights about gravity and motion were not the sudden epiphanies suggested by the story of the falling apple. Instead, the momentous idea that gravity is universal seems to have dawned on Newton gradually as he worked on revisions of the early drafts of *Principia.*

Previously, if scientists suspected that planets exerted a force of gravity, they believed the planets' gravity affected only their moons, but not other planets, as if each planet were a distinct world unto itself, with its own laws. Indeed, Newton himself had begun by investigating only whether the cause of things falling on earth might also explain the earth's pull on the moon, and not the pull of the sun on the planets.

It is a testament to Newton's creativity, his ability to think outside the box, that he eventually began to question that conventional thinking. He wrote to an English astronomer requesting data regarding the comets of 1680 and 1684, as well as the orbital velocities of Jupiter and Saturn as they approached each other. After performing some grueling calculations on that very accurate data and comparing the results, he became convinced that the same law of gravity applies everywhere—on the earth and also between all heavenly bodies. He revised the text of *Principia* to reflect this.

The strengths of Newton's laws did not lie solely in their revolutionary conceptual content. With them he could also make predictions of unprecedented accuracy and he could compare them with the results of experiments. For example, employing data on the distance of the moon and the radius of the earth and taking into account such minutiae as the distortion of the moon's orbit due to the pull of the sun, the centrifugal force due to the earth's rotation, and the deviations of the earth's shape from a perfect sphere, Newton concluded that, at the latitude of Paris, a body dropped from rest should fall fifteen feet and one-eighth of an inch in the first second. This, the ever-fastidious Newton reported, matched experiments to better than one part in three thousand. What's more, he painstakingly repeated the experiment with different materials—gold, silver, lead, glass, sand, salt, water, wood, and wheat. Every single body, he concluded, no matter its composition, and no matter whether on earth or in the heavens, attracts every other body, and the attraction always follows the same law.

* * *

By the time Newton had finished "getting to the bottom" of what he had begun, *De Motu Corporum in Gyrum* had grown from nine pages to three volumes—the *Principia*, or, in full, *Philosophiæ Naturalis Principia Mathematica* (Mathematical Principles of Natural Philosophy).

In the *Principia*, Newton no longer treated only the motion of bodies in orbit; he detailed a general theory of force and motion itself. At its core were the interrelationships of three quantities: force, momentum (which he called quantity of motion), and mass.

We've seen how Newton struggled to develop his laws. Now let's look at his three laws to see what they mean. The first was his refinement of Galileo's law of inertia, with the important added notion that force is the cause of change:

> *First Law: Every body perseveres in its state of being at rest or of moving uniformly straight forward, except insofar as it is compelled to change its state by forces impressed.*

Newton, like Galileo, identifies motion in which an object proceeds in a straight line at a constant speed as being the natural state of things. Because today we tend to think in Newtonian terms, it is difficult to appreciate how counterintuitive that idea was. But most motion we observe in the world does *not* proceed along the lines Newton described: things speed up as they fall, or slow down as they encounter the air, and they follow curved paths as they fall toward the earth. Newton maintained that all these are in some sense deviant motions, the result of invisible forces such as gravity or friction. If an object is left alone, he said, it will move uniformly; if its path is curved or its speed is changing, that is because a force is acting on it.

The fact that objects left alone will continue in their state of motion is what gives us the ability to explore space. An earthbound Ferrari, for example, can accelerate from zero to sixty miles per hour in less than four seconds, but it has to work hard to maintain its speed, due to air resistance and friction. A vehicle in outer space would encounter only about one stray molecule every 100,000 miles or so, so you wouldn't have to worry about friction or drag. That means that once you get a craft moving, it will continue to move in a straight line and at a constant speed without slowing down like the Ferrari. And if you keep your engines on, you can continue to accelerate it, without losing any energy to friction. If, say, your spacecraft had just the acceleration of the Ferrari, and you kept at it for a year instead of a second, you could reach more than half the speed of light.

There are, of course, a few practical problems, such as the weight of the fuel you'd have to carry and the effects of relativity, which we'll come to later. Also, if you wanted to reach a star, you'd need good aim: star systems are so sparse that if you pointed your spaceship randomly, on average it would travel farther than light has traveled since the Big Bang before it encountered another solar system.

Newton did not envision our visiting other planets, but, having asserted that force causes acceleration, in his second law he quantifies the relationship between amount of force, mass, and rate of acceleration (in modern terms, "change of motion" means change in momentum, i.e., it equals mass times acceleration):

> *Second Law: A change in motion is proportional to the motive force impressed and takes place along the straight line in which that force is impressed.*

Suppose you push a wagon with a child in it. The law says that, neglecting friction, if a push exerted over one second would get a 75-pound wagon-with-child moving at five miles per hour, then if your 150-pound teenager were in the wagon instead, you'd have to push twice as hard, or for twice as long, to achieve the same speed. The good news is that (again neglecting friction) you could accelerate a 750,000-pound jumbo jet to five miles per hour by pushing 10,000 times as hard, which is difficult, or 10,000 times as long, which merely requires patience. So if you could maintain your level of exertion for 10,000 seconds, which is only two hours and forty-seven minutes, you could give a whole jumbo jet full of passengers a wagon ride.

Today we write Newton's second law as $F = ma$—force equals mass times acceleration—but Newton's second law was not put into the form of an equation until long after Newton had died, and almost a hundred years after he stated it.

In his third law, Newton says that the total quantity of motion in the universe does not change. It can be transferred among objects, but it cannot be added to or subtracted from. The total quantity of motion that is here today was here when the universe began and will remain as long as there is a universe.

It is important to note that, in Newton's accounting, a quantity of motion in one direction added to an equivalent motion in the opposite direction produces a total quantity of motion equal to zero. Hence an object can be altered from a state of rest to a state of motion without violating Newton's third law as long as its motion is counteracted by a change in motion in the opposite direction by another body. Newton put it this way:

Third Law: To any action there is always an equal and opposite reaction.

This innocent-sounding sentence tells us that if a bullet flies forward, the gun moves back. If a skater pushes backward on the ice with her skate, she will move forward. If you sneeze, expelling air forward from your mouth, your head will fly backward (with, on average, an article in the journal *Spine* tells us, an acceleration three times greater than that due to the earth's gravity). And if a spaceship thrusts hot gases out its rear rockets, the ship will accelerate forward with a momentum that is equal in magnitude but opposite in direction to that of the hot gases it spewed out into the vacuum of empty space.

The laws Newton enunciated in the *Principia* were not just abstractions. He was able to give convincing proof of the fact that the mere handful of mathematical principles he enunciated could be used to account for countless real-world phenomena. Among the applications: he showed how gravity creates observed irregularities in the motion of the moon; he explained the marine tides; he calculated the speed of sound in air; and he showed that the precession of the equinoxes is an effect of the gravitational attraction of the moon on the earth's equatorial bulge.

These were astonishing accomplishments, and the world was indeed astonished. But in some ways, what was even more impressive was that Newton understood that there were certain limits to the practical

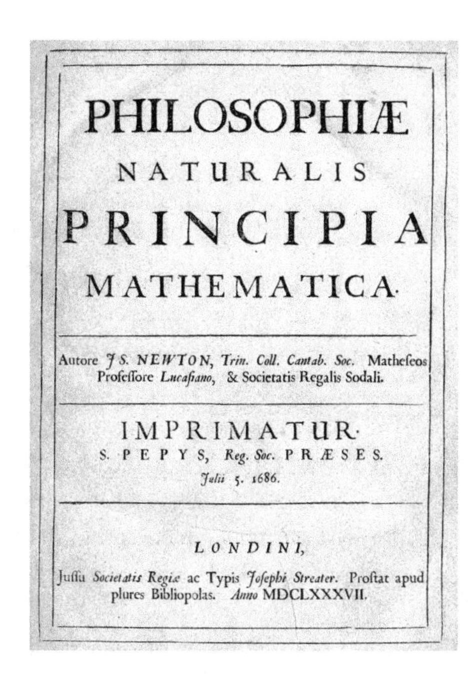

PHILOSOPHIÆ
NATURALIS
PRINCIPIA
MATHEMATICA.

Autore *J S.* NEWTON, *Trin. Coll. Cantab. Soc.* Mathefeos Profeffore *Lucafiano*, & Societatis Regalis Sodali.

IMPRIMATUR.
S. PEPYS, *Reg. Soc.* PRÆSES.
Julii 5. 1686.

LONDINI,

Juffu *Societatis Regiæ* ac Typis *Jofephi Streater*. Proftat apud plures Bibliopolas. *Anno* MDCLXXXVII.

applicability of his laws. He knew, for example, that though his laws of motion were, in general, excellent approximations of what we see happening around us, they held true in any absolute sense only in an idealized world, where there is no air resistance and no friction.

A great part of Newton's genius, like that of Galileo, was in his recognizing the myriad complicating factors that exist in our actual environment and then being able to strip those away to reveal the elegant laws that operate on a more fundamental level.

Consider free fall: Objects that are dropped speed up as Newton's laws dictate—but only at first. Then, unless the object is dropped in a vacuum, the medium it is falling through eventually acts to stop the acceleration. That's because the faster an object falls through a medium, the more resistance it experiences—because it encounters more molecules of the medium each second, and also because the collisions are more violent. Eventually, as the falling object gains speed, gravity and the resistance of the medium balance each other, and the object gains no more speed.

That maximum speed is what we now call the terminal velocity. The terminal velocity and the time of fall that it takes to achieve it depend upon the shape and weight of the object, and on the properties of the medium through which the object is falling. So while an object falling in a vacuum will gain 22 miles per hour in speed each second that it falls, a raindrop falling through the air will stop speeding up when it reaches about 15 miles per hour; for a Ping-Pong ball, the speed is 20; for a golf ball, 90; and for a bowling ball, 350 miles an hour.

Your own terminal velocity is about 125 miles per hour if you have your limbs fanned out, or 200 miles per hour if you roll yourself into a tight ball. And if you jump from a very high altitude, where the air is thin, you can fall fast enough to exceed the speed of sound, which is 761 miles per hour. An Austrian daredevil did just that in 2012—he jumped from a balloon at 128,000 feet and reached a speed of 843.6 miles per hour (American Alan Eustace jumped from a higher altitude in 2014 but didn't achieve as great a speed). Though Newton didn't know enough about the properties of air to derive such terminal velocities, in volume two of the *Principia* he did present the theoretical picture of free fall that I described above.

Shortly before Newton's birth, the philosopher and scientist Francis Bacon wrote, "The study of nature . . . [has met with] scanty success." A few decades after Newton's death, physicist-priest Roger Boscovich wrote, in contrast, that "if the law of forces were known, and the position, velocity and direction of all the points at any given instant," it is possible "to predict all the phenomena that necessarily followed from them." The powerful mind that accounted for the change of tune between these respective eras belonged to Newton, who gave such precise and profound answers to the chief scientific riddles of his day that for a hundred years, it was possible to make new advances only in subjects he himself had not touched.

* * *

On May 19, 1686, the Royal Society agreed to publish the *Principia*, but only if Halley paid the cost of printing. Halley had no choice but to agree. The Society was not a publisher. It had ventured into that business in 1685, and it had gotten burned, publishing a book called *The History of Fishes*, which, despite its exciting name, did not sell many copies. Its resources now thin, the Society could not even continue to provide Halley the fifty pounds per year he was due as its clerk; it paid him instead in copies of *The History of Fishes*. And so Halley accepted the Society's terms. The book would come out the following year.

By paying for publication, Halley was himself, essentially, Newton's publisher. He was also *Principia*'s informal editor and marketer. He sent copies of the *Principia* to all the leading philosophers and scientists of the time, and the book took Britain by storm. Word of it also spread quickly in coffeehouses and intellectual circles all around Europe. It soon became clear that Newton had written a book destined to reshape human thought—the most influential work in the history of science.

No one had been prepared for a work of such scope and profundity. Three of the leading Continental journals of opinion praised it in reviews, one saying that it offered "the most perfect mechanics one can imagine." Even John Locke, the great Enlightenment philosopher but no mathematician, "set himself to mastering the book." For all recognized that Newton had succeeded in finally overthrowing the age-old empire of Aristotle's qualitative physics and that his work would now be the template for how science should be done.

If there was a negative reaction to the *Principia*, it came mainly from those who complained that some of the central ideas it contained were not exclusive to Newton. The German philosopher and mathematician Gottfried Wilhelm Leibniz, who had independently, though slightly later, invented calculus, argued that Newton was trying to hog credit. He certainly was: the prickly Newton believed there could be only one decoder of divine knowledge on earth at any given time, and in his time, he was it. Meanwhile, Robert Hooke called the *Principia* "the most important discovery in nature since the world's creation"—and then went on to claim bitterly that Newton had stolen from him the important idea of the inverse square law. There was some justification for his claim, in that it does seem that the basic idea was Hooke's, even though it was Newton who worked out the mathematics of it.

Some also accused Newton of promoting supernatural or "occult powers," because his force of gravity acted at a distance, allowing massive bodies to affect faraway objects through the vacuum of space with no apparent means of transmitting their influence. This latter point would also puzzle Einstein—in particular the fact that the influence of Newtonian gravity was transmitted instantaneously. That aspect of Newtonian theory violated Einstein's theory of special relativity, which stated that nothing can travel faster than the speed of light. Einstein put his money where his mouth was and created his theory of gravity—general relativity—which remedied the issue and superseded Newton's theory of gravity. But those among Newton's contemporaries who criticized the idea of gravity acting at a distance could offer no alternative, and had to acknowledge the scientific power of Newton's accomplishment.

Newton's reaction to the criticism was far different from his response to the hostility that had greeted his work on optics in the early 1670s. Then, cowed by Hooke and others, he had withdrawn from the world and severed most of his connections. Now, having seen his research through to a conclusion and with a full grasp of the great significance of his own achievement, he entered fully into the fray. He met his critics with loud and fierce counterattacks that continued, in the case of the arguments over credit, until Hooke and then Leibniz died—and even afterward. As for the charge of occultism, Newton responded with a disclaimer: "These principles I consider, not as occult Qualities . . . but as general Laws of Nature . . . their truth appearing to us by Phaenomena, though their Causes be not yet discover'd."

The *Principia* changed Newton's life not only because it was acknowledged as a major milestone in intellectual history but because it thrust him into the public eye, and fame turned out to agree with him. He became more social and dropped, for the next twenty years, most of his radical efforts in theology. He also moderated, though he didn't terminate, his work in alchemy.

The changes started in March 1687, shortly after Newton finished his great work. Now bolder than he had ever been, he took part in a political battle between Cambridge University and King James II. The king, who was trying to turn England toward Roman Catholicism, tried to pressure the university into granting a Benedictine monk a degree without requiring the usual examinations and oaths to the Church of England. The university prevailed, and for Newton this was a turning point. His participation in the fray made him such a prominent political figure in Cambridge that when the university senate met in 1689, it voted to send him to Parliament as one of its representatives.

By all accounts, he didn't care much for the year he spent in Parliament, and spoke only to complain of the cold drafts. But he did grow to love London and to bask in the admiration of many of the leading intellectuals and financiers with whom he became acquainted. In 1696, after spending thirty-five years in Cambridge, Newton gave up the academic life to relocate.

In that transition, Newton was moving from a highly prestigious position to a relatively minor bureaucratic post in London: warden of the mint. But he'd been bitten by the London bug and, well past fifty by then, he felt his intellectual powers were beginning to wane. What's more, he'd grown weary of his academic salary. It might have once seemed generous, but as warden of the mint he'd be getting a big raise, to four hundred pounds. He also may have realized that, as England's leading intellectual, with the proper politicking he could progress to the higher-level position of master of the mint when it opened up, and so he did, in 1700. His income in the new position averaged 1,650 pounds, about seventy-five times the wage of a typical craftsman—and a level of compensation that made his prior Cambridge salary look paltry indeed. As a result, over the next twenty-seven years, he lived in the style of the better circles of London society, and he relished it.

Newton also rose to the very top of the organization that had published his magnum opus: in 1703, upon Hooke's death, he was elected president of the Royal Society. Age and success had not mellowed him, however. He ruled the Society with an iron fist, even ejecting members from meetings if they showed any sign of "levity or indecorum." He also became increasingly unwilling to share credit for any of his discoveries, using the power of his position to assert his primacy via various vindictive plots.

* * *

On March 23, 1726, the Royal Society recorded in its logbook: "The Chair being Vacant by the death of Sir Isaac Newton there was no Meeting this Day." Newton had died a few days earlier, at age eighty-four.

Isaac Newton in his youth and in middle age

Newton had been expecting to die for some time, for he suffered from a chronic and serious inflammation of the lungs. He had suffered from many other maladies as well, which was only to be expected in an alchemist whose hair, analyzed centuries later, would reveal levels of lead, arsenic, and antimony four times the normal level, and a level of mercury fifteen times normal. Newton's deathbed diagnosis, however, was a bladder stone. The pain was excruciating.

Newton's fate stands in stark contrast to that of Galileo. Over the years, in light of the successes of Newtonian science, the Church's opposition to new ideas from science had cooled enough that even the Catholic astronomers in Italy had earned the right not just to teach but to further develop Copernican theory—as long as they stated repeatedly, like Kansas schoolteachers are mandated to say of evolution, that "it is only a theory." Meanwhile, in England the potential of science to aid industry and improve people's lives had become clear. Science had evolved a coherent culture of experiment and calculation and grown into an enterprise that enjoyed enormous prestige, at least in the upper ranks of society. What's more, in Newton's later years, Europe was entering a period in which opposition to authority would be a theme of European culture, be it opposition to the ideas of ancient authorities like Aristotle and Ptolemy, or to the authority of religion and the monarchies.

Nothing could better exemplify the different receptions given Galileo and Newton than their respective burial rites. While Galileo was permitted only a quiet, private funeral and laid to rest in an obscure corner of the church where he'd asked to be buried, Newton's body lay in state in Westminster Abbey, and after he was buried there, a vast monument was erected, with a stone sarcophagus on a pedestal bearing his remains. On the sarcophagus were sculpted in bas-relief several boys holding instruments representing Newton's greatest discoveries, and on his tomb is inscribed:

> Here is buried Isaac Newton, Knight, who by a strength of mind almost divine, and mathematical principles peculiarly his own, explored the course and figures of the planets, the paths of comets, the tides of the sea, the dissimilarities in rays of light, and, what no other scholar has previously imagined, the properties of the colours thus produced. Diligent, sagacious and faithful, in his expositions of nature, antiquity and the holy Scriptures, he vindicated by his phi-

> losophy the majesty of God mighty and good, and expressed the simplicity of the Gospel in his manners. Mortals rejoice that there has existed such and so great an ornament of the human race! He was born on 25th December 1642, and died on 20th March 1726.

Newton's life and Galileo's had together spanned more than 160 years, and together they witnessed—and in many respects accounted for—most of what is called the scientific revolution.

In his long career, Newton was able to tell us a lot about our planet and our solar system using his laws of motion and the single law of force he had discovered—his law describing gravity. But his ambitions had reached far beyond that knowledge. He believed that force was the ultimate cause of *all* change in nature, from chemical reactions to the reflection of light off a mirror. What's more, he was confident that when, at some future time, we grew to understand the forces of attraction or repulsion that act at small distances between the tiny "particles" that constitute matter—his version of the age-old concept of atoms—his laws of motion would be sufficient to explain everything that could be observed in the universe.

Today, it is clear that Newton was prescient. His vision of what it would mean to understand the forces between atoms was very much on target. But that understanding would not come for another 250 years. And when it did, we would find that the laws governing the atom wouldn't fit into the framework of physics that he had constructed. Instead, they would reveal a new world beyond the experience of our senses, a new reality humans can envision only in their imagination, a reality whose architecture is so exotic that Newton's famous laws would have to be replaced wholesale by a new set of laws that to Newton would have appeared even more foreign than the physics of Aristotle.

8

What Things Are Made Of

When I reached my teen years, I found myself intrigued by two distinctly different kinds of scientific approaches to the secrets of the universe. I kept hearing strange rumors about what physicists did, about their discovery of quantum laws of physics that supposedly said I could be in two places at once. I doubted such claims could hold true in real life, and anyway I didn't have that many places I wanted to be. But I heard, also, about the more down-to-earth kinds of secrets that chemists pursued, violent and dangerous ones that seemed to have little to do with a master key to the universe but appealed to my sense of adventure and promised to give me a kind of power you don't normally have as a child. Soon I was mixing ammonia with tincture of iodine, potassium perchlorate with sugar, and zinc dust with the nitrate and chloride of ammonia, and I was blowing things up. Archimedes said that with a lever long enough, he could move the world; I believed that with the right household chemicals I could *explode* it. That's the power of understanding the substances around you.

The world's first scientific thinkers paved the way for these two lines of inquiry into the workings of the physical world. They asked what causes change; and they investigated what things are made of and how their composition determines their properties. Eventually Aristotle offered a road map for both, but the paths he offered proved to be dead ends.

Newton and his predecessors went a long way toward understanding the question of change. Newton also tried to understand the science of matter, but he wasn't anywhere near as great a chemist as he was a physicist. The problem wasn't that his intellect proved insufficient, or even that he went down the long, ultimately dead-end road of alchemy. What held him back was that, although the science of substance, chemistry, was evolving alongside physics, the science of change, it is a science of a very different character. It's dirtier and more complicated, and exploring it as thoroughly as he had explored change would require the development of a number of technological innovations, most of which had yet to be invented in Newton's time. As a result, Newton was stymied, and chemistry lacked a single towering figure to catapult it (and himself or herself) to greatness. Instead it developed more gradually, with several pioneers sharing the limelight.

The story of how humankind figured out what things are made of is close to my heart, because chemistry was my first love. I was raised in a small duplex apartment in Chicago, which had tight, crowded living quarters, but a large basement, where, left to my own devices, I was able to build my own Disneyland—an elaborate laboratory with shelf upon shelf of glassware, richly colored powders, and bottles of the most powerful acids and alkalis.

I had to obtain certain of the chemicals illegally or through the unwitting aid of my parents ("If I only had a gallon of muriatic acid, I could get that cat pee off the concrete"). Untroubled by having to resort to trickery, I found that by studying chemistry I could learn to create cool fireworks while fulfilling an inner curiosity about the world. And like Newton, I suppose, I also realized that it had a lot of advantages over trying to have a social life. Chemicals were easier to procure than friends, and when I wanted to play with them they never said they had to stay home to wash their hair or, less politely, that they didn't associate with weirdos. In the end, though, like many first loves, chemistry and I grew apart. I began to flirt with a new subject, physics. That was when I learned that different areas of science not only focus on different questions, but also have different cultures.

The difference between physics and chemistry shone most brightly through the various mistakes I made. I learned rather quickly, for instance, that if my physics calculation eventually boiled down to the equation "4 = 28," it did not mean that I had uncovered some profound and previously unnoticed truth, but rather that I had made some sort of error. But it was a harmless error, a mistake that existed only on paper. In physics, such boo-boos almost inevitably lead to benign, albeit frustrating, mathematical nonsense. Chemistry is different. My mistakes in chemistry tended to produce large quantities of smoke and fire, as well as acid burns in the flesh, and they left scars that persisted for decades.

My father characterized the differences between physics and chemistry according to the two people he knew who came closest to practicing them. The "physicist" was the man—really a mathematician—in the concentration camp who explained how to solve that math puzzle in exchange for bread. The man he called a "chemist" was someone he met in the Jewish underground, before he was deported to Buchenwald.*

My father had been part of a group planning to sabotage the railroad that went through his town, Częstochowa. The chemist said he could derail the train with an explosive placed strategically on the tracks, but he had to sneak out of the Jewish ghetto to procure some of his raw materials, which he insisted he could obtain through bribes and theft. This took several trips, but he didn't return from his last mission and was never heard from again.

The physicist, my father told me, was an elegant and quiet man who took refuge from the horrors of the camp in the way he knew best: by retreating into the world of his mind. The chemist had the character of a wild-eyed dreamer and a cowboy, and he threw himself into action to confront the chaos of the world head-on. That, my father asserted, was the difference between chemistry and physics.

It is indeed true that, unlike the early physicists, the early chemists had to have a certain amount of raw physical courage, for accidental explosions were a hazard of their work, as were poisonings, for they often tasted substances to aid in their identification. Perhaps the most famous of the early experimenters was Swedish pharmacist and chemist Carl Scheele. Scheele survived being the first chemist to produce the intensely corrosive and poisonous gas chlorine and, somewhat miraculously, also managed to accurately describe the taste of hydrogen cyanide, an extremely toxic gas, without being killed. But in 1786, Scheele succumbed, at age forty-three, to an illness suspiciously reminiscent of acute mercury poisoning.

On a more personal level, it struck me that the difference between chemist and physicist paralleled the difference between my father and

*I first learned that my father had been in the underground not from him, but when I came across his name mentioned in a book on the subject that I found in the university library. After reading about him, I began to question him on his experiences.

me. For after the disappearance of the chemist, he and four other plotters went ahead with their plan, using only hand tools—"all kinds of screwdrivers," he told me—rather than an explosive, in their attempt to loosen the railroad tracks. Things went awry when one of the saboteurs panicked and attracted the attention of nearby SS officers. As a result, only my father and one other saboteur escaped with their lives—by lying on the tracks, unseen, as a long freight train traveled over them. I, on the other hand, rarely take real action of any significance in the external world, but only calculate the consequences of things using equations and paper.

The gulf between physics and chemistry reflects both the origins and the cultures of the two fields. Whereas physics began with the mental theorizing of Thales, Pythagoras, and Aristotle, chemistry was born in the back rooms of tradespeople and the dark dens of the alchemist. Though practitioners of both fields were motivated by the pure desire to know, chemistry also had roots in the practical—sometimes in a desire to improve people's lives, sometimes in greed. There is nobility in chemistry, the nobility of the quest to know and conquer matter; but there has also always been the potential for great profit.

* * *

The three laws of motion that Newton uncovered were in some sense simple, even though hidden from plain view by the fog of friction and air resistance and the invisibility of the force of gravity. Chemistry, however, is not ruled by a set of laws analogous to Newton's three universal laws of motion. It is much more complicated, for our world offers us a bewilderingly diverse array of substances, and the science of chemistry had to gradually sort that all out.

The first discovery to be made was that some substances—the elements—are fundamental, while others are made up of various combinations of the elements. This was recognized intuitively by the Greeks. According to Aristotle, for example, an element is "one of those bodies into which other bodies can be decomposed and which itself is not capable of being divided into others." He named four elements: earth, air, water, and fire.

It is obvious that many substances are made from other substances. Salt plus freshwater produces salt water; iron in water forms rust; vodka and vermouth make a martini. On the flip side, you can decompose many substances into their components, often by heating them. For example, when heated, limestone turns to lime and a gas, carbon dioxide. Sugar yields carbon and water. This kind of naive observation, though, doesn't get you very far, because it isn't a universally applicable description of what happens. For example, if you heat water, it turns to gas, but that gas is not chemically different from the liquid; it has just taken a different physical form. Mercury, when heated, also doesn't break down into its components; rather, it does the opposite—it combines with the invisible oxygen in the air to form a compound known as calx.

And then there is combustion. Consider the burning of wood. When wood burns, it becomes fire and ash, but it would be wrong to conclude that wood is *made* from fire and ash. What's more, contrary to Aristotle's categorization, fire is not a substance, but rather the light and heat given off when other substances undergo a chemical reaction. What is really being emitted when wood burns is invisible gases—mainly carbon dioxide and water vapor, but more than a hundred gases in all—and the ancients had none of the technology that would have allowed them to collect, much less separate and identify, those gases.

Challenges of this sort made it difficult to sort out just what was made of two or more substances, and what was fundamental. As a result of such confusion, many ancients, like Aristotle, while misidentifying water, fire, and so on as fundamental elements, failed to recognize as elements the seven metallic elements—mercury, copper, iron, lead, tin, gold, and silver—that they *were* familiar with.

Just as the birth of physics relied on new mathematical inventions, the birth of true chemistry had to wait for certain technological inventions—equipment for accurately weighing substances, for measuring the heat absorbed or emitted in reactions, for determining whether a substance is an acid or an alkali, for capturing, evacuating, and manipulating gases, and for gauging temperature and pressure. Only with advances like these, in the seventeenth and eighteenth centuries, could chemists begin to unravel the twisted strands of their knowledge and develop fruitful ways of thinking about chemical reactions. It is a testament to human perseverance, though, that even before such technical advancements, those who practiced the trades that arose in the ancient cities gathered a large body of knowledge in a number of diverse fields such as dyeing, perfume making, glassmaking, metallurgy, and embalming.

* * *

Embalming was the first. In that realm, the beginnings of the chemical sciences can be traced all the way back to Çatalhöyük, because, although they didn't embalm their dead, they did develop a culture of death and a particular way of caring for their dead. By the time of ancient Egypt, the growing concern over the fate of the dead had led to the invention of mummification. This was believed to be the key to a happy afterlife; certainly there were no disgruntled customers coming back to say otherwise. And so there arose a demand for embalming agents. A new industry was born, one that sought, to borrow from DuPont, *better things for a better afterlife, through chemistry.*

The world has always had its dreamers and, among them, those happy individuals who achieve their dream, or at least make a living pursuing it. Those in the latter group are not necessarily distinguished by talent or knowledge, but they are inevitably set apart by their hard work. It must have been a dream of Egyptian entrepreneurs and innovators to grow rich by perfecting the embalming process, for they invested long and arduous hours in the attempt. Over time, through extensive trial and error, Egyptian embalmers eventually learned to employ a potent combination of sodium salts, resins, myrrh, and other preservatives that were successful at protecting corpses from decay—all discovered without any knowledge of either the chemical processes involved, or of what causes a body to rot.

Since embalming was a business, not a science, its discoveries were treated less like the theories of ancient Einsteins than the recipes of Einstein Bros. Bagels: they were closely guarded secrets. And since embalming was associated with the dead and the underworld, the practitioners of this art came to be thought of as sorcerers and magicians. Over time, other secretive professions evolved, producing knowledge of minerals, oils, extracts of flowers, plant pods and roots, glass, and metals. Here, in the proto-chemistry practiced by tradespeople, were the origins of alchemy's mysterious and mystical culture.

As a group, the practitioners of these fields built an extensive body of specialized but disconnected expertise. That diverse array of technical know-how finally began to jell as a unified field of study when Greece's Alexander the Great founded his Egyptian capital, Alexandria, at the mouth of the Nile in 331 B.C.

Alexandria was a lavish city, with elegant buildings and avenues a hundred feet wide. Several decades after its founding, the Greek king of Egypt, Ptolemy II, built its cultural crown jewel, the Museum. The Museum did not, like modern museums, display artifacts, but rather harbored more than one hundred scientists and scholars, who received state stipends and free housing and meals from the Museum's kitchen. Associated with it was a grand library of a half million scrolls, an observatory, dissection laboratories, gardens, a zoo, and other facilities for research. Here was a glorious center for the exploration of knowledge, a living, functional monument to the human quest to know. It was the world's first research institute, and would play a role like that of the university in later Europe, though, sadly, it was destined to be destroyed by fire in the third century A.D.

Alexandria soon became a mecca of culture and, within a couple of centuries, the largest and greatest city in the world. There, various Greek theories of matter and change intersected with the whole range of Egyptian chemical lore. That meeting of ideas changed everything.

Prior to the invasion of the Greeks, Egyptian knowledge of the properties of substances had for millennia been purely practical. But now Greek physics offered a theoretical framework to provide the Egyptians a context for their knowledge. In particular, Aristotle's theory of matter provided an explanation for the way substances change and interact. Aristotle's theory, of course, was not correct, but it inspired a more unified approach to the science of substance.

One aspect of Aristotle's theory was especially influential: his ideas on the transformation of substances. Take the process of boiling. Aristotle considered the element water to have two essential qualities: those of being wet and cold. He characterized the element air, on the other hand, as being wet and hot. Boiling, in his view, was thus a process in which the element fire acts to convert the coldness of water to heat, and hence transforms water to air. The Egyptians, smelling a potential for profit in this concept, sought to push the envelope: if water could be transformed into air, would it be possible to transform some lesser material into gold? Somewhat like my daughter, Olivia, who, when told she could get a dollar from the tooth fairy if she left her tooth under the pillow, immediately replied, "How much could I get for my nail clippings?"

The Egyptians noted that gold, like Aristotle's fundamental elements, seemed to have some essential qualities: it is a metal, it is soft, it is yellow. Gold alone has *all* those qualities, but they can be found, in differing combinations, in many substances. Could one find a way to transfer properties among substances? In particular, if the process of boiling allowed one to employ fire to alter a physical property of water, thereby turning it into air, perhaps there was an analogous process by which one could transmute a combination of metallic, soft, and yellow substances into gold.

As a result of such considerations, by 200 B.C. hints of real chemical understanding had mixed with ideas from Greek philosophy, and the old proto-chemistry of embalming, metallurgy, and other practical endeavors had spurred a unified approach to exploring chemical change. Thus the field of alchemy was born, with the central goal of producing gold, and eventually also an "elixir of life" that would grant eternal youth.

Historians debate exactly when the science of chemistry can be said to have sprouted, but chemistry is not alfalfa, and so the date of its sprouting is more a matter of opinion than a matter of precise fact. One thing that no one can argue with, though, is that alchemy served a useful purpose: chemistry, whenever it achieved its modern form, is the science that grew from the arts and mysticism of that ancient subject.

* * *

The first nudge directing the sorcery of alchemy toward the methods of science came from one of the odder characters in the history of human thought. Born in a small village in what is now Switzerland, Theophrastus Bombast von Hohenheim (1493–1541) was sent by his father at age twenty-one to study metallurgy and alchemy, but afterward he claimed to have attained a medical degree, and he adopted that profession. Then, while still in his early twenties, he renamed himself Paracelsus, meaning "greater than Celsus," a Roman physician of the first century A.D. Since the works of Celsus were very popular in the sixteenth century, with the change of moniker Paracelsus managed to go from someone named Bombast to someone whose name exhibited that quality. But there was more to the change than bombast: Paracelsus was trumpeting his scorn for the reigning approach to medicine. He demonstrated that disdain rather graphically when he joined students at their traditional bonfire one summer and tossed into the flames, along with handfuls of sulfur, the medical works of the revered Greek physician Galen.

Paracelsus's grievance against Galen was the same one that would be voiced by Galileo and Newton against Aristotle: his work had been invalidated by the observations and experience of later practitioners. In particular, Paracelsus was convinced that the conventional idea that illness is caused by an imbalance of mysterious bodily fluids called humours had not stood the test of time. Instead he was persuaded that outside agents cause disease—and that the causes can be addressed by administering the proper drugs.

It was the quest for those "proper drugs" that led Paracelsus to attempt to transform alchemy. The field had borne much fruit, such as the discovery of new substances—metallic salts, mineral acids, and alcohol—but Paracelsus wanted it to abandon its search for gold and focus on the more important goal of creating chemicals that have a place in the body's lab and can cure specific diseases. Just as important, Paracelsus aimed to reform alchemical methods, which were imprecise and sloppy. Being a marketer as well as a scholar, he invented a new name for his revamped version of alchemy. By replacing the Arabic prefix *al* (meaning "the") with the Greek word for "medicine," *iatro*, he created

Paracelsus, as depicted in a seventeenth-century copy of a lost original by Flemish artist Quentin Massys (1466–1529)

the term "iatrochemia." The word is a mouthful, which may be why it soon morphed into the shorter "chemia," which became the basis of the English word "chemistry."

Paracelsus's ideas would later influence both the great Isaac Newton and his rival Leibniz, and they would help lead alchemy toward a new identity as the science of chemistry. But though Paracelsus was a passionate crusader for his new approach, his effectiveness at personal persuasion was limited by personality issues. He could be quite offensive—and when I say "offensive," I mean "he acted like a raving lunatic."

Paracelsus was beardless and effeminate, with no interest in sex, but if the Olympics had awarded a gold medal in the sport of carousing, Paracelsus would have won the platinum. He was drunk a good part of the time and was said by one contemporary to "live like a pig." He was not subtle in promoting himself and was prone to making statements like "All the universities and all the old writers put together have less talent than my ass." And he seemed to enjoy irritating the establishment, sometimes as an end in itself. For example, when appointed as a lecturer at the University of Basel, he showed up to his first lecture wearing a leather lab apron instead of the standard academic robe; he spoke in Swiss German instead of the expected Latin; and, after announcing that he would reveal the greatest secret in medicine, he unveiled a pan of feces.

Those antics had the same effect that they would today: he alienated his physician and academic colleagues but became popular with many of his students. Still, when Paracelsus spoke, people listened, for some of his medicines really worked. For example, having discovered that opiates are far more soluble in alcohol than in water, he created an opium-based solution he called laudanum that was very effective at reducing pain.

In the end, though, it was economics that was perhaps the best engine for the spread of Paracelsus's ideas. The promise of new chemical cures for disease increased the income, social status, and popularity of apothecaries, which created a demand for knowledge of the field. Textbooks and classes in the subject sprang up, and as the terms and techniques of alchemy were translated into the new language of chemistry, they became both more precise and more standardized, just as Paracelsus had desired. By the early 1600s, though there were many who still practiced the old alchemy, Paracelsus's new style of alchemy—chemia—had also caught fire.

Like the Merton scholars of mathematics, Paracelsus was a transitional figure who helped transform his subject and laid a primitive foundation that later practitioners could build upon. The extent to which Paracelsus had a foot in both the old and new worlds of chemistry is clear from his own life: though he was so critical of traditional alchemy, he dabbled in it himself. He undertook, throughout his life, experiments aimed at creating gold, and once even claimed that he had found and drunk the elixir of life, and that he was destined to live forever.

Alas, in September 1541, while Paracelsus was staying at an establishment called the White Horse Inn, in Salzburg, Austria, God called his bluff. Paracelsus was walking back to the inn along a dark, narrow street one night when he either had a bad fall or was beaten up by thugs hired by local physicians he had antagonized—depending on which version you believe. Both stories lead to the same end: his. Paracelsus succumbed to his injuries a few days later, at age forty-seven. He was said, at the time of his death, to look decrepit well beyond his years, due to a life of late nights and heavy drinking. Had he survived another year and a half, he might have witnessed the publication of Copernicus's great work, *De Revolutionibus Orbium Coelestium* (*On the Revolutions of the Heavenly Spheres*), which is often considered the start of the scientific revolution, a movement of which Paracelsus would almost certainly have approved.

* * *

The century and a half after the death of Paracelsus was a period, as we have seen, in which pioneers like Kepler, Galileo, and Newton, building on earlier work, created a new approach to astronomy and physics. Over time, theories about a qualitative cosmos governed by metaphysical principles gave way to the concept of a quantitative and measurable universe obeying fixed laws. And an approach to knowledge that relied on scholarly authority and metaphysical argument gave way to the notion that we should learn about the laws of nature through observation and experiment, and articulate those laws through the language of mathematics.

As in physics, the intellectual challenge facing the new generations of chemists was not just to develop rigorous ways of thinking and experimentation, but also to shed the philosophy and ideas of the past. To begin to mature, the new field of chemistry had to both learn the lessons of Paracelsus and dethrone the dead-end theories of Aristotle—not his theories of motion, which Newton and other physicists and mathematicians were doing, but his theories of matter.

Before you solve a puzzle, you must identify the pieces, and in the puzzle of the nature of matter, the pieces are the chemical elements. As long as people believed that all is made of earth, air, fire, and water—or some analogous scheme—their understanding of material bodies would be based on fables, and their ability to create new and useful chemicals would remain a matter of trial and error, without the possibility of any true understanding. And so it came to pass that in the new intellectual atmosphere of the seventeenth century, as Galileo and Newton were finally banishing Aristotle from physics and replacing his ideas with observation-and-experiment-based theory, one of the men whose work on optics had helped inspire Newton stood up to banish Aristotle from chemistry. I speak of Robert Boyle, son of the first Earl of Cork in Ireland.

One road to being able to devote oneself to a life of science is to obtain a university appointment. Another is to be filthy rich. Unlike the university professors who pioneered physics, many of the champions of early chemistry were men of independent means who, in an age when laboratories were scarce, could afford to fund and set up their own. Robert Boyle was the son of an earl who was not just rich, but perhaps the richest man in Great Britain.

Little is known about Boyle's mother, other than that she was married at seventeen and proceeded to bear fifteen children in the next twenty-three years, then dropped dead of consumption, which by then must have come as a relief. Robert was her fourteenth child and seventh son. The earl seems to have liked making children more than rearing them, for shortly after birth they were each shipped off to be cared for by foster nurses, then to boarding school and college or abroad to be educated by private tutors.

Boyle spent his most impressionable years in Geneva. At the age of fourteen, he was awakened one night by a violent thunderstorm, and he swore that if he survived, he would dedicate himself to God. If everyone obeyed or even remembered the oaths they made under duress, this would be a better world, but as Boyle told it, that oath stuck. Whether or not the thunderstorm was the true cause, Boyle became deeply religious and, despite his great wealth, led an ascetic life.

The year after the life-changing thunderstorm, Boyle was visiting Florence when Galileo died in his exile nearby. Somehow Boyle got his hands upon Galileo's book on the Copernican system, his *Dialogue Concerning the Two Chief World Systems.* It was a serendipitous but notable incident in the history of ideas, for after reading the book, Boyle, then fifteen, fell in love with science.

It is not clear from anything in the historical record why Boyle chose chemistry, but he had been looking, since his conversion, for a way he could properly serve God, and he decided this was it. Like Newton and Paracelsus, he was celibate, and would become obsessive about his work, and like Newton he believed that the struggle to understand nature's ways was a path to discovering God's ways. Unlike Newton the physicist, however, Boyle the chemist also considered science important because it could be used to alleviate suffering and improve people's lives.

Boyle was in a sense a scientist because he was a philanthropist. He moved to Oxford in 1656, at the age of twenty-nine, and though the university didn't yet offer official instruction in chemistry, he set up a laboratory using his own funds and devoted himself to research—largely, but not exclusively, in chemistry.

Oxford had been a Royalist stronghold during the English Civil War and was home to many refugees from Parliamentarian London. Boyle didn't seem to have strong feelings either way, but he did join a group of the refugees who met weekly to discuss their common interest in the new experimental approach to science. In 1662, not long after the restoration of the monarchy, Charles II granted the group a charter, and it became the Royal Society (or, more precisely, the Royal Society of London for Improving Natural Knowledge), which played such an important role in Newton's career.

The Royal Society soon became a place where many of the great scientific minds of the day—including Newton, Hooke, and Halley—came together to discuss, debate, and critique one another's ideas, and to support those ideas and see to it that they made their way out into the world. The Society's motto, *Nullius in verba,* means roughly "Take nobody's word for it," but in particular it meant "Don't take Aristotle's word for it"—for the members all understood that to make progress it was crucial to move beyond the Aristotelian worldview.

Boyle also took skepticism as his own personal mantra, as reflected in the title of his 1661 book, *The Sceptical Chymist,* which was in great part an attack on Aristotle. For Boyle, like his peers, realized that in order to bring scientific rigor to the understanding of the subject that compelled him, he would have to reject much of the past. Chemistry might have had its roots in the laboratories of embalmers, glassmakers, dye makers, metallurgists, and alchemists and, since Paracelsus, in apothecaries, but Boyle saw it as a unified field worthy of study for its own sake, as necessary to a basic understanding of the natural world as astronomy and physics, and as worthy of an intellectually rigorous approach.

In his book, Boyle offered example after example of chemical processes that contradicted Aristotle's ideas regarding the elements. He discussed in great detail, for example, the burning of wood to produce ash. When you burn a log, Boyle observed, the water boiling out the ends "is far from being elementary water," and the smoke "is far from being air" but rather, when distilled, yields oil and salts. To say that fire converts the log to substances that are elemental—earth, air, and water—thus does not stand up to scrutiny. Meanwhile, other substances such as gold and silver seemed impossible to reduce to simpler components, and therefore *should* perhaps be considered elements.

Boyle's greatest work came in attacking the idea that air is an element. He supported his contention with experiments in which he was aided by a cranky young assistant, an Oxford undergraduate and ardent Royalist named Robert Hooke. Poor Hooke: later to be slighted by Newton, in many historical accounts Hooke also gets scant credit for the experiments he did with Boyle, though he probably made all of the equipment and carried out most of the work.

In one of their series of experiments, they explored respiration, trying to understand in what way our lungs interact with the air we inhale. They figured something important must be happening. After all, if there's not some kind of interaction taking place, then all this breathing we do is either a massive waste of time or, for some, just a way of keeping the lungs busy between cigars. To investigate, they performed breathing experiments on animals such as mice and birds. They observed that when the animals were placed in a sealed vessel, their respiration became labored and eventually stopped.

What did Boyle's experiments demonstrate? The most obvious lesson is that Robert Boyle was not a man you'd want to have house-sit if you owned a pet. But they also showed that when animals breathe, they are either absorbing some component of the air, which, if used up, causes death, or else they are expelling some gas, which, in high enough concentrations, proves fatal. Or both. Boyle believed it was the former, but either way, his experiments suggested that air is not elemental but rather is made up of different components.

Boyle also investigated the role of air in combustion, using a much-improved version of the vacuum pump Hooke had recently invented. He observed that once the pump drew all the air out of sealed vessels that contained burning objects, the fires were extinguished. So Boyle concluded that in combustion, as in respiration, there is some unknown substance in the air that is necessary for the process to occur.

The search for the identity of the elements was at the heart of Boyle's work. He knew that Aristotle and his successors were wrong, but, given the limitations of the resources available to him, he could make only incomplete progress in replacing their ideas with more accurate ones. Still, simply to be able to show that air is composed of different component gases was as effective a blow to Aristotle's theories as Galileo's observation that the moon has hills and craters and that Jupiter has moons. Through such work, Boyle helped free the emerging science from its reliance on the conventional wisdom of the past, replacing it with careful experimentation and observation.

* * *

There is something especially meaningful in the chemical study of air. To know saltpeter or the oxides of mercury tells us nothing about ourselves, but air gives us all life. Yet before Boyle, air was never a favorite substance of study. For to study gases was difficult, and severely constrained by the state of the available technology. That wouldn't change until the late eighteenth century, when the invention of new laboratory equipment such as the pneumatic trough enabled the collection of the gases produced in chemical reactions.

Unfortunately, since invisible gases are often absorbed or emitted in chemical reactions, without an understanding of the gaseous state, chemists were led to incomplete and often misleading analyses of many important chemical processes—in particular combustion. For chemistry to truly emerge from the Middle Ages, that had to change—the nature of fire had to be understood.

A century after Boyle, the gas necessary for combustion—oxygen—was finally discovered. It is an irony of history that the man who discovered it had his house burned down by an angry mob in 1791. The mob's provocation was the man's support of the American and French revolutions. Due to the controversy, Joseph Priestley (1733–1804) left his native England for America in 1794.

Priestley was a Unitarian and a famous and passionate advocate of religious freedom. He began his career as a minister, but in 1761 became a modern language teacher at one of the Nonconformist academies that played the role of the university for those who dissented from the Church of England. There, he was inspired by the lectures of a fellow teacher to write a history of the new science of electricity. His research on that topic would lead him to perform original experiments.

The stark contrast between the lives and backgrounds of Priestley and Boyle reflect a contrast in their times. Boyle died at the start of the Enlightenment, the period in the history of Western thought and culture between roughly 1685 and 1815. Priestley, on the other hand, worked at the height of that era.

The Enlightenment was an era of dramatic revolutions in both science and society. The term itself, in Immanuel Kant's words, represents "mankind's exit from its self-incurred immaturity." Kant's motto for enlightenment was simple: *Sapere aude*—"Dare to know." And indeed, the Enlightenment was distinguished by an appreciation of the advance of science, a fervor to challenge old dogma, and the principle that reason should trump blind faith and could bring practical social benefits.

Just as important, in Boyle's day (and Newton's), science had been the province of only a few elite thinkers. But the eighteenth century saw the beginnings of the industrial era, the continuing rise of the middle class, and a decline in the dominance of the aristocracy. As a result, by the second half of the century, science had become the concern of a relatively large educated echelon, a more diverse group that included members of the middle class, many of whom used learning as a way to improve their economic standing. Chemistry in particular profited from this new broader base of practitioners—people like Priestley—and the inventive and entrepreneurial spirit that came with them.

Priestley's book on electricity came out in 1767, but that same year, he turned his interest from physics to chemistry, and, in particular, gases. He didn't change fields because he'd had any great insight into that science or had come to believe that it was a more important area of study. Rather, he had moved next door to a brewery, where copious gas bubbled up furiously within wood barrels as their contents fermented, and this had piqued his curiosity. He eventually captured large volumes of the gas and, in experiments reminiscent of Boyle's, he found that if burning wood chips were placed in a sealed vessel that contained it, the fire would be extinguished, while a mouse placed in such a vessel would soon die. He also noticed that if it was dissolved in water, the gas created a sparkling liquid with a pleasant taste. Today we know that gas as carbon dioxide. Priestley had inadvertently invented a way to create carbonated beverages, but alas, since he was a man of modest means, he didn't commercialize his invention. That was done a few years later by one Johann Jacob Schweppe, whose soda company is still in business today.

That Priestley should have been brought to chemistry through his fascination with a commercial by-product is fitting, for it was now, with the coming of the industrial revolution in the late eighteenth century, that we find science and industry spurring each other on to ever greater accomplishments. Very little of immediate practical use had resulted from the great advances of the previous century's science, but the advances that began in the late eighteenth century utterly transformed daily life. The direct results of the collaborations of science and industry include the steam engine, advances in the harnessing of water power for use in factories, the development of machine tools, and, later, the appearance of railroads, the telegraph and the telephone, electricity, and the lightbulb.

Though in its earliest stages, around 1760, the industrial revolution depended on contributions from artisan inventors rather than on the discovery of new scientific principles, it nevertheless stimulated a movement among the wealthy toward supporting science as a means of improving the art of manufacturing. One wealthy patron who had such an interest in science was William Petty, Earl of Shelburne. In 1773, he gave Priestley a position as librarian and tutor of his children but also built him a laboratory and allowed him plenty of spare time to conduct his research.

Priestley was a clever and meticulous experimenter. In his new lab, he began to experiment with calx, which we now know as an oxide of mercury—in other words, mercury "rust." Chemists of the day knew that when they heated mercury to make calx, the mercury was absorbing something from the air, but they didn't know what. What was intriguing was that when calx was heated further, it turned back into mercury, presumably expelling whatever it had absorbed.

Priestley found that the gas expelled by calx had remarkable properties. "This air is of exalted nature," he wrote. "A candle burned in this air with an amazing strength of flame . . . But to complete the proof of the superior quality of this air, I introduced a mouse into it; and in a quantity in which, had it been common air, it would have died in about a quarter of an hour, it lived . . . a whole hour, and was taken out quite vigorous." He went on to sample the "exalted" air—which of course was oxygen: "The feeling of it to my lungs was not sensibly different from that of common air; but I fancied that my breast felt peculiarly light and easy for some time afterwards." Perhaps, he speculated, the mysterious gas would become a popular new vice for the idle rich.

Priestley did not go on to become an oxygen dealer to the wealthy. Instead, he studied the gas. He exposed it to samples of dark, clotted blood, and found that the blood turned bright red. He also noted that if dark blood was placed in a small sealed space to absorb the gas from the air, after the blood turned bright red, any animals present quickly suffocated.

Priestley took these observations to mean that our lungs interact with the air to revitalize our blood. He also experimented with mint and spinach and discovered that growing plants could restore the air's ability to support both respiration and fire—in other words, he was the first person to note the effects of what we today know as photosynthesis.

Though Priestley learned much about the effects of oxygen and is often said to be its discoverer, he did not understand its significance in the burning process. Instead, he subscribed to a popular but complicated theory of the day that held that objects burn not because they are reacting with something in the air, but because they are *releasing* something, called "phlogiston."

Priestley had performed revealing experiments, but he failed to see what they had revealed. It was left to a Frenchman named Antoine Lavoisier (1743–1794) to do the work that explained the true meaning of Priestley's experiments—that respiration and combustion were processes that involved absorbing something (oxygen) from the air, not releasing "phlogiston" into it.

* * *

It might seem a futile dream that the field that had begun as alchemy could rise to the precise mathematical rigor of Newtonian physics, but many eighteenth-century chemists believed it could. There was even speculation that the forces of attraction between the atoms that make up substances are essentially gravitational in nature and could be used to explain chemical properties. (Today we know they were right, except that the forces are electromagnetic.) Such ideas had originated with Newton, who had asserted that there are "agents in nature able to make the particles of bodies [i.e., atoms] stick together by very strong attractions. And it is the business of experimental philosophy to find them out." That was one of the growing pains of chemistry: the issue of just how literally Newton's ideas could be translated from physics to other sciences.

Lavoisier was one of those chemists greatly influenced by the Newtonian revolution. He viewed chemistry as it was then practiced as a subject "founded on only a few facts . . . composed of absolutely incoherent ideas and unproven suppositions . . . untouched by the logic of science." Still, he sought to have chemistry emulate the rigorous quantitative methodology of experimental physics, and not the purely mathematical systems of theoretical physics. It was a wise choice, given the knowledge and technical capabilities of his day. Eventually, theoretical physics *would* be able to explain chemistry through its equations, but that wouldn't happen until the development of quantum theory and, even better, high-speed digital computers.

Lavoisier's take on chemistry reflected the fact that he loved both chemistry *and* physics. He might have actually preferred the latter, but having grown up the son of a wealthy attorney in Paris, in a family that was intensely protective of its status and privilege, he came to consider it too acrimonious and controversial. Though Lavoisier's family encouraged his ambitions, they expected him to be socially deft as well as industrious, and they emphasized caution and restraint—qualities that weren't exactly natural to him.

That Lavoisier's true love was science must have been apparent to everyone who knew him. He had wild ideas and grand plans to carry them out. While still a teen, he sought to investigate the effects of diet on health by ingesting nothing but milk for an extended period, and proposed shutting himself in a dark room for six weeks to increase his ability to judge small differences in the intensity of light. (He was apparently talked out of it by a friend.) That same passion for scientific exploration would be reflected throughout his life in an enormous capacity to engage, like so many other pioneers of science, in long hours of tedious work in his pursuit of understanding.

Lavoisier was lucky in that money was never an issue for him: while still in his twenties, he received an advance on his inheritance worth upward of ten million dollars in today's currency. He invested it profitably, purchasing a share in an institution called the Company of General Farmers. The general farmers didn't cultivate asparagus; they gathered certain taxes, the collection of which the monarchy had decided to farm out.

Lavoisier's investment was hands-on—it brought with it the responsibility of overseeing the enforcement of tobacco regulations. In exchange for his efforts, the farm paid Lavoisier an average of the equivalent of about two and a half million dollars per year as his share of the profit. He used the money to build the finest private laboratory in the world, reportedly stocked with so much glassware that one imagines he enjoyed gazing at his collection of beakers as much as he enjoyed employing them. He also spent his money on a number of humanitarian efforts.

Lavoisier heard about Priestley's experiments in the fall of 1774 from Priestley himself, who landed in Paris while touring Europe with Lord Shelburne, acting as a kind of science guide. The three of them, along with some dignitaries of Parisian science, all dined together and afterward talked shop.

When Priestley told Lavoisier about the work he had been doing, Lavoisier immediately realized that Priestley's experiments with burning had something in common with experiments he had conducted on rusting; this surprised and delighted him. But he also felt that Priestley had little understanding of the theoretical principles of chemistry, or even of the implications of his own experiments. His work, Lavoisier wrote, was "a fabric woven of experiments that is hardly interrupted by any reasoning."

To excel at both the theoretical and experimental aspects of a science is, of course, a tall order, and I know few top scientists who could make that claim. Personally, I was identified early as a budding theorist, and so in college I was required to take only one physics lab. In it, I was to design and build a radio from scratch, a project that consumed the entire semester. In the end, my radio worked only when held upside down and shaken, and even then it picked up only one station, that of a Boston broadcaster who played discordant avant-garde music. And so I am grateful for the division of labor in physics, as are most of my friends, be they theorists or experimentalists.

Lavoisier was a master at both the theoretical and experimental aspects of chemistry. Having dismissed Priestley as a lesser intellect, and excited by the possibility of exploring the parallels between the processes of rusting and burning, he repeated Priestley's work with mercury and its oxide, calx, early the next morning. He improved upon Priestley's experiments, measuring and weighing everything meticulously. And then he gave an explanation of Priestley's discoveries that Priestley himself had never imagined: when mercury burns (to form calx), it combines with a gas that is a fundamental element of nature, and—his measurements showed—it gains an amount of weight equal to that of the gas with which it combines.

Lavoisier's careful measurements also demonstrated something else: when the reverse occurs—when calx is heated to form mercury—it gets lighter, presumably giving off the same gas it had absorbed and losing a quantity of weight that is precisely equal to the weight gained when mercury forms calx. Though Priestley is credited with discovering the gas that was absorbed and released in the course of these experiments, it was Lavoisier who explained its significance—and eventually gave it the name oxygen.*

Lavoisier later turned his observations into one of the most famous laws in science, the law of conservation of mass: the total mass of products produced in a chemical reaction must be the same as the mass of the initial reactants. This was perhaps the greatest milestone in the journey from alchemy to modern chemistry: the identification of chemical change as the combining and recombining of elements.

Lavoisier's association with the tax farm had funded the important scientific work he did. But it would also prove to be his undoing, for it brought him to the attention of the revolutionaries who overthrew the French monarchy. At any time and any place, tax collectors are about as welcome as a guy with tuberculosis and a bad cough. But these tax farmers were especially despised, because many of the taxes they were

* "Oxygen" means "acid generator," a name Lavoisier chose because oxygen was present in all the acids whose composition he was familiar with.

charged with gathering were seen as irrational and unfair, particularly in their impact on the poor.

Lavoisier had by all accounts carried out his duties fairly and honestly, and with some sympathy for those from whom he was collecting, but the French Revolution wasn't known for its nuanced judgments. And Lavoisier had given the revolutionaries plenty to hate.

His worst offense was a massive wall of heavy masonry that he'd had the government construct around the city of Paris, at a cost of several hundred million dollars in today's currency. No one could enter or leave the city except through one of the tollgates in the wall, which were patrolled by armed guards who measured all the goods coming in and out and kept records that were used in imposing taxes. Thus—to the chagrin of the public—Lavoisier brought his penchant for meticulous measurement out of the lab and into his work as a tax agent.

When the Revolution began, in 1789, Lavoisier's wall was the first structure attacked. He was arrested in 1793—along with the other tax farmers—under the Reign of Terror, and sentenced to death. He requested that his execution be delayed in order that he might complete the research he was working on. He was reportedly told by the judge, "The republic has no need of scientists." Perhaps not, but chemistry had had the need, and fortunately, during the course of his fifty years of life, Lavoisier had already been able to transform the field.

By the time of his execution, Lavoisier had identified thirty-three known substances as elements. He was correct about all but ten. He had also created a standard system of naming compounds according to the elements that constitute them, to replace the dizzying and unenlightening language of chemistry that had existed before him. I made much of the importance of mathematics as the language of physics, but a viable language is equally crucial in chemistry. Prior to Lavoisier, for example, the calx of hydrargyrum and the calx of quicksilver were two names for the same compound. In Lavoisier's terminology, that compound became "mercuric oxide."

Lavoisier did not go quite as far as inventing modern chemical equations like "$2Hg + O_2 \rightarrow 2HgO$," the equation describing the production of mercuric oxide, but he did set the groundwork for it. His discoveries created a revolution in chemistry and aroused great enthusiasm in industry, which in turn supplied future chemists with new substances to work with, and new questions to answer.

In 1789, Lavoisier had published a book synthesizing his ideas, *Elementary Treatise on Chemistry.* Today it is viewed as chemistry's first modern textbook, clarifying the concept that an element is a substance that cannot be broken down, denying the four-element theory and the existence of phlogiston, asserting the law of conservation of mass, and presenting his new rational nomenclature. Within a generation the book had become a classic, informing and inspiring numerous later pioneers. Lavoisier himself, by then, had been killed, his body discarded in a mass grave.

Lavoisier had spent his life serving science, but he also thirsted desperately for fame, and regretted never having isolated a new element himself (though he sought to share the credit for discovering oxygen). Finally, in 1900, a century after denying that France needed scientists, his country erected a bronze statue of him in Paris. The dignitaries at the unveiling remarked that he "merited the esteem of men" and was a "great benefactor of humanity," because he'd "established the fundamental laws which govern chemical transformations." One speaker proclaimed that the statue had captured Lavoisier "in all the luster of his power and intelligence."

It sounds like the recognition Lavoisier had craved, but it's doubtful he would have enjoyed the ceremony. As it turned out, the face on the statue was not Lavoisier's but that of the philosopher and mathematician the Marquis de Condorcet, who'd been the secretary of the Academy of Sciences during Lavoisier's last years. The sculptor, Louis-Ernest Barrias (1841–1905), had copied the head from a sculpture made by a different artist and had not identified its subject correctly. That revelation didn't seem to bother the French, and they left the erroneous bronze standing—a memorial to a man they'd guillotined, bearing the head of another.* In the end, the statue lasted only about as long as Lavoisier himself. Like him, it eventually fell victim to the politics of war—it was scrapped during the Nazi occupation and recycled for use as bullets. At least Lavoisier's ideas proved durable. They remade the field of chemistry.

* * *

*Ironically, in 1913 it was reported that a life-size marble bust of Condorcet that had been gifted to the American Philosophical Society in Philadelphia turned out to be not Condorcet, but Lavoisier!

Statue of Lavoisier,
with the head of Condorcet

People often refer to "the march of science," but science does not propel itself; people move it forward, and our forward progress is more like a relay race than a march. What's more, it is a rather odd relay race, for those who grab the baton often take off in a direction that the prior runner did not anticipate, and would not approve of. That's precisely what happened when the next great visionary of chemistry took over after the great run Lavoisier had achieved.

Lavoisier had clarified the role of elements in chemical reactions and promoted a quantitative approach to describing them. Today we know that to truly understand chemistry—and, in particular, if you want a *quantitative* understanding of chemical reactions—you need to understand the atom. But Lavoisier had nothing but scorn for the concept of the atom. It's not that he was closed-minded or shortsighted. Rather, he opposed the idea of thinking in terms of atoms for an entirely practical reason.

Ever since the Greeks, scholars had conjectured about atoms—though sometimes calling them other names, such as "corpuscles," or "particles of matter." Yet, because they are so small, over the course of nearly two dozen centuries, no one had ever thought of a way to relate them to the reality of observations and measurements.

To get an idea of just how small atoms are, imagine filling all the world's oceans with marbles. And then imagine shrinking each of those marbles down to the size of an atom. How much space would they take up? Less than a teaspoon. What hope could there be of observing the effects of anything that tiny?

As it turns out, plenty—and that miraculous achievement was first accomplished by a Quaker schoolteacher, John Dalton (1766–1844). Many of the great scientists of history were colorful people, but Dalton, the son of a poor weaver, was not among those. He was methodical in everything from his science to the way he took tea at five each afternoon, followed by a supper of meat and potatoes at nine.

The book Dalton is known for, *A New System of Chemical Philosophy*, is a meticulous three-part treatise, all the more astonishing for having been researched and written entirely in Dalton's spare time. Part one, published in 1810, when he was in his midforties, is a mammoth work of 916 pages. Of those 916 pages, just one chapter, barely five pages long, presents the epoch-making idea that he is known for today: a way to calculate the relative weights of atoms from measurements you can do in the laboratory. That is the excitement and power of ideas in science—five pages can reverse the misguided theories of two thousand years.

The idea came to Dalton, like many ideas, in a roundabout way, and though it was now the nineteenth century, Dalton's idea was inspired by the influence of a man born in the middle of the seventeenth century—again, here was the reach of Isaac Newton.

Dalton liked to take walks, and when he was young he lived in Cumberland, the wettest part of England, and became interested in meteorology. He was also a prodigy who, as a teen, studied Newton's *Principia*. The two interests proved a surprisingly potent combination, for they led him to an interest in the physical properties of gases—like the damp air in the Cumberland countryside. Intrigued by Newton's theory of corpuscles, which was essentially the ancient Greek concept of atoms updated by Newton's ideas on force and motion, Dalton was led to suspect that the different solubilities of gases are due to their atoms being of different sizes, and that, in turn, led him to consider the atoms' weights.

Dalton's approach was based on the idea that if one is careful to consider only pure compounds, then compounds must always be formed from their constituents in the exact same proportions. For example, there exist two different oxides of copper. If one examines those oxides separately, one finds that, for each gram of oxygen consumed, the creation of one of the oxides uses up four grams of copper, while the creation of the other oxide consumes eight grams. This implies that in the latter type of oxide, twice as many copper *atoms* are combined with each atom of oxygen.

Now suppose, for the sake of simplicity, that in the former case, each oxygen atom is combining with one copper atom, while in the latter case each oxygen atom combines with two. Then, since in the former case the oxide is formed from four grams of copper for every gram of oxygen, you can conclude that a copper atom weighs about four times as much as an oxygen atom. As it happens, that assumption is true, and it is the kind of reasoning Dalton employed to calculate the relative atomic weights of the known elements.

Since Dalton was calculating *relative* weights, he needed to start somewhere, so he assigned the lightest known element—hydrogen—a weight of "1" and calculated the weights of all the other elements relative to it.

Unfortunately, his assumption that elements combine in the simplest possible proportions didn't always work. For example, that assumption assigned to water the formula HO, rather than the more complicated H_2O, which we know it to be today. Therefore, when he calculated the weight of the oxygen atom relative to that of hydrogen, his result was half what it should have been. Dalton was well aware of the uncertainty, and with regard to water, he recognized both HO_2 and H_2O as good alternative possibilities. The relative weights would have been far more difficult to decipher if common compounds had formulas like $H_{37}O_{22}$, but luckily that is not the case.

Dalton knew that his estimates were provisional, and needed to be based on data from a large number of compounds in order to reveal inconsistencies that might point out errors in the assumed formulas. That difficulty would plague chemists for another fifty years, but that it took time to work out the details didn't diminish its impact on the field, for Dalton's version of atomism was one that finally made practical sense in that it could be related to laboratory measurements. What's more, building on Lavoisier's work, Dalton used his ideas to create, for the first time, a quantitative language for chemistry—a new way to understand the experiments chemists carried out in terms of the exchange of atoms among molecules. In the modern version, for example, to describe the creation of water from oxygen and hydrogen, a chemist (or high school student) would write, "$2H_2 + O_2 \rightarrow 2H_2O$."

The new language of chemistry revolutionized chemists' ability to understand and reason about what they were observing and measuring when they created chemical reactions, and his ideas have been central to the theory of chemistry ever since. Dalton's work made him world famous, and though he shunned public honors, he received them, including a membership in the Royal Society that was bestowed over his vehement objections. When he died, in 1844, the funeral that he had hoped would be modest drew more than forty thousand mourners.

Through Dalton's efforts, human thinking regarding the nature of substance had progressed from the theories proposed by ancient mystical lore to the beginnings of an understanding of matter on a level far beyond the reach of our senses. But if each element is distinguished by the weight of its atoms, how is that atomic property related to the chemical and physical characteristics we observe? That is the next leg of the relay, and indeed the last of the profound questions about chemistry that could be answered without moving beyond Newtonian science. Deeper insights would come, but they would have to await the quantum revolution in physics.

* * *

Stephen Hawking, having survived for decades despite being paralyzed with an illness that was supposed to kill him in just a few years, once told me that he considered stubbornness to be his greatest trait, and I believe he may be right. Though it makes him at times difficult to work with, he knew that it's his stubbornness that both keeps him alive and gives him the strength to continue to do his research.

The finished theories of science may seem almost self-evident once they have been formulated, but the struggle to create them can usually be won only through magnificent perseverance. Psychologists speak of a quality called "grit," an attribute related to perseverance and stubbornness, but also to passion, all characteristics we've seen much of in these pages. Defined as "the disposition to pursue long-term goals with sustained interest and effort over time," they, not surprisingly, have found grit related to success in everything from marriage to the Army's Special Operations Forces. Perhaps that's why so many of the characters we've met thus far in our story were headstrong, even arrogant. Most great innovators are. They have to be.

Our next pioneer, Dmitri Mendeleev (1834–1907), a Russian chemist known for his tantrums and fits of rage (and for having his hair and beard trimmed just once each year), fits comfortably in that pantheon of stubborn mules. So strong was his personality, in fact, that his wife eventually learned to avoid him by living at their country estate—except when he showed up there, at which time she'd grab the kids and move to their city residence.

Mendeleev, like Hawking, was a survivor. In his late teens, he was hospitalized for tuberculosis, but he not only survived—he found a nearby laboratory and spent his days of recuperation performing chemistry experiments there. Later, after obtaining his teaching credential, he angered an official at the Ministry of Education and as a result was assigned to teach at a high school in faraway Crimea. The year was 1855, and when Mendeleev arrived, he found that not only was the high school

Dmitri Mendeleev

in a war zone, but it had long ago been closed. Undeterred, he returned home, gave up on his prospects of a high school teaching career, and took a job as a *privatdozent*—one who lectures for tips—at the University of St. Petersburg. He eventually became a professor there.

That Mendeleev was a chemist, or even educated at all, he owed entirely to his mother. He was born to a poor family in western Siberia, the youngest of either fourteen or seventeen children—accounts vary. He did poorly in school but enjoyed performing makeshift science experiments. His mother believed in his intellect, and when he was fifteen, after his father died, she set off on the road with him, looking for a university that would take him.

It proved to be a journey of fourteen hundred miles, involving much hitchhiking on horse-drawn wagons, but in the end he got a small scholarship at the Central Pedagogical Institute, in St. Petersburg, where the director was an old friend of his late father. His mother died soon afterward. Thirty-seven years later, he dedicated a scientific treatise to her memory, quoting what he called her "sacred" dying words: "Refrain from illusions, insist on work and not on words. Patiently seek divine and scientific truth." Mendeleev, like so many great scientists before him, would live his life by those words.

In one sense, Mendeleev was lucky to have been born when he was. Virtually every great discovery and innovation has arisen from a combination of human insight and happy circumstance. Einstein was lucky to have begun his career shortly after the formulation of the modern theory of electromagnetism, which implied that the speed of light is constant, an idea that would become the essence of his theory of relativity. Steve Jobs was likewise lucky to have begun his career at a time when technology was just reaching the stage at which a useful personal computer could be developed. On the other hand, Armenian American inventor and businessman Luther Simjian held many patents, but he had his best idea just a decade or so too early: he conceived of the automated teller machine (ATM), which he called the Bankograph in 1960. He persuaded a New York City bank to install a few, but people did not trust them to accept deposits, so the only ones who used them were prostitutes and gamblers who didn't want to deal with tellers face-to-face. A decade later, times had changed, and ATMs took off, but using someone else's design.

Unlike Simjian, Mendeleev had the spirit of the times on his side.

He reached adulthood at a moment when chemistry was ripe for advancement—the idea that the elements could be organized into families was in the air all over Europe in the 1860s. It hadn't gone unnoticed, for example, that fluorine, chlorine, and bromine—classified as "halogens" by Swedish chemist Jöns Jakob Berzelius in 1842—seem to belong together: they are all extremely corrosive gases that become tamed when united with sodium, forming harmless saltlike crystals. (Table salt, for example, is sodium chloride.) Nor was it hard to detect the similarities among the alkali metals, like sodium, lithium, and potassium. These are all shiny, soft, and highly reactive. In fact, the members of the alkali metal family are so alike that if you swap potassium for sodium in table salt, what you get is close enough to sodium chloride that it is used as a salt substitute.

Chemists inspired by Carl Linnaeus's scheme for classifying biological organisms sought to develop a comprehensive family system of their own to explain the relationships among the elements. But not all the groupings were obvious, nor was it known how they related to one another, or what property of atoms was responsible for the family resemblances. Those issues attracted thinkers all over Europe. Even a sugar refiner got into the act, or at least the house chemist at a sugar refinery. But though a handful of thinkers knocked on the door of the answer, just one man—Mendeleev—crashed right through it.

You'd think that if the idea of organizing the elements was "in the air," then the person who succeeded at it would get a hearty cheer, but you wouldn't necessarily conclude that that person would be considered one of the greatest geniuses ever to work in his field—which Mendeleev is. What puts him in the class with titans like Boyle, Dalton, and Lavoisier?

The "periodic table" that Mendeleev developed isn't the chemist's version of a field guide to the birds; rather, it is chemistry's version of Newton's laws, or at least it's as close to that magical achievement as chemistry can hope to come. For it is not simply a table listing families of elements; it is a veritable Ouija board allowing chemists to understand and predict the properties of any element, even those yet unknown.

Looking back, it is easy to credit Mendeleev's breakthrough to his asking the right question at the right time, and to his work ethic, passion, stubbornness, and extreme self-confidence. But as has often been the case in discovery and innovation—and in many of our own lives—

just as important as his intellectual qualities was the role of chance, or at least an unrelated circumstance that set the stage for those qualities to achieve their triumph. In this case, it was Mendeleev's chance decision to write a chemistry textbook.

The decision to write the text came in 1866, after Mendeleev received his appointment as professor of chemistry in St. Petersburg, at age thirty-two. St. Petersburg had been created a century and a half earlier by Peter the Great and was finally emerging as an intellectual center of Europe. Its university was the best in Russia, but Russia was behind the rest of Europe, and, scanning Russia's chemistry literature, Mendeleev found no decent and up-to-date book to use in his teaching. Hence, his decision to write his own. The text would take years to complete and was destined to be translated into all the major languages and used in universities far and wide for decades to come. It was unorthodox, laden with anecdotes, speculation, and eccentricity. It was a work of love, and his drive to make it the best book possible was what forced him to focus on the issues that would lead him to his great discovery.

The first challenge Mendeleev faced in writing his book was how to organize it. He decided to treat the elements and their compounds in groups, or families, defined by their properties. After the relatively easy task of describing the halogens and alkali metals, he confronted the issue of which group to write about next. Should the order be arbitrary? Or might there be some organizing principle that would dictate it?

Mendeleev struggled with the problem, looking deep into his vast chemical knowledge for clues to how the different groups might be related. One Saturday, he became so focused that he worked through the entire night and into the morning. He got nowhere, but then something possessed him to write the names of the elements in the oxygen, nitrogen, and halogen groups—twelve elements in all—on the back of one of his envelopes, in ascending order according to their atomic weights.

Now he suddenly noticed a striking pattern: the list began with nitrogen, oxygen, and fluorine—the lightest member of each group—and then continued with the second-lightest member of each group, still in turn, and so on. The list, in other words, formed a repeating, or "periodic," pattern. Only in the case of two of the elements did the pattern not hold.

Mendeleev made his discovery more apparent by arranging the elements of each group into a row and writing the rows atop one another to make a table. (Today we write the groups in columns.) Was there really something to this? And if these twelve elements really did form a meaningful pattern, would the other fifty-one known at the time fit his scheme?

Mendeleev and his friends used to play a card game called patience, in which they laid out a deck and had to arrange the cards in a certain way. The cards formed a table that, he would later recount, looked much like the table of twelve elements he made that day. He decided to write the names and atomic weights of *all* the known elements on cards and try to make a table of that, playing what he now called "chemical patience." He began moving the cards around, trying to arrange them in a way that made sense.

There were serious problems with Mendeleev's approach. For one, it wasn't clear which group some elements ought to belong to. The properties of others were not well understood. There was also disagreement about the atomic weights of certain elements, and, as we now know, the weights assigned to some elements were just plain wrong. Perhaps most serious, there were elements yet to be discovered, and this had the effect of making his sequencing appear not to work.

All these issues served to make Mendeleev's task difficult, but there was something else, something more subtle: there was no reason to believe that a scheme based on atomic weight *should* work, for no one at that time understood what chemical aspect of an atom its weight reflected. (Today we know that it is the number of protons and neutrons in the atom's nucleus, and the contribution to the weight from the neutrons is unrelated to the atom's chemical properties.) Here, especially, is where Mendeleev's stubbornness supported his passion to pursue his idea: he continued based on intuition and faith alone.

Mendeleev's work reveals, more literally than most, how the scientific process is an activity of puzzle solving. But it also illustrates important differences, because unlike the jigsaw you would buy in the store, the pieces in Mendeleev's puzzle did not fit. Part of science, and all innovation, is to sometimes ignore issues that seem to indicate that your approach cannot possibly work, in the belief that you'll eventually find a workaround, or that the issues will prove irrelevant after all. In this case, through both remarkable brilliance and extraordinary persistence, Mendeleev created his picture by remaking some pieces of the puzzle and completely fabricating others.

In hindsight, it is easy to characterize Mendeleev's accomplishment in a heroic light, as I suppose I have done. Even if your ideas sound crazy, if they work, we make you a hero. But there is a flip side, for there have been many crazy schemes throughout the ages that have proved wrong. Those schemes that work, in fact, are vastly outnumbered by those that don't. The wrong ones are quickly forgotten, the hours and days and years of work their believers put into them having ultimately been wasted. And often we call the proponents of those schemes failures and crackpots. But heroism is about taking risks, and so what's really heroic about research, whether it pans out or doesn't, is that risk we scientists and other innovators take—those long hours and days, months or even years, of intense intellectual struggle, which may or may not lead to a fruitful conclusion or product.

Mendeleev certainly put in the time. And when an element didn't fall into place as he had wanted, he refused to accept that his scheme was wrong. Instead, he stuck to his guns and concluded that those who had measured the atomic weight were wrong—and he boldly crossed out the measured weight and filled in the value that would make the element fit.

His most audacious pronouncements came when his table left him with a gap somewhere—a spot in which no element existed that possessed the properties that were required. Rather than abandoning his ideas, or attempting to alter his organizing principle, he steadfastly insisted that those gaps represented undiscovered elements. He even predicted the new elements' properties—their weight, their physical characteristics, which other elements they combine with, and the kinds of compounds they make—based solely on the spot in which the gap appeared.

There was a gap next to aluminum, for example. Mendeleev filled it with an element he called eka-aluminum, and went on to predict that when some chemist eventually discovered eka-aluminum, it would be a shiny metal that conducted heat very well and had a low melting point, and that a cubic centimeter of the stuff would weigh precisely 5.9 grams. A few years later, a French chemist named Paul-Émile Lecoq de Boisbaudran discovered an element in ore samples that fit the bill—except that he found that it weighed 4.7 grams per cubic centimeter. Mendeleev immediately sent Lecoq a letter telling him that his sample must have been impure. Lecoq repeated his analysis with a new sample that he made sure to rigorously purify. This time, it weighed in at exactly what

Mendeleev had predicted: 5.9 grams per cubic centimeter. Lecoq named it gallium, after the Latin name for France, Gallia.

Mendeleev published his table in 1869, first in an obscure Russian journal, and then in a respected German publication, under the title "On the Relationship of the Properties of the Elements to their Atomic Weights." In addition to gallium, his table included placeholders for other yet unknown elements—today called scandium, germanium, and technetium. Technetium is radioactive and so rare that it wasn't dis-

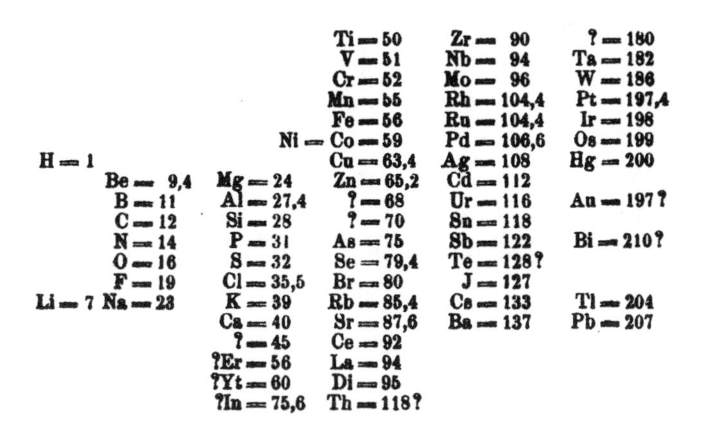

			Ti = 50	Zr = 90	? = 180
			V = 51	Nb = 94	Ta = 182
			Cr = 52	Mo = 96	W = 186
			Mn = 55	Rh = 104,4	Pt = 197,4
			Fe = 56	Ru = 104,4	Ir = 198
			Ni = Co = 59	Pd = 106,6	Os = 199
H = 1			Cu = 63,4	Ag = 108	Hg = 200
	Be = 9,4	Mg = 24	Zn = 65,2	Cd = 112	
	B = 11	Al = 27,4	? = 68	Ur = 116	Au = 197?
	C = 12	Si = 28	? = 70	Sn = 118	
	N = 14	P = 31	As = 75	Sb = 122	Bi = 210?
	O = 16	S = 32	Se = 79,4	Te = 128?	
	F = 19	Cl = 35,5	Br = 80	J = 127	
Li = 7	Na = 23	K = 39	Rb = 85,4	Cs = 133	Tl = 204
		Ca = 40	Sr = 87,6	Ba = 137	Pb = 207
		? = 45	Ce = 92		
		?Er = 56	La = 94		
		?Yt = 60	Di = 95		
		?In = 75,6	Th = 118?		

hydrogen 1 H 1.0079																		helium 2 He 4.0026
lithium 3 Li 6.941	beryllium 4 Be 9.0122												boron 5 B 10.811	carbon 6 C 12.011	nitrogen 7 N 14.007	oxygen 8 O 15.999	fluorine 9 F 18.998	neon 10 Ne 20.180
sodium 11 Na 22.990	magnesium 12 Mg 24.305												aluminium 13 Al 26.982	silicon 14 Si 28.086	phosphorus 15 P 30.974	sulfur 16 S 32.065	chlorine 17 Cl 35.453	argon 18 Ar 39.948
potassium 19 K 39.098	calcium 20 Ca 40.078		scandium 21 Sc 44.956	titanium 22 Ti 47.867	vanadium 23 V 50.942	chromium 24 Cr 51.996	manganese 25 Mn 54.938	iron 26 Fe 55.845	cobalt 27 Co 58.933	nickel 28 Ni 58.693	copper 29 Cu 63.546	zinc 30 Zn 65.39	gallium 31 Ga 69.723	germanium 32 Ge 72.61	arsenic 33 As 74.922	selenium 34 Se 78.96	bromine 35 Br 79.904	krypton 36 Kr 83.80
rubidium 37 Rb 85.468	strontium 38 Sr 87.62		yttrium 39 Y 88.906	zirconium 40 Zr 91.224	niobium 41 Nb 92.906	molybdenum 42 Mo 95.94	technetium 43 Tc [98]	ruthenium 44 Ru 101.07	rhodium 45 Rh 102.91	palladium 46 Pd 106.42	silver 47 Ag 107.87	cadmium 48 Cd 112.41	indium 49 In 114.82	tin 50 Sn 118.71	antimony 51 Sb 121.76	tellurium 52 Te 127.60	iodine 53 I 126.90	xenon 54 Xe 131.29
caesium 55 Cs 132.91	barium 56 Ba 137.33	57-70 *	lutetium 71 Lu 174.97	hafnium 72 Hf 178.49	tantalum 73 Ta 180.95	tungsten 74 W 183.84	rhenium 75 Re 186.21	osmium 76 Os 190.23	iridium 77 Ir 192.22	platinum 78 Pt 195.08	gold 79 Au 196.97	mercury 80 Hg 200.59	thallium 81 Tl 204.38	lead 82 Pb 207.2	bismuth 83 Bi 208.98	polonium 84 Po [209]	astatine 85 At [210]	radon 86 Rn [222]
francium 87 Fr [223]	radium 88 Ra [226]	89-102 * *	lawrencium 103 Lr [262]	rutherfordium 104 Rf [261]	dubnium 105 Db [262]	seaborgium 106 Sg [266]	bohrium 107 Bh [264]	hassium 108 Hs [269]	meitnerium 109 Mt [268]	ununnilium 110 Uun [271]	unununium 111 Uuu [272]	ununbium 112 Uub [277]		ununquadium 114 Uuq [289]				

*Lanthanide series	lanthanum 57 La 138.91	cerium 58 Ce 140.12	praseodymium 59 Pr 140.91	neodymium 60 Nd 144.24	promethium 61 Pm [145]	samarium 62 Sm 150.36	europium 63 Eu 151.96	gadolinium 64 Gd 157.25	terbium 65 Tb 158.93	dysprosium 66 Dy 162.50	holmium 67 Ho 164.93	erbium 68 Er 167.26	thulium 69 Tm 168.93	ytterbium 70 Yb 173.04
* * Actinide series	actinium 89 Ac [227]	thorium 90 Th 232.04	protactinium 91 Pa 231.04	uranium 92 U 238.03	neptunium 93 Np [237]	plutonium 94 Pu [244]	americium 95 Am [243]	curium 96 Cm [247]	berkelium 97 Bk [247]	californium 98 Cf [251]	einsteinium 99 Es [252]	fermium 100 Fm [257]	mendelevium 101 Md [258]	nobelium 102 No [259]

Mendeleev's original periodic table, as published in 1869, and the periodic table today

covered until 1937, when it was synthesized in a cyclotron—a kind of particle accelerator—some thirty years after Mendeleev died.

The Nobel Prize in Chemistry was first awarded in 1901, six years before Mendeleev's death. It is one of the great misjudgments of Nobel history that he never received the prize, for his table of elements was the central organizing principle of modern chemistry, the discovery that made possible our mastery of the science of substance, and the capstone to two thousand years of work that had begun in the laboratories of embalmers and alchemists.

But Mendeleev did eventually become a member of a more select club. In 1955, scientists at Berkeley produced just over a dozen atoms of a new element, again in a cyclotron, and in 1963 they designated it mendelevium, in honor of his magnificent achievement. The Nobel Prize has been awarded to more than eight hundred individuals, but only sixteen scientists have an element named for them. And Mendeleev is one of them, giving him his own spot on his table, where he appears as element 101, just a short distance from the likes of einsteinium and copernicium.

9

The Animate World

Though scholars since antiquity had speculated that material objects are made of fundamental building blocks, no one had guessed that so are living things. And so it must have come as quite a surprise when, in 1664, our old friend Robert Hooke sharpened his penknife until it was "as keen as a razor," shaved a thin slice from a piece of cork, peered at it through his homemade microscope, and became the first human to see what he would call "cells." He chose that name because they reminded him of the tiny bedrooms assigned to monks in their monasteries.

One can think of cells as the atoms of life, but they are more complex than atoms, and—even more shocking to those who first perceived them—they are themselves alive. A cell is a vibrant living factory that consumes energy and raw materials, and produces from them many diverse products, mainly proteins, which carry out almost every crucial biological function. It takes a lot of knowledge to perform the functions of a cell, so although cells don't have brains, they do "know" things—they know how to make the proteins and other materials we need to grow and function, and, perhaps most crucial, they know how to reproduce.

The most important single product of a cell is a copy of itself. As a result of that ability, we humans start from a single cell and, through a series of forty-plus cell doublings, we eventually come to be made of about thirty trillion cells—a hundred times more cells than there are stars in the Milky Way. It is a great wonder that the sum of our cells' activities, the interaction of a galaxy of unthinking individuals, adds up to a whole that is us. Just as staggering a thought is the notion that we could untangle how that all works, like computers that, unbidden by any programmer, analyze themselves. That is the miracle of biology.

The miracle appears even greater when you consider that most of the world of biology is invisible to us. That's partly due to the minuteness of cells and partly due to the magnificent diversity of life. If you exclude creatures like bacteria and count only living things with cells that have a nucleus, then, scientists estimate, there are roughly ten million species on our planet, of which we have discovered and classified only about 1 percent. There are at least 22,000 species of ants alone, and somewhere between one and ten million individual ants for every person on earth.

We are all familiar with a medley of backyard insects, but a scoop of good soil contains more types of creatures than we could ever count—hundreds or even thousands of invertebrate species, several thousand microscopic roundworms, and tens of thousands of types of bacteria. The presence of life on earth is so pervasive, in fact, that we are continually ingesting organisms that we'd probably rather not eat. Try buying peanut butter that's free of insect fragments: you can't. The government recognizes that producing insect-free peanut butter is impractical, so regulations allow for up to ten insect fragments per thirty-one-gram serving. Meanwhile, a serving of broccoli may contain sixty aphids and/or mites, while a jar of ground cinnamon may contain four hundred insect fragments.

That all sounds unappetizing, but it's good to remember that even our own bodies are not free of foreign life—we are, each of us, an entire ecosystem of living things. Scientists have identified, for example, forty-four genera (species groups) of microscopic organisms that live on your forearm, and at least 160 species of bacteria that live in people's bowels. Between your toes? Forty species of fungi. In fact, if you bother to total it up, you find that there are far more microbial cells in our bodies than human cells.

Our body parts each form a distinct habitat, and the creatures in your intestines or between your toes have more in common with the organisms in those regions of *my* body than with the creatures on your own forearm. There is even an academic center called the Belly Button Biodiversity project at North Carolina State University, set up to study the life that exists in that dark, isolated landscape. And then there are the infamous skin mites. Relatives of ticks, spiders, and scorpions, these creatures are less than a third of a millimeter long and live on your face—in hair follicles and glands connected to hair follicles—mainly near the nose, eyelashes, and eyebrows, where they suck the innards out of your juicy cells. But don't worry, they normally cause no ill effects, and if you're an optimist, you can hope you're among the half of the adult population that is free of them.

Given the complexity of life, its diversity in size, shape, and habitat, and our natural disinclination to believe that we are "mere" products of physical law, it is not surprising that biology lagged behind physics and chemistry in its development as a science. Like those other sciences, for biology to grow it had to overcome the natural human tendencies to feel that we are special and that deities and/or magic govern the world. And, as in those other sciences, that meant overcoming the God-centric doctrine of the Catholic Church and the human-centric theories of Aristotle.

Aristotle was an enthusiastic biologist—almost a quarter of his surviving writings pertain to that discipline. And while Aristotle's physics has our earth at the physical center of the universe, his biology, more personal, exalts humans, and males in particular.

Aristotle believed that a divine intelligence designed all living beings, which differ from the inanimate in that they have a special quality or essence that departs or ceases to exist when the living thing dies. Among all those blueprints for life, Aristotle argued, humans represent the high point. On this point Aristotle was so vehement that when he described a characteristic of a species that differs from the corresponding human characteristic, he referred to it as a deformity. Similarly, he viewed the human female as a deformed or damaged male.

The erosion of such traditional but false beliefs set the stage for the birth of modern biology. One of the important early victories over such ideas was the debunking of a principle of Aristotle's biology called spontaneous generation, in which living things were said to arise from inanimate matter such as dust. Around the same time, by showing that even simple life has organs as we do, and that we, like other plants and animals, are made of cells, the new technology of the microscope cast doubt on the old ways of thinking. But biology could not begin to really mature as a science until the discovery of its great organizing principle.

Physics, which concerns how objects interact, has its laws of motion; chemistry, which concerns how elements and their compounds interact, has its periodic table. Biology concerns itself with the ways in which species function and interact, and to succeed, it needed to understand why those species have the characteristics they do—an explanation other than "Because God made them that way." That understanding finally came with Darwin's theory of evolution based on natural selection.

* * *

Long before there was biology, there were observers of life. Farmers, fishermen, doctors, and philosophers all learned about the organisms of the sea and the countryside. But biology is more than what is detailed in catalogs of plants or field guides to the birds, for science doesn't just sit quietly and describe the world; it jumps up and screams ideas that explain what we see. To explain, though, is much more difficult than to describe. As a result, before the development of the scientific method, biology, like the other sciences, was plagued by explanations and ideas that were reasonable—but wrong.

Consider the frogs of ancient Egypt. Each spring, after the Nile inundated the surrounding lands, it left behind nutrient-rich mud, the kind of land that, through the farmers' diligent toil, would soon feed the nation. The muddy soil also yielded another crop that did not exist on drier land: frogs. The noisy creatures appeared so suddenly, and in numbers so vast, that they seemed to have arisen from the mud itself—which is precisely how the ancient Egyptians believed they came into being.

The Egyptian theory was not the product of flabby reasoning. Assiduous observers through most of history have reached the same conclusion. Butchers noted that maggots "appeared" on meat, farmers found mice "appearing" in bins in which wheat was stored. In the seventeenth century, a chemist named Jan Baptist van Helmont even went so far as to recommend a recipe to create mice from everyday materials: just put a few grains of wheat in a pot, add dirty underwear, and wait twenty-one days. The recipe reportedly often worked.

The theory behind van Helmont's concoction was spontaneous generation—that simple living organisms can arise spontaneously from certain inanimate substrates. Ever since the time of ancient Egypt, and probably before, people believed that some sort of life force or energy exists in all living creatures. Over time, a by-product of such views was the conviction that life energy could somehow become infused into inanimate matter, creating new life, and when that doctrine was synthesized into a coherent theory by Aristotle, it gained special authority. But just as certain key observations and experiments in the seventeenth century represented the beginning of the end of Aristotle's physics, so too did the rise of science in that century finally bring his ideas about biology under potent attack. Among the most memorable challenges was a test of spontaneous generation performed by Italian physician Francesco Redi in 1668. It was one of biology's first truly scientific experiments.

Redi's method was simple. He procured some widemouthed jars and placed in them samples of fresh snake meat, fish, and veal. Then he left some of the jars uncovered while covering others with a gauzelike material or paper. He hypothesized that if spontaneous generation really occurred, flies and maggots should appear on the meat in all three situations. But if maggots arose instead, as Redi suspected, from tiny invisible eggs laid by flies, they should appear on the meat in the uncovered jars but not in the jars covered by paper. He also predicted that maggots would appear on the gauze covering in the remaining jars, which was as close to the meat as the hungry flies could get. That was exactly what occurred.

Redi's experiment met with a mixed reception. To some, it appeared to debunk spontaneous generation. Others chose to ignore it, or to find fault. Many probably fell in the latter category simply because they were biased toward maintaining their prior beliefs. After all, the issue had theological implications—some felt that spontaneous generation preserved a role for God in creating life. But there were also scientific reasons for doubting Redi's conclusion—for example, it could be wrong to extrapolate the validity of his experiment beyond the creature he'd studied. Perhaps all he had demonstrated was that spontaneous generation does not apply to flies.

To his credit, Redi himself kept an open mind—he even found other cases where he suspected that spontaneous generation *did* occur. Ultimately, the issue would be argued about for another two hundred years, until Louis Pasteur, in the late nineteenth century, put it to rest for good with his careful experiments showing that not even microorganisms are generated spontaneously. Still, though not definitive, Redi's work was gorgeous. It stands out especially because anyone could have carried out a similar test, and yet no one thought to.

People often think of great scientists as having extraordinary intelligence, and in society, and especially in business, we tend to shun people who don't blend well with others. But it is those who are different who often see what others do not. Redi was a complex man—a scientist but also a man of superstition who greased himself with oil to ward off disease, a physician and naturalist but also a poet who wrote a classic in praise of Tuscan wines. With regard to spontaneous generation, only Redi was odd enough to think outside the box, and in an age before scientific reasoning was commonplace, he reasoned and acted as scientists do. In doing so, he not only cast doubt on an invalid theory, but he poked a stick at Aristotle, and pointed conspicuously to a new approach to answering the questions of biology.

* * *

Redi's experiment was in great part a reaction to microscope studies that had recently revealed that minute creatures are complex enough to have reproductive organs—for the belief that "lower animals" are too simple to reproduce was one of Aristotle's arguments for spontaneous generation.

The microscope had actually been invented decades earlier—around the time of the telescope—though no one knows for certain exactly when, or by whom. We do know that at first the same Latin word, *perspicillum*, was employed for both, and Galileo even used the same instrument—his telescope—to gaze both inward and outward. "With this tube," he told a visitor in 1609, "I have seen flies which look as big as lambs."

By revealing the details of a realm of nature that could never have been imagined by the ancients—or accounted for in theories—the microscope, like the telescope, eventually helped to open scholars' minds to different ways of thinking about their subject, creating a thread of intellectual progress that would reach a high point with Darwin. But, like the telescope, the microscope also initially met with strong opposition. Medieval scholars were wary of "optical illusions" and distrusted any device that stood between them and the objects they perceived. And while the telescope had its Galileo, who quickly stood up to the critics and adopted the device, it took half a century for the champions of the microscope to make their mark.

One of the greatest champions was Robert Hooke, who did his microscopic studies at the behest of the Royal Society, and thus contributed to the roots of biology, just as he had contributed to chemistry and physics. In 1663, the Royal Society assigned Hooke the task of presenting at least one novel observation at each meeting. Despite an eye infirmity that made it both difficult and painful to stare into a lens for a long time, he lived up to the challenge and made a long series of extraordinary observations employing improved instruments that he himself had designed.

In 1665, the thirty-year-old Hooke published a book titled *Micrographia*, or "Small Drawings." It was a bit of a hodgepodge of Hooke's work and ideas in several fields, but it made a splash by revealing a strange new microworld through fifty-seven amazing illustrations that Hooke himself drew. They exposed to human perception for the first time the anatomy of a flea, the body of a louse, the eye of a fly, and the stinger of a bee, all blown up to full-page images, some even presented as foldouts. That even simple animals have body parts and organs as we do was not just a striking revelation to a world that had never seen an insect magnified; it was also a direct contradiction of Aristotelian doctrine, a revelation similar to Galileo's discovery that the moon has hills and valleys just like the earth does.

The year *Micrographia* was published was the year the Great Plague, which would kill one in seven Londoners, reached its summit. The next year, London was engulfed by the Great Fire. But despite all that chaos and suffering, people read Hooke's book, and it became a best seller. So engrossed was Samuel Pepys, the famous diarist and naval administrator and later a member of Parliament, that he sat up until two in the morning devouring it, then called it "the most ingenious book that ever I read in my whole life."

Though Hooke excited a new generation of scholars, he also drew the ridicule of doubters who found it difficult to accept his sometimes gro-

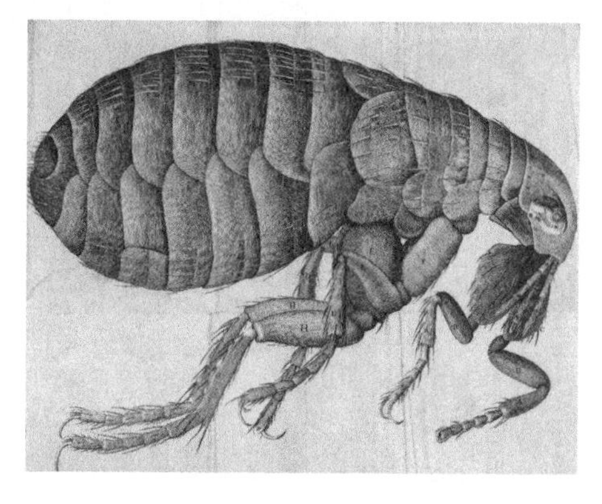

Hooke's *Micrographia*

tesque portrayals, based on observations employing an instrument they did not trust. The low point came when, attending a satire of contemporary science written by English playwright Thomas Shadwell, Hooke was humiliated upon realizing that the experiments being mocked on the stage before him were mostly his own. They had been drawn from his beloved book.

One man who didn't doubt Hooke's claims was an amateur scientist named Anton van Leeuwenhoek (1632–1723). He was born in Delft, Holland. His father made baskets in which the famous blue-and-white Delft pottery was packaged for shipment around the world; his mother came from a family engaged in another Delft specialty—brewing beer. At age sixteen, young Anton took a job as cashier and bookkeeper for a cloth merchant, and in 1654 he opened his own business, selling fabric, ribbons, and buttons. To that he would soon add another, unrelated, occupation: seeing to the maintenance and upkeep of Delft's town hall.

Leeuwenhoek never attended college and did not know Latin—then the language of science. And though he would live past ninety, he left the Netherlands just twice—once to visit Antwerp, in Belgium, and once to go to England. But Leeuwenhoek did read books, and one that inspired him was Hooke's best seller. The book changed his life.

In its preface, *Micrographia* explains how to build simple microscopes, and as a fabric merchant, Leeuwenhoek probably had some experience grinding lenses, for he would have needed them to examine samples of linen. But after reading *Micrographia*, he became a fanatical lens maker, devoting hour after hour to creating new microscopes and making observations employing them.

In his early work, Leeuwenhoek simply repeated Hooke's experiments, but he soon eclipsed them. Hooke's microscopes were, for their time, technologically superior, and he had wowed the Royal Society with his magnifications of twenty to fifty power. So one can only imagine the astonishment when, in 1673, the secretary of the Society, Henry Oldenburg, received a letter informing him that an uneducated custodian and fabric purveyor in the Netherlands had "devised microscopes which far surpass those which we have hitherto seen." In fact, the forty-one-year-old Leeuwenhoek was reaching magnifications ten times those achieved by Hooke.

It was superior craftsmanship rather than clever design that made Leeuwenhoek's microscopes so powerful. They were in fact simple devices, made from just a single lens ground from select fragments of glass, or even grains of sand, and mounted on plates made from gold and silver that he sometimes extracted from the ore himself. He fixed each specimen permanently and made a new microscope for every study, perhaps because achieving the proper positioning was as difficult as creating the lens. Whatever the reason, he didn't share it with anyone and was generally very secretive about his methods, because, like Newton, he wanted to avoid "contradiction or censure from others." Over his long life, he produced more than five hundred lenses, but to this day no one knows exactly how he created them.

When word of Leeuwenhoek's achievements arrived, the English and Dutch navies were firing cannons at each other in the Anglo-Dutch Wars, but being at war with Leeuwenhoek's country didn't stop Oldenburg: he encouraged Leeuwenhoek to report his findings, and the Dutchman did. In his first letter, Leeuwenhoek, intimidated by the attention of the famous Royal Society, apologized if they noted shortcomings in his work. It was, he wrote, "the outcome of my own unaided impulse and curiosity alone; for beside myself in our town there are no philosophers who practice this art; so pray take not amiss my poor pen and the liberty I have taken in setting down my random notions."

Leeuwenhouk's "notions" were an even greater revelation than Hooke's. For where Hooke had seen in detail the body parts of tiny insects, Leeuwenhoek saw entire creatures that were too small to see with the naked eye, complete societies of organisms whose existence no one had previously suspected, some a thousand or even ten thousand times tinier than the smallest animal visible to the naked eye. He called them "animalcules." Today we call them microorganisms.

If Galileo reveled in viewing the landscape of the moon and spying the rings of Saturn, Leeuwenhoek took equal delight in observing through his lenses new worlds of bizarre and minute creatures. In one letter, he wrote of the world that existed in a drop of water: "I now saw very plainly that these were little eels, or worms, lying all huddled up together and wriggling . . . the whole water seemed to be alive with these multifarious animalcules . . . I must say, for my part, that no more pleasant sight has ever come before my eye than these thousands of living creatures, seen all alive in a little drop of water."

But if Leeuwenhoek sometimes reported a God's-eye overview of entire worlds, in other reports he told of magnifying individual creatures enough to describe many new species in great detail. For example, he described one creature as having "stuck out two little horns which were continually moved, after the fashion of a horse's ears . . . [with a roundish body], save only that it ran somewhat to a point at the hind end; at which pointed end it had a tail." Over a period of fifty years, Leeuwenhoek never attended a meeting of the Royal Society, but he wrote it hundreds of letters, and the majority have been preserved. Oldenburg had them edited and translated into English or Latin, and the Royal Society published them.

Leeuwenhoek's work created a sensation. The world was stunned to learn that there were whole universes of creatures in every drop of pond water, and entire classes of life entirely hidden from our senses. What's more, when Leeuwenhoek turned his microscopes upon human tissue such as sperm cells and blood capillaries, he helped to reveal our own construction, and how it is unexceptional, in that we have much in common with other life forms.

Like Hooke, Leeuwenhoek had doubters who believed he was making everything up. He countered them by providing signed testimonials from respected eyewitnesses, notaries public, and even the pastor of the congregation in Delft. The majority of scientists believed him, and Hooke was even able to reproduce some of Leeuwenhoek's research. As the word spread, visitors from everywhere showed up at Leeuwenhoek's shop, asking to gaze at his tiny beasts. Charles II, the founder and patron of the Royal Society, asked Hooke for a showing of one of the Leeuwenhoek experiments he had replicated, and Peter the Great of Russia visited Leeuwenhoek himself. Not bad for a guy who ran a fabric store.

In 1680, Leeuwenhoek was elected, in absentia, a fellow of the Royal Society, and he kept working till his end, at age ninety-one, some forty years later. No comparable microbe hunter would come onto the scene for another 150 years.

As Leeuwenhoek lay dying, the last thing he did was ask a friend to translate his final two letters into Latin and send them to the Society. He had also prepared them a gift: a black-and-gilt cabinet filled with his best microscopes, some of which he had never before shown anyone. Today, only a handful of his microscopes remain intact; in 2009, one sold at auction for 312,000 pounds.

In his long life, Leeuwenhoek helped establish many aspects of what would become biology—microbiology, embryology, entomology, histology—causing one twentieth-century biologist to call Leeuwenhouk's letters the "most important series of communications that a scientific society has ever received." Just as important—like Galileo in physics and Lavoisier in chemistry—Leeuwenhoek helped to establish a scientific tradition in the field of biology. As the pastor of the New Church at Delft wrote to the Royal Society on Leeuwenhoek's death, in 1723, "Anton van Leeuwenhoek considered that what is true in natural philosophy can be most fruitfully investigated by the experimental method, supported by the evidence of the senses; for which reason, by diligence and tireless labor he made with his own hand certain most excellent lenses, with the aid of which he discovered many secrets of Nature, now famous throughout the whole philosophical World."

* * *

If Hooke and Leeuwenhoek were in a sense the Galileos of biology, its Newton was Charles Darwin (1809–1882). Fittingly, he is buried just a few feet away from Newton in Westminster Abbey, his pallbearers having included two dukes and an earl as well as past, present, and future presidents of the Royal Society. Though Darwin's burial in an abbey may seem incongruous to some, "it would have been unfortunate," said the Bishop of Carlisle in his memorial sermon, "if anything had occurred to give weight and currency to the foolish notion . . . that there is a necessary conflict between a knowledge of Nature and a belief in God." The interment was a glorious end for a man whose principal scientific achievement was met, at first, with little more than a yawn, and then with much venom and skepticism.

One of those initially underwhelmed was Darwin's own publisher, John Murray, who had agreed to release the book in which Darwin elaborated on his theory but gave it an initial press run of just 1,250 copies. Murray had good reason to worry, for those who had seen Darwin's book in advance were unenthusiastic. One early reviewer even recommended that Murray not publish it at all—it was "an imperfect and comparatively meager exposition of his theory," he wrote. And then the reviewer suggested that Darwin write a book on pigeons instead, and include in *that* book a brief statement of his theory. "Everybody is interested in pigeons," the reviewer advised. "The book would . . . soon be on every table." The advice was passed on to Darwin, but he declined it.

Not that he himself was confident that his book would sell. "God knows what the public will think," he remarked.

Darwin need not have worried. *On the Origin of Species by Means of Natural Selection; or, the Preservation of Favoured Races in the Struggle for Life* would become biology's *Principia*. Published on November 24, 1859, all 1,250 copies were immediately snatched up by eager booksellers, and it has been in print ever since. (The book did not, however, sell out on its publication day, as legend has it.) It was gratifying validation for the man who had had the passion and patience to spend twenty years accumulating evidence for his ideas—an effort so monumental that just one of its many by-products was a 684-page monograph on barnacles.

Darwin's predecessors had learned many descriptive details about life forms from bacteria to mammals, but they hadn't had a clue about the more fundamental question of what drove species to have the characteristics they possess. Like physicists before Newton or chemists before the periodic table, pre-Darwinian biologists gathered data but didn't know how it fit together. They couldn't, for before Darwin the young field of biology was shackled by the conviction that the origins and interrelationships of different forms of life were beyond science—a conviction that arose from the literal acceptance of the biblical story of creation, which held that the earth and all life forms were created in six days and that, in the time since, species had not changed.

It's not that there hadn't been thinkers who'd pondered the idea that species evolve—there were, going back to the Greeks, and these included Darwin's own grandfather, Erasmus Darwin. But pre-Darwinian evolution theories were vague and not much more scientific than the religious doctrine they would be replacing. As a result, though there was talk of the idea of evolution before Darwin, most people, including scientists, accepted that humans rested atop a pyramid of more primitive species whose characteristics were fixed and had been designed by a creator to whose thinking we could never be privy.

Darwin changed that. If before him there existed a grove of speculations about evolution, his theory towered over the other trees, a majestic specimen of careful science. For every argument or piece of evidence his precursors supplied, he offered a hundred. More important, he discovered the *mechanism* behind evolution—natural selection—and thus made evolution theory testable, and scientifically respectable, freeing biology from its reliance on God and allowing it to become a true science, one rooted, like physics and chemistry, in physical law.

* * *

Born at his family's home in Shrewsbury, England, on February 12, 1809, Charles was the son of Robert Darwin, the town physician, and Susannah Wedgwood, whose father had founded the pottery firm of that name. The Darwins were a well-to-do and illustrious family, but Charles was a poor student who hated school. He would later write that he had a bad memory for rote learning and "no special talents." He was selling himself short, for he also recognized that he had a "great curiosity about facts, and their meaning," and "energy of mind shown by vigorous and long-continued work on the same subject." These latter two traits are, for a scientist—or any innovator—indeed special talents, and they would serve Darwin well.

Darwin's curiosity and determination are well illustrated by an incident that occurred when he was in college at Cambridge, obsessed with a hobby of collecting beetles. "One day," he wrote, "on tearing off some old bark, I saw two rare beetles & seized one in each hand; then I saw a third & new kind, which I could not bear to lose, so I popped the one which I held in my right hand into my mouth." Only from a boy of that character can emerge a man with the tenacity to put together 684 pages on the topic of barnacles (though before he was finished he would write, "I hate a barnacle as no man ever did before").

It took many years for Charles to find his calling. His journey began in the fall of 1825 when his father sent him, at age sixteen, not to Cambridge but to the University of Edinburgh—to study medicine, as both he and Charles's grandfather had done. It proved to be a bad decision.

For one, Charles was famously squeamish, and this was an era in which operations featured copious splashes of blood and screaming patients, cut into without the benefit of anesthetic. Still, squeamishness wouldn't stop Charles, years later, from dissecting dogs and ducks as he searched for evidence supporting his theory of evolution. Probably what proved fatal to his medical studies was the lack of both interest and motivation. As he would later write, he had become convinced that his father would leave him enough property "to subsist on with some comfort," and this expectation was "sufficient to check any strenuous effort to learn medicine." And so, in the spring of 1827, Charles left Edinburgh without a degree.

Cambridge was his second stop. His father sent him there with the idea that he should study divinity and then embark upon a clerical career.

This time Charles completed his degree, ranking tenth out of 178 graduates. His high ranking surprised him, but it reflected, perhaps, that he had developed a genuine interest in geology and natural history—as evidenced by his beetle collecting. Still, he seemed headed for a life in which science would be at best a hobby, while his professional energies would be devoted to the church. But then, on returning home from a postgraduation geological walking tour of North Wales, Darwin found a letter that presented a different option: the chance to sail around the world on the HMS *Beagle*, under one Captain Robert Fitzroy.

The letter was from John Henslow, a Cambridge professor of botany. Despite his high ranking, Darwin hadn't stood out to many at Cambridge; Henslow, however, had seen potential in him. He once remarked, "What a fellow that Darwin is for asking questions"—a seemingly bland compliment, but it says that in Henslow's mind, Darwin had the soul of a scientist. Henslow befriended the curious student, and when he was asked to recommend a young man for the position of naturalist on the voyage, he recommended Darwin.

Henslow's letter to Darwin was the culmination of a series of unlikely events. It all began when the *Beagle*'s previous captain, Pringle Stokes, shot himself in the head and, after the bullet didn't kill him, died of gangrene. Fitzroy, Stokes's first lieutenant, brought the ship home, but it wasn't lost on him that Stokes's depression had been spurred by the loneliness of a multiyear sea voyage in which the captain was forbidden to socialize with his crew. Fitzroy's own uncle had slit his throat with a razor a few years earlier, and some four decades later Fitzroy himself would follow suit, so he must have sensed that his captain's fate was one he should do his best to avoid. As a result, when the twenty-six-year-old Fitzroy was offered the opportunity to succeed Stokes, he decided he needed a companion. It was the custom at the time for the ship's doctor to double as its naturalist, but Fitzroy instead put word out that he sought a young "gentleman-naturalist" of high social standing—a person to serve, essentially, as his hired friend.

Darwin was not Fitzroy's first choice for the position—it had previously been offered to a number of others. Had any of them accepted it, Darwin most likely would have proceeded to his quiet life in the church and never created his theory of evolution—just as Newton would probably never have completed and published his great work had Halley not stopped to see him and ask about the inverse square law. But the position Fitzroy was offering paid nothing—the compensation was to come from the later sale of specimens collected on shore visits along the way—and none of those asked were willing or able to spend years at sea, self-financed. As a result, the choice finally fell to the twenty-two-year-old Darwin, offering him a chance at adventure—and to avoid starting a career in which he'd preach that the earth had been created on the night preceding October 23, 4004 B.C. (as claimed in a biblical analysis published in the seventeenth century). Darwin seized the opportunity. It would change both his life and the history of science.

The *Beagle* set off in 1831 and wouldn't return until 1836. It was not a comfortable trip. Darwin resided and worked in the ship's tiny poop cabin, in the roughest-riding section of the ship. He shared the room with two others and slept in a hammock slung over the chart table. "I have just room to turn round & that is all," he reported in a letter to Henslow. Not surprisingly, he was racked by seasickness. And though Darwin forged a friendship of sorts with Fitzroy—he was the only member of the ship to have any intimacy with the captain, and they usually dined together—they quarreled often, especially over slavery, which Darwin despised but which they observed repeatedly in their time ashore.

Still, the discomforts of the voyage were offset by the unparalleled excitement of the shore visits. During those periods, Darwin participated in the Carnival in Brazil, watched a volcano erupt outside Osorno, Chile, experienced an earthquake and walked through the ruins it left in Concepción, and observed revolutions in Montevideo and Lima. All the while, he collected specimens and fossils, packaged them, and shipped them in crates back to Henslow in England for storage.

Darwin would later consider the voyage to have been the main formative event of his life, both for the impressions it left on his character and for the new appreciation it gave him of the natural world. It was *not* during the voyage, however, that Darwin made his famous discoveries regarding evolution, or even grew to accept that evolution occurred. He in fact ended the voyage as he had begun it—with no doubts regarding the moral authority of the Bible.

Yet his plans for the future did change. As the voyage ended, he wrote to a cousin who'd made a career in the church, "Your situation is above envy; I do not venture to even frame such happy visions. To a person fit to take the office, the life of a Clergyman is . . . respectable & happy."

Despite those encouraging words, Darwin had decided that he himself was not suited to that life, and he chose instead to make his way in the world of London science.

* * *

Back in England, Darwin found out that the observations he had detailed in casual letters to Professor Henslow had received some scientific notice—in particular, those regarding geology. Soon Darwin was lecturing at the prestigious Geological Society of London on topics like "the Connexion of Certain Volcanic Phenomena and the Formation of Mountain-Chains and the Effects of Continental Elevations." Meanwhile, he enjoyed financial independence thanks to a stipend of four hundred pounds per year from his father. Coincidentally, it was the same amount Newton had earned when he began at the mint, but in the 1830s, according to the British National Archives, it was "only" five times the wage of a craftsman (though still enough to buy twenty-six horses or seventy-five cows). The money allowed Darwin to devote time to turning his *Beagle* diary into a book, and to sorting through the many animal and plant specimens he'd collected. It was this effort that would change our ideas about the nature of life.

Since Darwin hadn't had any great epiphanies about biology during his voyage, he probably expected the scrutiny of the specimens he'd sent home to result in a solid but not revolutionary body of work. There were soon signs, however, that his investigations might be more exciting than expected—he had given some of his specimens to specialists to analyze, and many of their reports astonished him.

One group of fossils, for example, suggested "a law of succession"—that extinct South American mammals had been replaced by others of their kind. Another report, on the mockingbirds of the Galápagos Islands, informed him that there were three species, not the four he had believed, and that they were island specific, as were the giant tortoises found there. (The story about his being inspired to a eureka moment by observing differences in the beaks of the finches on different islands of the Galápagos is apocryphal. He did bring back finch specimens, but he hadn't been trained in ornithology and had actually misidentified them as a mix of finches, wrens, "Gross-beaks," and blackbird relatives—and they were not labeled by island.)

Perhaps the most striking of the experts' reports concerned a specimen of rhea, or South American ostrich, that Darwin and his team had cooked and eaten before realizing its possible significance and shipping its remains home. That specimen turned out to belong to a new species, which, like the common rhea, had its own principal range but which also competed with the common rhea in an intermediate zone. This contradicted the conventional wisdom of the time, which held that every species is optimized for its particular habitat, leaving no room for ambiguous regions in which similar species compete.

As these provocative studies were coming in, Darwin's own thoughts on the role of God in creation were evolving. One major influence was Charles Babbage, who held Newton's former position as Lucasian professor of mathematics at Cambridge, and is best known for inventing the mechanical computer. Babbage hosted a series of soirees attended by freethinkers and was himself writing a book proposing that God worked through physical laws, not fiat and miracle. That idea, which provided the most promising basis for the coexistence of religion and science, appealed to the young Darwin.

Gradually, Darwin became convinced that species were not unchanging life forms designed by God to fit into some grand scheme, but rather had somehow adapted themselves to fit into their ecological niche. By the summer of 1837—the year after the *Beagle* ended her voyage—Darwin had become a convert to the idea of evolution, though he was still far from formulating his particular theory of it.

Soon Darwin was rejecting the notion that humans are superior, or indeed that any animal is superior to another, and in its place he now held the conviction that each species is equally marvelous, a perfect or nearly perfect fit for its environment and its role within it. None of this, to Darwin, precluded God from playing a hands-on role: he believed that God had designed the laws governing reproduction to allow species to alter themselves as needed to adapt to environmental change.

If God created laws of reproduction that enabled species to become tailored to their environment, what are those laws? Newton understood God's plan for the physical universe through his mathematical laws of motion; so, too, did Darwin—at least initially—seek the mechanism of evolution, thinking it would explain God's plan for the living world.

Like Newton, Darwin began to fill a series of notebooks with his thoughts and ideas. He analyzed the relationships among the species and fossils he had observed in his travels; he studied an ape, an orangutan, and monkeys in the London Zoo, taking notes on their human-like emotions; he examined the work of pigeon, dog, and horse breeders and pondered how great a variation in traits could be produced through their method of "artificial selection"; and he speculated in a grand manner regarding how evolution had an impact on metaphysical questions and human psychology. And then, around the month of September 1838, Darwin read T. R. Malthus's popular *Essay on the Principle of Population.* It put him on the path to finally discovering the process through which evolution occurred.

Malthus had not written a pleasant book. Misery, in his view, was the natural and eventual state of humankind, because population increases invariably lead to violent competition for food and other resources. Due to limits on land and production, he argued, those resources can only increase "arithmetically," as in the series 1, 2, 3, 4, 5, and so on, while population increases each generation according to the series 1, 2, 4, 8, 16, and so on.

Today we know that a single squid can produce three thousand eggs in one season. If each egg grew into a squid and reproduced, by the seventh generation the volume of squid would fill a hollowed-out earth; in fewer than thirty generations, the eggs alone would fill the observable universe.

Darwin didn't have that particular data, and he wasn't good at math, but he knew enough to realize that the Malthusian scenario doesn't happen. Instead, he reasoned, of the prodigious number of eggs and offspring nature produces, competition leaves only a few—on average, those best adapted—to survive. He called the process "natural selection," to emphasize the comparison with the artificial selection exercised by breeders.

Later, in his autobiography, Darwin described having an epiphany: "It at once struck me that under these circumstances favorable variations would tend to be preserved, and unfavorable ones to be destroyed." But new ideas rarely jump into the discoverer's mind that suddenly, or neatly formed, and Darwin's description seems to have been a distortion of happy hindsight. Examinations of the notebooks he was keeping at the time reveal a different story: at first he only sniffed the trace of an idea, and it would be several years before he perceived it clearly enough to set it down on paper.

One reason the idea of natural selection took some time to develop is that Darwin recognized that weeding out the unfit in every generation may hone a species's traits, but it will not create *new* species—individuals that are so dissimilar from the original species that they can no longer interbreed and produce fertile offspring. For that to happen, the culling of existing traits must be complemented by a source of *new* traits. That, Darwin eventually concluded, arises from pure chance.

Bill color in zebra finches, for example, normally varies from paler to stronger red. Through careful breeding, one might be able to create populations favoring one or the other, but a zebra finch with a new beak color—say, blue—can occur only through what we now call a mutation: a chance alteration in the structure of a gene that results in a novel, variant form of the organism.

Now Darwin's theory could finally jell. Together, random variation and natural selection create individuals with new traits, and give those traits that are advantageous an increased chance of being propagated. The result is that, just as breeders produce animals and plants with the traits they desire, so, too, does nature create species that are well adapted to their environment.

The realization that randomness plays a role represented an important milestone in the development of science, for the mechanism Darwin had discovered made it difficult to reconcile evolution with any substantive idea of divine design. Of course, the concept of evolution in itself contradicts the biblical story of creation, but Darwin's *particular* theory now went further, making it difficult to rationalize the Aristotelian and traditional Christian views that the unfolding of events is driven by purpose rather than indifferent physical law. In that respect, Darwin did for our understanding of the living world what Galileo and Newton had done for the inanimate: he divorced science from its roots in both religious questioning and the ancient Greek traditions.

* * *

Darwin, like Galileo and Newton, was a man of religious faith, so his evolving theory presented him with contradictions in his belief system. He tried to avoid the clash by accepting both the theological and scientific views, each in their own context, rather than actively attempting to reconcile them. Yet he couldn't avoid the issue entirely, for in January 1839 he married his first cousin, Emma Wedgwood, a devout Christian who was disturbed by his views. "When I am dead," he once wrote her,

Annie Darwin (1841–1851)

"know that many times I have . . . cryed over this." Despite their differences, their bond was a strong one, and they remained throughout their lives a devoted couple, producing ten children.

Though much has been written regarding the question of reconciling evolution with Christianity, it was the death, years later, of the Darwins' second child, Annie, at age ten, that, as much as his work on evolution, finally destroyed Darwin's faith in Christianity. The cause of Annie's death is unclear, but she suffered at the end for more than a week, with high fever and severe digestive issues. Afterward Darwin wrote, "We have lost the joy of the Household, and the solace of our old age:—she must have known how we loved her; oh that she could now know how deeply, how tenderly we do still & shall ever love her dear joyous face."

The couple's first child had come in 1839. By then Darwin, just thirty, had begun to suffer debilitating bouts of a (to this day) mysterious illness. For the rest of his life, the joy he experienced from his family and scientific work would be punctuated by frequent eruptions of painful disability that at times left him unable to work for months on end.

Darwin's symptoms pointed everywhere, like the plagues in the Bible: stomach pain, vomiting, flatulence, headaches, heart palpitations, shivering, hysterical crying, tinnitus, exhaustion, anxiety, and depression. The attempted cures, some of which the desperate Darwin signed on for against his better judgment, were equally diverse: vigorous rubbing with cold wet towels, footbaths, ice rubs, freezing-cold showers, faddish electrotherapy employing shocking belts, homeopathic medicines, and that Victorian standard, bismuth. Nothing worked. And so the man who at twenty was a rugged adventurer had become, by age thirty, a frail and reclusive invalid.

With the new child, his work, and the illness, the Darwins began to withdraw, giving up parties and their old circles. Darwin's days became quiet and routine, as alike "as two peas." In June 1842, Darwin finally completed a thirty-five-page synopsis of his evolution theory; that September, he persuaded his father to lend him the money to buy a fifteen-acre retreat in Down, Kent, a parish of about four hundred inhabitants, sixteen miles from London. Darwin referred to it as "the extreme verge of the world." His life there was like that of the prosperous country parson he had once intended to be, and by February 1844 he had used his quiet time to expand the work into a 231-page manuscript.

Darwin's manuscript was a scientific will, not a work intended for immediate publication. He entrusted it to Emma with a letter that she was to read it in the event of his "sudden-death," which, due to his illness, he feared might be imminent. The letter instructed her that it was his "most solemn and last request" that after his demise the manuscript be made public. "If it be accepted even by one competent judge," Darwin wrote, "it will be a considerable step in science."

Darwin had good reason not to want his views published in his lifetime. He had earned a stellar reputation in the highest circles of scientific society, but his new ideas were bound to attract criticism. What's more, he had many clerical friends—not to mention a wife—who supported the creationist status quo.

Darwin's reasons for hesitation seemed to be validated in the fall of that year, when a book called *Vestiges of the Natural History of Creation* appeared, published anonymously.* The book did not present a valid theory of evolution, but it did weave together several scientific ideas, including the transmutation of species, and it became an international

*Robert Chambers, an Edinburgh publisher of popular periodicals, was officially named as the book's author in 1884, thirteen years after his death, but Darwin had guessed that Chambers had written the book after a meeting with him in 1847.

best seller. The religious establishment, however, railed against its unknown author. One reviewer, for example, accused him of "poisoning the foundations of science, and sapping the foundations of religion."

Some in the scientific community were not much kinder. Scientists have always been a tough crowd. Even today, with easy communication and travel enabling more cooperation and collaboration than ever before, presenting new ideas can open you to rude attack, for, along with a passion for their subject and their ideas, scientists sometimes also exhibit fervor in opposing work they deem misguided, or simply uninteresting. If the talk being given by a visitor describing his work at a research seminar did not prove worthy of his attention, one famous physicist I knew would pull out a newspaper, open it wide, and start reading, displaying conspicuous boredom. Another, who liked to sit near the front of the room, would stand in the midst of the talk, announce his negative opinion, and walk out. But the most interesting display I've seen came from yet another well-known scientist, a fellow familiar to generations of physicists because he'd written the standard graduate text on electromagnetism.

Sitting up front in a seminar room in which the rows of chairs were only about a dozen deep, this professor raised his Styrofoam coffee cup high over his head and rotated it back and forth slightly so that all those behind him—but not the puzzled speaker in front of him—could see that he had written on it, in large block letters, the message THIS TALK IS BULLSH-T! And then, having made his contribution to the discourse, he got up and walked out. Ironically, the talk was on the topic of "The Spectroscopy of Charm-Anticharm Particles." Though the word "charm" here is a technical term unrelated to its everyday meaning, I think it is fair to say that this professor belonged to the category "anticharm." However, if that is the reception given an idea judged dubious in a field as arcane as that, one can imagine the brutality exhibited toward "big ideas" that challenge received wisdom.

The fact is, though much is made of the opposition of advocates of religion to new ideas in science, there is also a strong tradition of opposition from scientists themselves. That is usually a *good* thing, for when an idea is misguided, scientists' skepticism serves to protect the field from racing in the wrong direction. What's more, when shown the proper evidence, scientists are quicker than just about anyone else to change their minds and accept strange new concepts.

Still, change is difficult for all of us, and established scientists who've devoted a career to furthering one way of thinking sometimes react quite negatively to a contradictory mode. As a result, to propose a startling new scientific theory is to risk exposing yourself to attack for being unwise, misguided, or just plain inadequate. There aren't many foolproof ways to foster innovation, but one way to kill it is to make it unsafe to challenge the accepted wisdom. Nevertheless, that is often the atmosphere in which revolutionary advances must be made.

In the case of evolution, Darwin had plenty to fear, as evidenced, for example, by the reaction to *Vestiges* by his friend Adam Sedgwick, a distinguished professor at Cambridge who had taught Darwin geology. Sedgwick called *Vestiges* a "foul book" and wrote a scathing eighty-five-page review. Before opening himself up to such attacks, Darwin would amass a mountain of authoritative evidence to support his theory. That effort would occupy him for the next fifteen years, but in the end, it would be responsible for his success.

* * *

Through the 1840s and '50s, Darwin's family grew. His father died in 1848 and left him that considerable sum that Darwin had speculated about decades earlier while studying medicine—it came to about fifty thousand pounds, the equivalent of millions of dollars today. He invested wisely and became very prosperous, easily able to care for his large family. But his stomach issues continued to plague him, and he became yet more reclusive, missing even his father's burial service due to his own illness.

All the while, Darwin continued to develop his ideas. He examined and experimented on animals, such as the pigeons his colleague would suggest he write about and, of course, those barnacles. He also experimented with plants. In one series of investigations, he tested the common notion that viable seeds could not reach distant oceanic islands. He attacked the question from many angles: he tested garden seeds that had been steeped for weeks in brine (to mimic seawater); he searched for seeds on birds' legs and in their droppings; and he fed seed-stuffed sparrows to an owl and an eagle in the London Zoo and then examined their pellets. All his studies pointed to the same conclusion: seeds, Darwin found, are hardier and more mobile than people had thought.

Another issue Darwin spent considerable time on was the question of diversity: Why had natural selection produced such a wide variation between species? Here he took inspiration from the economists of his day, who spoke often of the concept of "division of labor." Adam Smith had shown that people are more productive if they specialize, rather than each trying to create a complete article. The idea triggered Darwin to theorize that a given tract of land could support more life if its inhabitants were each highly specialized to exploit different natural resources.

Darwin expected that, if his theory were true, he would find more diverse life in areas where there was massive competition for limited resources, and he sought evidence to support or contradict that idea. Such thinking was typical of Darwin's novel approach to evolution: while other naturalists looked for evidence of evolution in the development, over time, of family trees linking fossils and living forms, Darwin looked for it in the distribution and relationships among species in his own time.

To examine the evidence, Darwin had to reach out to others. So, though physically isolated, he solicited input from many and, like Newton, depended on the mail service—in particular a new, cheap "penny post" program that enabled him to build an unparalleled network of naturalists, breeders, and other correspondents who supplied him with information on variation and heredity. The arm's-length back-and-forth allowed Darwin to test his ideas against their practical experience without exposing himself to ridicule by revealing his ultimate purpose. It also allowed him to gradually sort out who among his colleagues might be sympathetic to his views—and to eventually share, with that select group, his unorthodox ideas.

By 1856 Darwin had divulged his theory in detail to a few close friends. These included Charles Lyell, the foremost geologist of the day, and biologist T. H. Huxley, the world's leading comparative anatomist. His confidants, especially Lyell, were encouraging him to publish, lest he be scooped. Darwin was then forty-seven and had been working on his theory for eighteen years.

In May 1856, Darwin began work on what he intended to be a technical treatise aimed at his peers. He decided to call it *Natural Selection*. By March 1858 the book was two-thirds done and running to 250,000 words. And then in June, Darwin received by mail a manuscript and a congenial cover letter from an acquaintance working in the Far East, Alfred Russel Wallace.

Wallace knew that Darwin was working on a theory of evolution and hoped that he would agree to pass to Lyell the manuscript—a paper outlining Wallace's independently conceived theory of natural selection. Like Darwin, he'd been inspired to his theory by Malthus's views on overpopulation.

Darwin panicked. The worst of what his friends had warned him about seemed to have come true: another naturalist had reproduced the most important aspect of his work.

When Newton heard claims of similar work he turned nasty, but Darwin was a far different man. He agonized about the situation and seemed to have no good alternative. He could bury the paper or rush to publish first, but those options were unethical; or he could help Wallace get it published and give up credit for his own life's work.

Darwin sent the manuscript to Lyell with a letter on June 18, 1858:

> [Wallace] has to-day sent me the enclosed, and asked me to forward it to you. It seems to me well worth reading. Your words have come true with a vengeance—that I should be forestalled. . . . I never saw a more striking coincidence; if Wallace had my MS. sketch written out in 1842, he could not have made a better short abstract! Even his terms now stand as heads of my chapters. Please return me the [manuscript], which he does not say he wishes me to publish, but I shall, of course, at once write and offer to send to any journal. So all my originality, whatever it may amount to, will be smashed, though my book, if it will ever have any value, will not be deteriorated; as all the labour consists in the application of the theory. I hope you will approve of Wallace's sketch, that I may tell him what you say.

* * *

As it turned out, the key to who would be credited for the theory rested in Darwin's observation that the value of his book lay in the applications he'd detailed. Not only had Wallace not made an exhaustive study of the evidence for natural selection, as Darwin had, but he had also failed to duplicate Darwin's detailed analysis of how change can be of such magnitude as to generate new species, rather than merely new "varieties," which we today call subspecies.

Lyell replied with a compromise: he and another of Darwin's close friends, botanist Joseph Dalton Hooker, would read to the prestigious Linnean Society of London both Wallace's paper and an abstract of

Darwin's ideas, and the two would be simultaneously published in the society's *Proceedings.* As Darwin agonized over the plan, the timing could hardly have been worse. Not only was Darwin sick with his usual maladies, but his old friend biologist Robert Brown had recently died, and his tenth and youngest child, Charles Waring Darwin, just eighteen months old, was gravely ill with scarlet fever.

Darwin left the matter for Lyell and Hooker to handle as they saw fit, and so on July 1, 1858, the secretary of the Linnean Society read Darwin's and Wallace's papers to the thirty-odd fellows present. The readings drew neither hoots nor applause, only stony silence. Then came the reading of six other scholarly papers, and, in case anyone was left awake after the first five, saved for last was a lengthy treatise describing the vegetation of Angola.

Neither Wallace nor Darwin attended. Wallace was still in the Far East and unaware of the proceedings in London. When later informed, he graciously accepted that the matter had been handled fairly, and in future years he would always treat Darwin with respect and even affection. Darwin was ill at the time, so he probably wouldn't have traveled to the meeting in any case, but as it turned out, while the meeting transpired, he and his wife, Emma, were burying their second deceased child, Charles Waring, in the parish churchyard.

With the presentation at the Linnean Society, after twenty years of hard work to develop and back up his theory, Darwin had finally exposed his ideas to the public. The immediate reaction was, to say the least, anticlimactic. That everyone present had missed the significance of what they'd heard was perhaps best reflected in comments by the society's president, Thomas Bell, who on his way out lamented, as he later put it, that the year had not "been marked by any of those striking discoveries which at once revolutionize, so to speak, [our] department of science."

After the presentation at the Linnean Society, Darwin moved quickly. In less than a year, he reworked *Natural Selection* into his masterpiece, *On the Origin of Species.* It was a shorter book, and targeted at the general public. He finished the manuscript in April 1859. By then he was exhausted and, in his own words, "weak as a child."

Always cognizant of the need to nurture a consensus in his favor, Darwin arranged for Murray, his publisher, to distribute a great many complimentary copies of the book, and Darwin personally mailed self-deprecating letters to many of the recipients. But in writing his book, Darwin had actually been careful to minimize any theological objections. He argued that a world governed by natural law is superior to one governed by arbitrary miracles, but he still believed in a distant deity, and in *Origin of Species* he did everything he could to create the impression that his theory was not a step toward atheism. Rather, he hoped to show that nature worked toward the long-term benefit of livings things by guiding species to progress toward mental and physical "perfection" in a manner consistent with the idea of a benevolent creator.

"There is grandeur in this view of life . . . ," he wrote, "having been originally breathed into a few forms or into one . . . whilst this planet has gone cycling on according to the fixed law of gravity, from so simple a beginning endless forms most beautiful and most wonderful have been, and are being, evolved."

* * *

The reaction to *Origin of Species* was not muted. His old mentor Professor Sedgwick of Cambridge wrote, for example, "I have read your book with more pain than pleasure . . . parts I read with absolute sorrow, because I think them utterly false and grievously mischievous."

Still, by presenting a superior theory, better supported by evidence, and in a somewhat mellower time, *On the Origin of Species* didn't stimulate as much ire as *Vestiges* had. Within a decade, the debate among scientists was largely over, and by the time of Darwin's death, ten years after that, evolution had become almost universally accepted and a dominant theme of Victorian thought.

Darwin had already been a respected scientist, but with the publication of his book he became, like Newton after *Principia*, a public figure. He was showered with international recognition and honors. He received the prestigious Copley Medal from the Royal Society; he was offered an honorary doctorate by both Oxford and Cambridge; he was granted the Order of Merit by the king of Prussia; he was elected a corresponding member of both the Imperial Academy of Sciences, in St. Petersburg, and the French Academy of Sciences; and he was granted an honorary membership by the Imperial Society of Naturalists, in Moscow, as well as by the Church of England's South American Missionary Society.

Like Newton's, Darwin's influence stretched far beyond his scien-

Darwin in the 1830s, 1850s, and 1870s

tific theories to encompass new ways of thinking about totally unrelated aspects of life. As one group of historians wrote, "Everywhere Darwinism became synonymous with naturalism, materialism, or evolutionary philosophy. It stood for competition and co-operation, liberation and subordination, progress and pessimism, war and peace. Its politics could be liberal, socialist, or conservative, its religion atheistic or orthodox."

From the point of view of science, though, Darwin's work, like Newton's, was only a beginning. His theory proposed a fundamental principle governing the way the characteristics of species change over time in response to environmental pressures, but scientists of the day remained completely in the dark with regard to the mechanics of how heredity functions.

Ironically, at the very time Darwin's work was being presented to the Linnean Society, Gregor Mendel (1822–1884), a scientist and monk in a monastery in Brno—now part of the Czech Republic—was in the midst of an eight-year program of experiments that would suggest a mechanism for heredity, at least in the abstract. He proposed that simple characteristics are determined by two genes, one contributed by each parent. But Mendel's work was slow to catch on, and news of it never reached Darwin.

In any case, an understanding of the material realization of Mendel's mechanism would require the advances of twentieth-century physics—especially quantum theory and its products, such as X-ray diffraction techniques, the electron microscope, and transistors enabling the creation of digital computers. Those technologies eventually revealed the detailed structure of the DNA molecule and the genome, and allowed genetics to be studied on the molecular level, finally enabling scientists to begin to understand the nuts and bolts of how inheritance and evolution occur.

Even that, though, is just a beginning. Biology seeks to understand life on all its levels, all the way down to the structures and biochemical reactions within the cell—the attributes of life that are the most direct results of the genetic information we carry. That grand goal, no less than the reverse engineering of life, is no doubt—like the physicists' unified theory of everything—far in our future. But no matter how well we grow to understand the mechanisms of life, the central organizing principle of biology will probably always remain that nineteenth-century epiphany, the theory of evolution.

Darwin himself was not the fittest specimen, but he survived to old age. In his late years, his chronic health issues improved, though he became constantly tired. Still, he kept working until the end, publishing his last paper, *The Formation of Vegetable Mould through the Action of Worms*, in 1881. Later that year, he began to experience chest pains upon exercise, and at Christmastime he had a heart attack. The following spring, on April 18, he had another heart attack and was barely brought back to consciousness. He muttered that he was not afraid to die, and hours later, around four the following morning, he did die. He was seventy-three. In one of his last letters, written to Wallace, he had said, "I have everything to make me happy and contented, but life has become very wearisome."

Part III

Beyond the Human Senses

It's the best possible time to be alive, when almost everything you thought you knew is wrong.

—Tom Stoppard, *Arcadia*, 1993

10

The Limits of Human Experience

Two million years ago, we humans created our first great innovation when we learned to turn stone into cutting tools. That was our initial experience in harnessing nature to serve our needs, and virtually no discovery since represents as great an epiphany or has resulted in a larger change in the way we live. But one hundred years ago, another discovery was made that was of equal power and significance. Like the use of stone, it concerned something that is present everywhere, something that had been before our very eyes, albeit invisible to them, since the beginning of time. I am speaking of the atom—and the strange quantum laws that govern it.

A theory of the atom is obviously the key to understanding chemistry, but the insights gained through studying the atomic world also revolutionized physics and biology. And so, when scientists accepted the reality of the atom and began to decipher the workings of its laws, they arrived at grand insights that transformed society, shedding light on subjects ranging from the fundamental forces and particles of nature to DNA and the biochemistry of life, while enabling the creation of the new technologies that have shaped the modern world.

People speak of the technology revolution, the computer revolution, the information revolution, and the nuclear age, but ultimately these all boil down to just one thing: turning the atom into a tool. Today our ability to manipulate atoms is what makes possible everything from our televisions to the fiber-optic cables that carry the signals they display, from our phones to our computers, from the technology of the Internet to our MRI machines. We even use our knowledge of atoms to create light—our fluorescent bulbs, for example, emit light when the electrons in atoms, having been excited by an electric current, make "quantum jumps" to lower energy states. Today, even the most seemingly mundane of our appliances—ovens, clocks, thermostats—have components that depend for their design on an understanding of the quantum.

The great revolution that led to our understanding of the functioning of the atom and the quantum laws of the atomic world dates to the beginning of the twentieth century. Years earlier, it had been noticed that what we refer to today as "classical physics" (physics based on Newton's laws of motion rather than the quantum laws) failed to explain a phenomenon called blackbody radiation, which we now know depends on the quantum properties of the atom. That isolated failure of Newtonian theory was not immediately seen as a red flag. Instead, it was thought that physicists were somehow confused about how to apply Newtonian physics to the problem, and that when they figured out how to do so, blackbody radiation would be understandable within the classical framework. But physicists eventually discovered other atomic phenomena that also resisted explanation under Newtonian theory, and ultimately they realized they had to overthrow much of Newton, just as earlier generations had had to dispose of Aristotle before him.

The quantum revolution occurred over a period of struggle that lasted twenty years. That this revolt was completed in a matter of a couple of decades, rather than centuries or millennia, is testament to the vastly greater number of scientists working on the problem, and not an indication that the new way of thinking was any easier to accept. In fact, the new philosophy behind quantum theory is still, in some quarters, a lively topic of discussion. For the picture of the world that arose after those twenty years is heresy to anyone who, like Einstein, disdains the role of chance in the outcome of events or who believes in the usual laws of cause and effect.

* * *

The thorny issue of causality in the quantum universe didn't arise until near the end of the quantum revolution, and we will get to it later. But there was another issue, one that was both philosophical and practical, that got in the way from the very beginning: atoms were far too small to be seen, or even measured individually—it wouldn't be until the late twentieth century that scientists first "saw" an image of a molecule. As a result, in the nineteenth century, all experiments regarding atoms were able to reveal only phenomena that were due, at best, to the average behavior of immense numbers of those minute unseen objects. Was it legitimate to consider unobservable objects real?

Despite Dalton's work on the atom, few scientists thought so. Even those chemists who employed the concept because it was useful in helping them understand phenomena that they could observe and measure tended to employ it as a mere working hypothesis: i.e., chemical reactions proceed *as if* they are due to the reshuffling of the atoms that make up the compounds. Others thought of atoms as appropriate for philosophy but not science, and sought to ban the idea altogether. Said German chemist Friedrich Wilhelm Ostwald: they are "hypothetical conjectures that lead to no verifiable conclusions."

The hesitation is understandable, for over the centuries science had parted ways with philosophy on precisely the issue of whether conceptions of nature must be supported through experimentation and observation. By insisting on verifiability as the criterion for acceptance of any hypothesis, scientists were able to discard ancient speculations as either untestable or—as proved to be the case when many of Aristotle's theories were tested—wrong. In their place came mathematical laws that yielded accurate quantitative predictions.

The existence of atoms was *not* directly testable, but the hypothesis of their existence *did* lead to testable laws, and those laws proved valid—for example, the concept of atoms could be used to derive, mathematically, the relation between temperature and pressure in gases. So what to make of the atom? This was the meta-question of that age. The answer was unclear, and as a result, for most of the nineteenth century the atom maintained an existence as a ghostly spirit on the shoulders of physicists, an intangible that whispered into their ears the secrets of nature.

The question of the atom was eventually answered so powerfully that today it is no question at all: if science is to progress, we know it must direct its focus beyond our direct sensory experience. By early in the twenty-first century our acceptance of the unseen world had gone so far that no one flinched when the discovery of the famous "Higgs particle" was announced, even though no one had laid eyes on a Higgs particle or even observed the tangible result of Higgs particles interacting with some apparatus that could make them visible *indirectly*—the way, say, a fluorescent screen makes electrons "visible" by glowing when they strike it.

Instead, the evidence that the Higgs particle exists is mathematical, inferred from certain signature numerical characteristics of electronic data. That data was generated by debris—such as radiation—that resulted from more than three hundred trillion proton-proton collisions and analyzed, statistically, long after the fact, employing almost two hundred computing facilities in three dozen countries. Today, that is what a physicist means when he or she says, "We saw a Higgs particle."

Physics laboratories for the study of the elementary particles of matter, in 1926 and today (the position of the accelerator ring, seventeen miles in circumference and a few hundred feet underground, is indicated by the white circle)

With the Higgs and all the other subatomic particles scientists have "seen" in a similar manner, the once indivisible atom now seems more like a whole universe of objects, a billion billion such universes in every drop of water, tiny worlds not only invisible to us but separated by *several degrees* from direct human observation. So forget explaining the theory of the Higgs boson to a nineteenth-century physicist; you'd have trouble even explaining what you meant when you said we "saw" one.

The new style of observation, detached from human sensory experience, created new demands on the scientist. Newton's science was based on that which could be perceived through the senses, aided perhaps by a microscope or telescope, but still with a human eye at one end of the device. Twentieth-century science would remain dedicated to observation, but would accept a far broader definition of "seeing," one that includes indirect statistical evidence such as that for the Higgs. Due to the new attitude toward what it means to "see," twentieth-century physicists had to develop mental pictures corresponding to theories that involve bizarre avant-garde concepts like the quantum, concepts that are far beyond our human experience and rooted instead in abstract mathematics.

The new way of doing physics was reflected by a growing division among physicists. The increasing role of arcane mathematics in the theories of physics, on one hand, and the burgeoning technical sophistication of experiments, on the other, encouraged the growing apart of the formal specializations of experimental and theoretical physics. At roughly the same time, the visual arts were evolving in a manner that was comparable, causing a split between traditional representational artists and pioneers of cubism and abstraction, such as Cézanne, Braque, Picasso, and Kandinsky, who, like the new quantum theorists, also "saw" the world in radically new ways.

In music and literature, too, a new spirit was challenging the entrenched norms of rigid nineteenth-century Europe. Stravinsky and Schoenberg questioned the assumptions of traditional Western tone and rhythm; Joyce and Woolf and their counterparts on the Continent experimented with new narrative forms. In 1910, philosopher, psychologist, and educator John Dewey wrote that critical thinking often involves "the willingness to endure a condition of mental unrest and disturbance." That's not only true of critical thinking but applies to creative efforts as well. Whether practicing art or science, none of these pioneers had it easy.

* * *

The picture I just painted of early twentieth-century science was laid down with the advantages of hindsight. The physicists studying the atom in the late nineteenth century didn't realize what was coming. In fact, it is astounding to look back and note that, despite the time bomb of the atom resting on their doorstep, those physicists viewed their field as more or less settled, and were advising young students to avoid physics because there was nothing exciting left to do.

The department head at Harvard, for example, was famous for shooing away prospective students by telling them that everything important had already been discovered. Across the ocean, the head of the physics department at the University of Munich warned in 1875 that the field was not worth going into anymore, because "physics is a branch of knowledge that is just about complete." As far as prescience goes, this advice was on par with the pronouncement by the builder of the *Titanic* that the ship was "as nearly perfect as human brains can make her." Physics around 1900, like the *Titanic*, was considered unassailable, and yet the version practiced in that era was destined to sink.

One of the fellows who got the bum advice from the head of the physics department in Munich was Max Planck (1858–1947). A thin, almost gaunt young man who, even in his early years, had a receding hairline and glasses, Planck radiated an air of seriousness that belied his age. Born in Kiel, Germany, he was the product of a long line of pastors, scholars, and jurists, and he fit perfectly the template of the nineteenth-century physicist: diligent, dutiful, and, in his own words, "disinclined to questionable adventures." Hardly the words you'd expect from a man whose work would one day upend Newton, but Planck didn't plan to start a revolution. In fact, for many years he didn't even support the movement his own discovery had sparked.

Though disinclined to adventures, Planck did begin his career by taking a chance—he ignored the advice of the department head and enrolled in the physics program. He had been inspired to study physics by a high school instructor who imparted to him a passion to "investigate the harmony that reigns between the strictness of mathematics and the multitude of natural laws," and he believed in himself enough to pursue his passion. Years later he would tell one of his own students, "My maxim is always this: consider every step carefully in advance, but then, if you believe you can take responsibility for it, let nothing stop you." That statement doesn't have the swagger of Nike's classic "Just do it" ad campaign, or the bold pronouncements we are used to hearing from sports figures, but in his own way, the quiet and conventional Planck was giving voice to the same inner strength.

Having decided upon physics, Planck had to choose the topic of his doctoral research. Here again, he made a bold and critical choice. He chose thermodynamics—the physics of heat. It was then a rather obscure area of physics, but it was the one that, in high school, had initially inspired him, and Planck again chose to follow his interests rather than work on what was fashionable.

At the time, a handful of scientists who had accepted the atom had begun to understand the mechanism behind thermodynamics as the statistical result of the motion of individual atoms. For instance, if at some time there is a cloud of smoke confined to a small region in a room, thermodynamics tells us that at a later time it will be more spread out, rather than even more concentrated. That process defines what physicists call an "arrow of time"—the future is the direction in time in which smoke disperses, the past is the direction in which it concentrates. That is puzzling, because the laws of motion applying to each *individual* atom of the smoke (and the air) give no indication regarding which direction of time is the future and which is the past. But the phenomenon can be explained employing a statistical analysis of the atoms: the "arrow of time" is apparent only when you observe the cumulative effect of many atoms.

Planck did not like arguments of that sort. He viewed the atom as a fantasy and made it the goal of his Ph.D. research to extract concrete and testable results from the principles of thermodynamics without employing the concept of the atom—in fact, without making *any* assumptions about the internal structure of substances. "Despite the great success that the atomic theory has so far enjoyed," he wrote, "ultimately it will have to be abandoned in favor of the assumption of continuous matter."

A clairvoyant, Planck was not. What would ultimately be abandoned was not the atomic theory, but his resistance to it. In fact, in the end, his work could be taken as evidence *for* and not *against* the existence of atoms.

Since my name is so hard to spell and pronounce, when I make a restaurant reservation, I often do it under the name Max Planck. It's very rare that the name is recognized, but one time when it was, I was asked if I was related to the "guy who invented quantum theory." I said, "I *am* that man." The maître d', in his early twenties, didn't believe me. He said I was too young. "Quantum theory was invented around 1960," he said. "It was during World War II, as part of the Manhattan Project."

We didn't talk further, but the issue I would have liked to chat about was not his fuzziness regarding history, but the fuzziness of what it means to "invent" a theory in physics. The word "invent" means to create something that did not before exist. To discover, on the other hand, means to become aware of something that was previously not known. One can look at theories either way—as mathematical structures scientists invent to describe the world, or as the expression of laws of nature that exist independently of us, and which scientists discover.

In part this is a metaphysical question: To what extent should we take the pictures painted by our theories as literal reality (that we discover), as opposed to mere models (that we invent) of a world that could equally well be modeled in other ways, say, by people (or aliens) who think differently than we do? But philosophy aside, the distinction between invention and discovery has another dimension, having to do with process: we make discoveries through exploration, and often by accident; we make inventions through planned design and construction, and accident plays less of a role than trial and error.

Certainly Einstein, when he came up with relativity, knew what he was setting out to do and did it, so one might call relativity an invention. But quantum theory was different. In the steps leading to the development of quantum theory, more often than not, "discover" or even "stumble on" would be a better term than "invent," and what the (many) discoverers stumbled on was often, as in Planck's case, just the opposite of what they were hoping and expecting to find—as if Edison had set out to invent artificial light and instead discovered artificial darkness. What's more, as was Planck's fate, they sometimes didn't quite understand the meaning of their own work, and when others interpreted it for them, they argued against it.

Planck didn't succeed at supporting either the existence or nonexistence of atoms with his 1879 Ph.D. dissertation on thermodynamics. Even worse, it got him nowhere professionally. His professors in Munich did not understand it; Gustav Kirchhoff, an expert in thermodynamics at Berlin, judged it wrong; and two of the other founding pioneers of the field, Hermann von Helmholtz and Rudolf Clausius, declined to read it. Planck, after receiving no answer to two letters, even went so far as to travel to Bonn to call on Clausius at his home, but the professor refused to receive him. Unfortunately, other than those few physicists, when it came to thermodynamics, as one of Planck's colleagues put it, "nobody . . . had any interest whatever."

The lack of interest didn't bother Planck, but it did result in a series of bleak years during which he lived at his parents' home and worked at the university as an unpaid lecturer, eking out a small living as Mendeleev had done, by, collecting fees directly from the students who attended his classes.

Whenever I've mentioned that to anyone, I've gotten looks of surprise. For some reason, people expect artists to have such love for their art that they will make any kind of sacrifice, live in the shabbiest of garrets or even, worst of all, with their parents, to keep working at it; but people don't see physicists as being that passionate. In graduate school, though, I knew two fellow students who faced failure as Planck did. One, sadly, attempted suicide. The other convinced the Harvard physics department to let him work at a desk in a crowded office, without pay. (A year later, they hired him.) A third fellow, whom I didn't know, had flunked out some years earlier and had since submitted his personal pet (and wildly wrong) theories to various faculty members and had them ignored, then showed up one day determined to convince them—by brandishing a knife. He was intercepted by security and never came back. Popular lore tells no famous tales of lonely, underappreciated physicists cutting off an ear, but in my three years as a graduate student in Berkeley, here were three stories, each fueled by a passion for physics.

Planck, like my jobless graduate student friend who landed at Harvard, would do good enough research during his "volunteer" period that he would eventually find a paying job. It took five years. Finally, through sheer perseverance, luck, and—some say—the intervention of his father, he managed to get a professorship at the University of Kiel. Four years after that, his work would be deemed impressive enough that he was called to the University of Berlin, where he was named full professor in 1892, making him a member of the small cadre of thermodynamics elite. But that was only the beginning.

* * *

In Berlin, Planck's research passion remained squarely focused on understanding thermodynamics in a context in which one did not have to "resort" to the concept of atoms—that is, in which substances were considered to be "endlessly divisible" rather than constituted from discrete building blocks. The question of whether that could be done was, in his mind, the most pressing issue in all of physics, and, being in the academic world, Planck had no boss who could tell him—at least directly—otherwise. That's a good thing, because his thinking was so far from the mainstream of physics that in the summer of 1900, just months before he would announce his world-shattering breakthrough, the official historian at the international physics meeting in Paris expressed the opinion that, besides Planck, there were at most only three other people in the world who thought this to be a question worth considering. Very little, it seemed, had changed in the twenty-one years since he had submitted his dissertation.

In science, as in all fields, there are plenty of ordinary people asking ordinary questions, and many of them will do just fine in life. But the most successful researchers are often the ones who ask the odd questions, questions that haven't been thought of or that weren't deemed interesting by others. For their trouble, these individuals will be considered odd, eccentric, maybe even crazy—until the time comes when they're considered geniuses.

Of course, a scientist who asks, "Does the solar system rest on the

Max Planck, circa 1930

back of a giant moose?" would also be an original thinker—as was, I suppose, the knife brandisher I mentioned above. So when looking at a group of freethinkers, one must be choosy, and therein lies the problem: the people whose ideas are *only* weird are often not easy to distinguish from people whose ideas are weird but *true*. Or weird but destined to lead, perhaps after a long time and many missteps, to something true. Planck was an original thinker who asked questions that didn't even seem interesting to his fellow physicists. But they would prove to be precisely the questions that classical physics could not answer.

Chemists of the eighteenth century had found in the study of gases a kind of Rosetta Stone, a key that led to the deciphering of important scientific principles. Planck sought his own Rosetta Stone in blackbody radiation, the thermodynamic phenomenon that Gustav Kirchhoff had identified and named in 1860. Today, "blackbody radiation" is a familiar term among physicists: it is the form of electromagnetic radiation emitted by a body that is literally black, and held at some fixed temperature.

"Electromagnetic radiation" sounds complicated, not to mention dangerous, like something that drones fire at Al Qaeda camps. But it refers to a whole family of energy waves—for example, microwaves, radio waves, visible and ultraviolet light, X-rays, and gamma radiation—that, when harnessed, have a range of practical effects, some of which are lethal but all of which are very much part of the world we have come to take for granted.

In Kirchhoff's day, the concept of electromagnetic radiation was still new and mysterious. The theory that described it—in the context of the Newtonian laws—came from the work of Scottish physicist James Clerk Maxwell. Maxwell is still a physics hero today, and on college campuses it is not uncommon to find his face or his equations gracing the T-shirts of physics majors. The reason for all that adoration is that in the 1860s, he accomplished the greatest unification in the history of physics: he explained electric and magnetic forces as manifestations of the same phenomenon, the "electromagnetic field," and revealed that light and other forms of radiation are waves of electromagnetic energy. To a physicist, to elucidate deep connections between different phenomena, as Maxwell did, is one of the most exciting things a person can do.

That a Maxwell would someday arise had been Newton's hope and dream, for Newton knew that his theory was not complete. He had provided *laws of motion* explaining how objects react to force, but in order for those laws to be used, they had to be supplemented with separate *laws of force*, laws that describe whatever force is acting on the objects under consideration. Newton provided the laws of one kind of force—gravity—but he knew that other kinds of force must exist.

In the centuries following Newton, two other forces of nature did gradually reveal themselves to physics: electricity and magnetism. By creating a quantitative theory of those forces, Maxwell had in a sense completed the Newtonian (i.e., "classical") program—in addition to Newton's laws of motion, scientists now had theories of all the forces that manifest themselves in our everyday existence. (We would, in the twentieth century, discover two additional forces, the "strong" and "weak" forces, whose effects are *not* apparent in our daily affairs but which act within the tiny regions of the atomic nucleus.)

Previously, by employing Newton's law of gravity along with his laws of motion, scientists could describe only gravitational phenomena, such as the orbits of the planets and the trajectories of artillery shells. Now, by employing Maxwell's theory of the electric and magnetic forces in combination with Newtonian laws of motion, physicists could analyze a vast new range of phenomena, such as radiation and its interaction with matter. In fact, physicists believed that, with the addition of Maxwell's theory to their arsenal, they could in principle explain every single natural phenomenon we can observe in the world—hence the rampant optimism of late-nineteenth-century physics.

Newton had written that there are "certain forces by which the particles of bodies, by some causes hitherto unknown, are either mutually impelled towards one another, and cohere in regular figures, or are repelled and recede from one another." These, he believed, cause "local motions which on account of the minuteness of the moving particles cannot be detected . . . [but] if any one shall have the good fortune to discover all these, I might almost say that he will have laid bare the whole nature of bodies." What physicists had discovered with electromagnetism fulfilled that dream of understanding the forces that act among the minute particles of bodies—atoms—but Newton's dream that his theory would thus be able to explain the properties of material objects would never become reality. Why? Because, though physicists had discovered the laws of the electric and magnetic *forces*, when those laws were applied to atoms, Newton's laws of *motion* would be found to fail.

Though no one realized it at the time, the shortcomings of Newtonian physics showed themselves most dramatically in precisely the phenomenon Planck had chosen to study: blackbody radiation. For when physicists applied Newtonian physics to calculate how much radiation the black material should give off at different frequencies, those calculations did not just prove wrong, they yielded an absurd result: that a blackbody would emit an infinite amount of high-frequency radiation.

If those calculations had been correct, the phenomenon of blackbody radiation would mean that sitting in front of a warm fireplace or opening the door of your hot oven would result not only in your being able to bask in the warmth of low-frequency infrared radiation, or the soothing glow of somewhat higher-frequency red light, but also in your being bombarded with dangerous high-frequency ultraviolet rays, X-rays, and gamma rays. And the lightbulb, newly invented at that time, would have proved to be not so much a useful instrument of artificial illumination but, due to the radiation that resulted from its elevated operating temperature, a weapon of mass destruction.

When Planck began his work in the field, though everyone knew the blackbody calculations were wrong, no one knew why. Meanwhile, as most physicists interested in the problem were scratching their heads, a few focused on concocting various ad hoc mathematical formulas to describe the experimental observations. These formulas yielded, for each frequency, the intensity of radiation emitted by a blackbody of any given temperature, but they were merely descriptive, created to yield the necessary data, not derived from a theoretical understanding. And none were accurate for all frequencies.

Planck began working on the challenge of providing an accurate description of the radiation emitted by blackbodies in 1897. Like the others, he didn't suspect that the problem suggested that there was anything wrong with Newtonian physics, but rather that the physical description used for the material of the blackbody must be fundamentally flawed. After several years, he had gotten nowhere.

Finally, he decided to work backward and, like those applied physicists, simply find a formula that worked. He focused on two ad hoc formulas—one that was an accurate description of the low-frequency light emitted as blackbody radiation and another that was accurate for high frequencies. After considerable trial and error, he managed to "sew" them together into his own ad hoc formula, an elegant mathematical expression he concocted simply to combine the correct features of both.

You might think that if you spend years working on a problem, you deserve to make an important discovery at the end—like the microwave oven, or at least a new way of making popcorn. All Planck came away with was a formula that seemed—for unknown reasons—to work pretty well, though Planck didn't have enough data to put the predictive powers of his equation to a thorough test.

Planck announced his formula on October 19, 1900, at a meeting of the Berlin Physical Society. As soon as the meeting ended, an experimentalist named Heinrich Rubens went home and started plugging in numbers to check the formula against his own voluminous data. What he found astonished him: the Planck formula was more accurate than it had any right to be.

Rubens grew so excited that he worked through most of the night, painstakingly cranking through the math of Planck's equation for different frequencies and comparing the predictions with the records of his observations. The next morning, he raced to Planck's home to give him the shocking news: the agreement was freakishly good, and for *all* frequencies. Planck's formula was too accurate to be merely an ad hoc guess. It had to *mean* something. The only problem was that neither Planck nor anyone else knew what it meant. It seemed like magic, a formula that presumably had deep and mysterious principles behind it but had been "derived" purely by guessing.

* * *

Planck had chosen to work on the theory of blackbody radiation with the aim of explaining it without resorting to the concept of atoms. In a sense he had done that. But he had pulled his formula from thin air, so he still felt compelled to answer the question *Why did it work?* His success must have been exciting, but his ignorance must have been frustrating.

Ever the patient scientist, Planck turned—perhaps out of sheer desperation—to research by the grand advocate of the atom, Austrian physicist Ludwig Boltzmann (1844–1906). Boltzmann had been fighting for decades to accomplish exactly the opposite of what Planck sought to prove—that atoms should be taken seriously—and in the process Boltzmann had made much progress in developing the techniques of what is now called statistical physics (though he'd made little headway in convincing people of the importance of his work).

Planck's willingness to turn, however reluctantly, to Boltzmann's research is an act worth taking time to appreciate: here was an evangelist for doing physics without the atom, seeking intellectual salvation in the work of the man who championed the very theory he had long opposed. That kind of openness to ideas that contradict one's own preconceptions is the way science is supposed to be done, and it is one reason Einstein would later have great admiration for Planck; but it is not the way science is *usually* done. Indeed, it's not the way most human enterprises operate. For example, when the Internet, smartphones, and the other new media were on the rise, like the established physicists who had trouble accepting the atom or the quantum, established companies—Blockbuster Video, the music labels, the major bookstore chains, the established media outlets—resisted accepting the new way of life and business. They were therefore superseded by younger people and companies with more mental flexibility, such as Netflix, YouTube and Amazon. What Planck himself would later say about science, in fact, also seems to apply to any revolutionary new human idea: "A new scientific truth does not triumph by convincing its opponents and making them see the light, but rather because its opponents eventually die, and a new generation grows up that is familiar with it."*

Reading Boltzmann's work, Planck noted that in the Austrian's statistical description of thermodynamics, he had found it necessary to employ a mathematical trick in which energy is treated as if it came in discrete portions, as, say, eggs do, in contrast to flour, which seems to be endlessly divisible. That is, while you can have only an integral number of eggs, like one or two or two hundred, you can measure 2.7182818 ounces of flour, or any amount you like. At least that's how it seems to the chef, though flour actually is not endlessly divisible, but rather consists of discrete building blocks—fine individual grains—which you can see under a microscope.

Boltzmann's trick was a mere calculational expedient; at the end, he always made the portion size approach zero, meaning that energy could again come in any amount, rather than just discrete quantities. To his

*Planck is often misquoted as having said, instead, the pithier version: "Science advances one funeral at a time."

great surprise, though, Planck discovered that by applying Boltzmann's methods to the blackbody problem, he could derive his formula, but *only if he skipped that last step* and left the energy as a quantity that could only be doled out, like eggs, in multiples of a certain fundamental (very tiny) portion. Chef Planck called the basic portion of energy a "quantum," from the Latin for "how much."

That, in a nutshell, was the origin of the quantum concept. Quantum theory arose not from the relentless efforts of a scientist pursuing a deep principle to its logical conclusions, or from the drive to discover a new philosophy of physics, but from a man who was like a chef peering for the first time through a microscope and discovering to his surprise that flour is like eggs after all, made of discrete individual units, and can be doled out only in multiples of those tiny portions.

Planck found that the size of the portion, or quantum, is different for different frequencies of light—which, for visible light, correspond to different colors. In particular, Planck discovered that a quantum of light energy is equal to the frequency multiplied by a proportionality factor, which Planck called h, and which today is known as Planck's constant. Had Planck taken Boltzmann's last step, in essence setting h equal to zero, then energy would have been posited as endlessly divisible. By not doing that, and instead fixing h by comparing his formula with experimental data, Planck was asserting that—at least as far as blackbody radiation is concerned—energy comes in tiny, fundamental packages and cannot take on just any value.

What did his theory mean? Planck had no idea. In a way, all he had succeeded in doing was to create an enigmatic theory to explain his enigmatic guess. Still, Planck announced his "discovery" at the December 1900 meeting of the Berlin Physical Society. Today we call that announcement the birth of quantum theory, and indeed, his new theory would win him the 1918 Nobel Prize and eventually turn the field of physics inside out. But no one, including Planck, knew it then.

To most physicists, it seemed that Planck's long study of blackbody radiation had only made his theory even more elusive and mysterious, and what was the good of that? Planck himself, however, had learned something important from the experience. He had finally "understood" blackbody radiation using a picture in which the black material was revealed to be made of tiny oscillators, like springs, which he eventually came to believe were atoms or molecules—so he finally became con-

Ludwig Boltzmann, circa 1900

vinced that atoms were real. Still, neither he nor anyone else at the time realized that the quanta he described might be a fundamental characteristic of nature.

Some of Planck's contemporaries thought that there would eventually be found a route to Planck's blackbody formula that didn't require the quantum. Others thought that the quantum would someday be explained not as a fundamental principle of nature, but as a result of some yet unknown characteristic of materials that would be fully consistent with physics as they knew it—for instance, a mundane mechanical property resulting from the internal structure of atoms, or the way atoms interact. And some physicists simply dismissed Planck's work as nonsense, despite its agreement with the experimental data.

In attacking Planck, for example, Sir James Jeans, a well-known physicist who had worked on the problem but unlike Planck had not been able to derive the full formula, wrote, "Of course, I am aware that Planck's law is in good agreement with experiment . . . while my own law, obtained [from Planck's] by putting $h = 0$, cannot possibly agree with experiment. This does not alter my belief that the value $h = 0$ is the only value which it is possible to take." Yeah, those pesky experimental observations are such a hassle—better to ignore them. Or, as Robert

Frost wrote in 1914, "Why abandon a belief / Merely because it ceases to be true."

The bottom line is that, other than annoying James Jeans, Planck's work just didn't create much of a stir. Whether they believed his work to be nonsense, or thought it would have a mundane explanation, those in the community of physicists were simply unexcited, like the fans at a rock festival where they enforce the drug laws. Those drugs wouldn't arrive for quite some time. Over the next five years, in fact, there wouldn't be another piece of research that carried his ideas forward—not by him or anyone else. That wouldn't happen until 1905.

* * *

I said above that when Planck proposed the quantum idea, no one realized it was a fundamental principle of nature. But soon after, there entered the field one new player who would have a very different attitude. An unknown, and just out of college when Planck made his announcement, he would see Planck's work on the quantum as profound, even troubling. "It was as if the ground had been pulled out from under us, with no firm foundation to be seen anywhere," he would later write.

That man, who assimilated Planck's work on the quantum and showed what it was worth, is not, in popular culture, known for that work, but instead for eventually taking the opposite stance and, in the tradition of Jeans, disapproving of an idea despite the many observations that seemed to reveal it as true. He is Albert Einstein (1879–1955).

Einstein was twenty-five and had not yet completed his Ph.D. dissertation when he took Planck's quantum idea and ran with it. By the time he had reached the age of fifty, however, he came to oppose what he'd wrought. Einstein's reasons for changing his mind about quantum theory were ultimately philosophical, or metaphysical, rather than scientific. The ideas he proposed at age twenty-five had concerned "merely" a new way of understanding light, as energy that is made of quantum particles. The quantum ideas that arose—and that he rejected—in his later years concerned, by contrast, a fundamentally new way of looking at *reality*.

That is, as quantum theory evolved, it became clear that if one is to accept it, one must adopt a new view of what it means to exist, what it means to exist at some particular place, and even what it means for one event to cause another. The new quantum worldview would be an even greater break from our intuitive Newtonian worldview than the mechanical Newtonian view was from the purpose-driven perspective of Aristotle, and Einstein, willing as he had been to revise *physics*, would go to his grave without accepting the radical revision of *metaphysics* that arose out of his own work.

By the time I was exposed to quantum theory, a mere couple of decades after Einstein's death, I was of course taught the modern formulation, complete with all the radical ideas that Einstein hadn't liked. In my college courses these were presented as ho-hum albeit odd aspects of a now well-developed and well-tested theory. The "quantum strangeness" people sometimes talk about—like the possibility of something being, essentially, in two places at once—was by then considered a long-established fact. It sometimes made for some interesting drinking discussions, but nothing we college students ever lost sleep over. Still, Einstein was one of my heroes, and so it bothered me that he had had such trouble accepting ideas that I had allowed into my own mind without a fight. I knew I was no Einstein, so what wasn't I seeing?

While I was grappling with that issue, my father told me a story. It was prewar Poland, and he and some friends had come upon a deer lying on the road, having been killed by a car or truck. Food was scarce at the time, so they took the deer home and ate it. My father told me that they saw nothing wrong with eating "roadkill" but that Americans—like me—find it disgusting, because we were raised to believe it is disgusting. I realized then that you don't have to turn to deep questions of the cosmos or to strongly held moral beliefs to find ideas people have trouble accepting. Such ideas are everywhere, and most depend simply on the fact that people tend to continue to believe what they have always believed.

The metaphysical implications of quantum theory were like Einstein's version of roadkill. Having grown up believing in traditional notions of causality, he would have been loath to accept an idea so profoundly different in its implications. But had he been born eight decades later and been my classmate, he'd have grown up with the strangeness of quantum theory and probably reacted to it in the same matter-of-fact way that I—and all the other students—did. By then it would have been just part of the accepted intellectual environment, so although one might recognize the novelties of the quantum world, in the absence of experiments that contradict it, one would not think of turning back.

* * *

Though Einstein would eventually fight to maintain central aspects of the Newtonian worldview, he was never a conventional thinker, nor was he one to give undue credence to figures in authority. In fact, that willingness to think differently and challenge authority was so pronounced that it got him into trouble when he was still a teenager, enrolled as a stu-

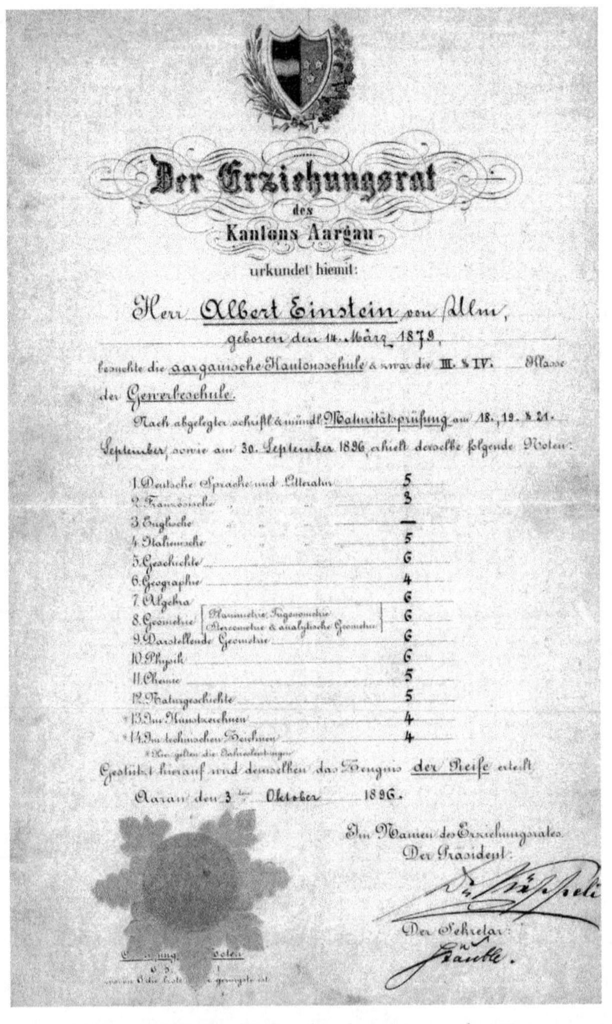

Der Erziehungsrat
des
Kantons Aargau
urkundet hiemit:

Herr Albert Einstein von Ulm,
geboren den 14. März 1879,
besuchte die aargauische Kantonsschule & zwar die III. & IV. Klasse
der Gewerbeschule.

Nach abgelegter schriftl. & mündl. Maturitätsprüfung am 18., 19. & 21. September, sowie am 30. September 1896, erhielt derselbe folgende Noten:

1. Deutsche Sprache und Litteratur	5
2. Französische " "	3
3. Englische " "	—
4. Italienische " "	5
5. Geschichte	6
6. Geographie	4
7. Algebra	6
8. Geometrie (Planimetrie, Trigonometrie, Stereometrie & analytische Geometrie)	6
9. Darstellende Geometrie	6
10. Physik	6
11. Chemie	5
12. Naturgeschichte	5
*13. Im Kunstzeichnen	4
*14. Im technischen Zeichnen	4

*Hier gelten die Jahresleistungen

Gestützt hierauf wird demselben das Zeugnis der Reife erteilt.

Aarau den 3ten Oktober 1896.

Im Namen des Erziehungsrates.
Der Präsident:
Der Sekretär:

Einstein's Swiss high school report card, 1896.
The grades are on a scale from 1 to 6,
with 6 being the best mark.

dent in a Munich *gymnasium*—the German equivalent of high school. At the age of fifteen, he was informed by one of his teachers that he would never amount to anything, and he was subsequently either forced or "politely encouraged" to leave the school, because he had shown his teachers disrespect and was viewed as a negative influence on other students. He would later call the *gymnasium* an "education machine," by which he did not mean that it performed useful work, but rather that it belched out mind-stifling pollution.

Luckily for physics, Einstein's desire to understand the universe trumped his aversion to formal education, so after being expelled from high school, he applied to the Swiss Federal Institute of Technology, in Zurich. He failed the entrance exam, but after a brief remedial stint in a Swiss high school, he was admitted to the Federal Institute in 1896. He liked it no better than the *gymnasium*, and didn't attend many lectures, but he did manage to graduate, by cramming for exams using notes taken by a fellow student he befriended. Marcel Grossmann, Einstein later wrote, was "the irreproachable student, I myself disorderly and a dreamer. He, on good terms with the teachers and understanding everything; I a pariah, discontented and little loved." Meeting Grossmann was not just a lucky break for Einstein's college career: Grossmann would also later become a mathematician and teach Einstein the exotic geometry he needed to complete his theory of relativity.

Einstein's college degree did not set him on an easy path to success. In fact, one of his university professors spitefully wrote him bad references. At least in part due to that, Einstein was unable to find a traditional job after graduating from college in Zurich—he had wanted a university appointment in physics and mathematics—and he instead became a private tutor for two *gymnasium* boys.

Not long after accepting the position, Einstein suggested to his employer that the two boys be pulled from school altogether, to avoid its destructive influence. His beef with the educational system was its extraordinary focus on preparing students for tests, thereby snuffing out any genuine curiosity or creativity. Ironically, about a century later, with the No Child Left Behind program promulgated by President George W. Bush, a test-driven curriculum focused on the ability to recite facts became the centerpiece of official American educational policy. Everyone knew that Bush was no Einstein, but apparently, when it came to having a politician's ability to sway people to accept your views, Einstein was no Bush: his employer, to whom he decried the fatal influence of the *gymnasium*, fired him.

Of his struggles at this time, Einstein's father wrote, "My son is deeply unhappy with his current state of unemployment. Day by day the feeling grows in him that his career is off the track . . . the awareness weighs on him that he is a burden to us, people of small means." The letter was sent to Leipzig physicist Friedrich Wilhelm Ostwald, to whom Albert had sent a reprint of his first paper, along with a request for a job. Neither Albert nor his father received a reply. Ten years later, Ostwald would be the first to propose Albert for the Nobel Prize. But in 1901, no one was sufficiently impressed with Einstein's intellect to give him a job that was in any way suited to his abilities.

Einstein's professional life finally stabilized in 1902, when Marcel Grossmann's father introduced him to the director of the Swiss patent office in Bern, who invited him to take a written exam. Einstein did well enough on it that the director offered him a position. The job was reading highly technical patent applications and translating them into language simple enough that his less intelligent superiors could understand them. He began work there, on a trial basis, that summer.

Einstein was apparently good at his job, though in 1904 he applied for a promotion from patent clerk third class to patent clerk second class and was turned down. Meanwhile, his work in physics, though he found it rewarding, was undistinguished. His first two papers, written in 1901 and 1902, dealt with a hypothesis regarding a universal force among molecules and were, by his own later account, worthless. These were followed by three more papers of mixed quality, which also had little impact in the world of physics. After that came a year in which his first son was born, but he did not publish a single physics paper.

Chronic money problems and a stagnating physics career must have been discouraging, but Einstein enjoyed his job, finding it intellectually stimulating and commenting that it left him, after work, "eight hours of idleness" during which he could exercise his passion and think about physics. He also supplemented his after-hours research by stealing time from his job at the patent office, hastily shoving his calculations into a drawer when anyone came near. All that work finally paid off in a most spectacular fashion: in 1905, he turned out three disparate and revolutionary papers that would propel him from third-class patent clerk to first-class physicist.

Each of those papers was worthy of a Nobel Prize, though only one of them won it for him. One can perhaps understand why the Nobel Committee would hesitate to award multiple prizes to the same individual, but over the years it has unfortunately become famous for many less understandable oversights. Among physicists alone, the committee has blundered by passing up scientists like Arnold Sommerfeld, Lise Meitner, Freeman Dyson, George Gamow, Robert Dicke, and Jim Peebles.*

Denying the prize to Meitner was especially egregious, because for thousands of years women had been almost universally barred from advanced education and from employment opportunities that would have allowed them to contribute to our understanding of the world. That began to change only about a hundred years ago, a social change that is still in progress. Meitner, a pioneer both as a scientist and as a woman, was only the second of her gender ever to receive a physics doctorate from the University of Vienna. After graduating, she convinced Max Planck to allow her to study with him, even though he had previously never even allowed a woman to sit in on one of his lectures. Eventually she began to collaborate with a young Berlin chemist named Otto Hahn. Together they made many breakthroughs, the most important being the discovery of nuclear fission. Sadly, for that, Hahn, but not Meitner, received the 1944 Nobel Prize in Chemistry.†

* * *

One of the intoxicating attractions of theoretical physics is the potential for your ideas to have a great impact on the way we think and even live. Yes, it takes years to understand and assimilate the subject, and to understand its techniques and issues. Yes, many problems you attack turn out to be insoluble. And yes, the majority of your ideas turn out to be nonsense, and in most cases it takes you months to make even a tiny contribution to a much larger body of work. Certainly, if you are going

*Sommerfeld was an important quantum pioneer; Meitner, as I say, made many discoveries, including fission; Dyson was instrumental in the quantum theory of electromagnetism; and Gamow, Dicke, and Peebles explained and predicted the cosmic microwave background radiation, but the prize for it was awarded to Arno Penzias and Robert Wilson, who accidentally detected it and had no idea what they had found.

†Like Mendeleev, though, Meitner was recognized by the International Union of Pure and Applied Chemistry, which in 1997 named element 109 meitnerium. Meitner had died in 1968.

to be a theoretical physicist, you had better be stubborn and persistent, and get a thrill out of even minor discoveries, bits of mathematics that seem to magically work out and tell you a secret of nature that, until you publish, only you know. But there is always another possibility: that you will think of or stumble across an idea so powerful that it is far more than a little secret of nature, but rather something that changes the way your colleagues—or even all of humanity—views the universe. It was that kind of idea that Einstein churned out three times, in the course of a single year at the patent office.

Of those three groundbreaking theories, the one for which Einstein is best known is relativity. His work in this area revolutionized our concepts of space and time, showing that they are intimately related, and that measurements of those quantities are not absolute but depend on the state of the observer.

The issue Einstein aimed to resolve with relativity was a paradox that had emerged from Maxwell's theory of electromagnetism, which implied that all observers who measure the speed of light will find the same result, regardless of their own speed relative to the light source.

In the spirit of Galileo, we can employ a simple thought experiment to understand why the above statement contradicts our everyday experience. Imagine a food vendor standing on a platform at the train station as a train whizzes past. A ball (or any material object) thrown forward by a passenger on that moving train will appear to the vendor to be moving faster than a ball that the vendor might hurl with equal gusto. That's because, from the point of view of the vendor, the ball on the train will move at the speed with which the passenger on the train threw it, *plus* the speed of the train. Yet a flash of light shone from a moving train, Maxwell's theory said, will *not* travel any faster. It will appear to both the vendor and the passenger to propagate at the same speed. To physicists, who wish to reduce everything to a matter of principle, that demands an explanation.

What principle distinguishes light from matter? For years, this was the question physicists addressed, the most popular approach holding that it had to do with an as-yet-undetected medium through which light waves propagate. But Einstein had other ideas. The explanation does not lurk in some unknown property of light propagation, he realized, but in our understanding of *speed.* Since speed is distance divided by time, Einstein reasoned, by saying that the speed of light is fixed, Maxwell's theory is telling us that in measuring distance and time, there can be no universal agreement. There are no universal clocks or universal meter sticks, Einstein showed, but rather all such measures are dependent on the motion of the observer—in just the manner required so that all observers measure the same speed for light. What we each observe and measure is therefore no more than our own personal view, not a reality upon which everyone can agree. That is the essence of Einstein's special theory of relativity.

Relativity didn't require the replacement of Newtonian theory, but rather a modification: Newton's laws of motion had to be altered, rebuilt to rest comfortably upon Einstein's new framework of space and time, in which the outcome of measurements depend on one's motion. For objects and observers moving at relatively slow speeds with respect to each other, Einstein's theory is essentially equivalent to Newton's. It is only when the speeds in question approach that of light that the effects of relativity become noticeable.

Since the novel effects of relativity are apparent only in extreme circumstances, it has far less importance to everyday existence than quantum theory, which explains the very stability of the atoms that constitute us. But no one at the time knew the far-reaching implications the quantum would have; meanwhile, relativity hit the physics community like an earthquake: Newton's worldview had shaped science for more than two hundred years, and now here was the first crack in its structure.

Newton's theory was based on there being a single objective reality. Space and time formed a fixed framework, a stage on which the world's events transpired. Observers could watch, and no matter where they were or how they were moving, they would all see the same play, like God observing us all from the outside. Relativity contradicted that view. By asserting that there is no single play—that, as in our daily lives, the reality we each experience is personal and, depends on our place and our motion, Einstein had begun the demolition of Newton's world, just as Galileo began the job of dismantling Aristotle's.

Einstein's work had important implications for the culture of physics: it emboldened generations of new thinkers and made it easier for them to consider challenging old ideas. It was, for example, a book on relativity that Einstein wrote for high school kids that inspired Werner Heisenberg, whom we will meet shortly, to go into physics, and it was Einstein's approach to relativity that gave Niels Bohr, whom we will also meet shortly, the courage to imagine that the atom might follow laws that are radically different from those of our everyday existence.

Ironically, of all the great physicists who assimilated and understood Einstein's theory of relativity, it was Einstein himself who was the least impressed. In his view, he was not advocating the overthrow of a major aspect of the Newtonian worldview, but merely supplying some corrections—corrections that had little effect on most experimental observations of the day, but were important in that they fixed a defect in the theory's logical structure. What's more, the mathematical alterations necessary to make Newton's theory compatible with relativity were fairly easy to make. And so, while Einstein would later consider quantum theory to be a dismantling of Newtonian physics, in the words of physicist and biographer Abraham Pais, he "considered relativity theory no revolution at all." To Einstein, relativity was the least important of his 1905 papers. Far more profound, in his eyes, were his other two papers, on the atom and on the quantum.

Einstein's paper on the atom analyzed an effect called Brownian motion, discovered by Darwin's old friend Robert Brown in 1827. The "motion" refers to the mysterious, random meandering of tiny particles, such as the grains found in pollen dust when it is suspended in water. Einstein explained it as the result of submicroscopic molecules bombarding the floating particle from all sides, and with a very high frequency. Although individual collisions are too gentle to budge the particle, Einstein showed statistically that the magnitude and frequency of the observed particle jiggling could be explained by the rare occasions in which, by pure chance, a great many more molecules hit the particle on one side than on points opposite, imparting enough oomph to move it.

The paper was an immediate sensation, so compelling that even the atom's archenemy Friedrich Wilhelm Ostwald remarked that after reading Einstein's work, he was convinced that atoms were real. The grand champion of the atom, Boltzmann, on the other hand, inexplicably never got word of Einstein's work, or of the change in attitudes it brought about. In part because he was despondent over the reception of his own ideas, he committed suicide the following year. That's especially sad because, with Einstein's paper on Brownian motion and another paper he wrote in 1906, physicists were finally persuaded of the reality of objects they could neither touch nor see—precisely the idea that Boltzmann had been preaching, without much success, since the 1860s.

Within three decades, scientists employing new equations that describe the atom would begin to be able to explain the underlying principles of chemistry—finally providing explanation and proof for Dalton's and Mendeleev's ideas. They would also start work on Newton's dream of understanding the properties of materials on the basis of the forces that act among their constituent particles, i.e., their atoms. By the 1950s scientists would go further, putting their knowledge of the atom in the service of a more profound understanding of biology. And in the latter half of the twentieth century, the theory of the atom would usher in the technology revolution, the computer revolution, and the information revolution. What started as an analysis of the motion of pollen dust would grow into a tool that shaped the modern world.

The laws on which all those practical endeavors depended, the equations that describe the properties of the atom, would not, however, come from the classical physics of Newton, not even from its amended "relativistic" form. To describe the atom would instead demand new laws of nature—the quantum laws—and it was the quantum idea that was the subject of Einstein's other revolutionary paper of 1905.

In that paper, which bore the title "On a Heuristic Viewpoint Concerning the Production and Transformation of Light," Einstein took Planck's ideas and turned them into deep physical principles. Einstein was aware that, like the theory of relativity, the quantum theory was a challenge to Newton. But at that point, quantum theory gave no hint of the scope of that challenge, or of the disturbing philosophical implications that would arise when it was further developed, so Einstein did not know what he had wrought.

Because the "viewpoint" that Einstein presented in his paper involved treating light as a quantum particle rather than as a wave—as Maxwell's very successful theory had described it—this paper was not embraced as were as his other groundbreaking works of 1905. In fact, it would take more than a decade for the physics community to accept his ideas. Regarding Einstein's own feelings on the matter, it's telling to look back at a letter he wrote to a friend in 1905, in advance of all three papers. Of his paper on relativity, Einstein remarked that part of it "will interest you." Meanwhile, he described his paper on the quantum as "very revolutionary." And indeed, it was that work that eventually had the greatest impact, and it was that work, in particular, that would earn him the Nobel Prize in 1921.

* * *

It was no accident that Einstein picked up the quantum where Planck had left off. Like Planck, he had started his career working on issues related to the role of atoms in that backwater field, thermodynamics. Unlike Planck, though, he was an outsider, out of touch with most of contemporary physics. And with regard to atoms, Einstein and Planck had precisely opposite goals, for while Planck's Ph.D. research had been aimed at ridding physics of the atom, Einstein said that his aim in his first papers, written between 1901 and 1904, was "to find facts which would guarantee as much as possible the existence of atoms of definite finite size," a goal he finally accomplished with his revolutionary 1905 analysis of how the random movement of atoms cause Brownian motion.

But though Einstein helped physicists finally come to terms with atoms, in his work on Planck's quantum idea, Einstein introduced a *new* "atom-like" theory of light that physicists would find even more difficult to swallow. He had been led to it after considering Planck's research on blackbody radiation. Not satisfied with Planck's analysis, he developed his own mathematical tools to analyze the phenomenon. And though he came to the same conclusion—that blackbody radiation could be explained only through the concept of quanta—his explanation had a crucial, if seemingly technical, difference: Planck had assumed that the discrete character of the energy he analyzed was due to the way the atoms or molecules in the blackbody oscillated when emitting the radiation; Einstein saw the discrete nature as an inherent property of *radiation* itself.

Einstein saw blackbody radiation as evidence of a radical new principle of nature: that *all* electromagnetic energy comes in discrete packets, and that radiation is made of particles akin to atoms of light. It was with that insight that Einstein became the first to realize that the quantum principle is revolutionary—a fundamental aspect of our world, and not just an ad hoc mathematical trick employed to explain blackbody radiation. He called the particles of radiation "light quanta," and in 1926 his light quanta would be given their modern name: *photons*.

Had he left it at that, Einstein's theory of photons would have been merely an alternate model concocted, like Planck's, to explain blackbody radiation. But if the photon idea was truly fundamental, it should elucidate the nature of phenomena other than just the one it had been constructed to explain. Einstein found one of those in a phenomenon called the photoelectric effect.

The photoelectric effect is a process in which light impinging on a metal causes it to emit electrons. These can be captured as an electric current to be employed in various devices. The technique would become important in the development of the television, and it is still used in gadgets such as smoke detectors and the sensors that keep elevator doors from shutting on you as you walk in. In the latter application, a beam of light is shone across the doorway, impinging on a photoelectric receptor on the other side, which generates a current; if you step into the elevator, you break the beam and hence the current, and elevator manufacturers arrange things so that a break in the current causes the doors to open.

That light shone on metals can create electric currents was discovered in 1887 by German physicist Heinrich Hertz, who was the first to deliberately produce and detect the electromagnetic waves emitted by accelerating electric charges, and for whom the unit of frequency, the hertz, is therefore named. But Hertz couldn't explain the photoelectric effect, for the electron hadn't been discovered. That development, from the laboratory of British physicist J. J. Thomson, came in 1897—three years after Hertz died, at age thirty-six, from a rare disease that causes inflammation of the blood vessels.

The existence of the electron offered a simple explanation of the photoelectric effect: when waves of light energy hit a metal, they excite electrons within it, causing them to fly off into space and show themselves as sparks, rays, and currents. Inspired by Thomson's work, physicists began to study the effect in more detail. But the long and difficult experiments eventually uncovered aspects of the photoelectric effect that didn't fit the theoretical picture.

For example, when you increase the intensity of the beam of light, it results in more electrons being given off by the metal but has no effect on the energy of those ejected electrons. This contradicts the prediction of classical physics, because more intense light carries more energy, and so, on being absorbed, it ought to produce faster, more energetic electrons.

Einstein had pondered the issues for several years, and then in 1905 he finally made the quantum connection: the data could be explained if light is made of photons. Einstein's picture of the photoelectric effect was this: each light photon that strikes the metal transfers its energy to some particular electron. The energy that each photon carries is proportional to the frequency, or color, of the light, and if a photon carries enough energy, it will send the electron flying free. Light of higher frequency consists of photons of higher energy. On the other hand, if only the *intensity* of the light is increased (and not the frequency), the light will consist of more photons, but not more energetic ones. As a result, more intense light will result in more electrons being emitted, but the energy of the electrons will be unchanged—exactly as had been observed.

The proposal that light is made of photons—particles—contradicted Maxwell's highly successful theory of electromagnetism, which said that light travels in waves. Einstein suggested—correctly—that the classical "Maxwellian" wavelike qualities of light could arise when we make optical observations involving the net effects of a very large number of photons, which under ordinary circumstances is the case.

A one-hundred-watt lightbulb, for example, emits roughly a billion photons each billionth of a second. By contrast, the quantum nature of light is apparent when working with very low-intensity light—or in the case of certain phenomena (such as the photoelectric effect) whose mechanism depends on the discrete nature of photons. But Einstein's speculation was not enough to convince others to accept his radical ideas, and they were met with great and almost universal skepticism.

One of my favorite comments on Einstein's work was the 1913 recommendation written jointly by Planck and several other leading physicists for Einstein's induction into the prestigious Prussian Academy of Sciences: "In sum, one can say that there is hardly one among the great problems in which modern physics is so rich to which Einstein has not made a remarkable contribution. That he may sometimes have missed the target in his speculations, as, for example, in his hypothesis of light quanta, cannot really be held too much against him, for it is not possible to introduce really new ideas even in the most exact sciences without sometimes taking a risk."

* * *

Ironically, it was one of the initial opponents of the photon theory, Robert Millikan, who would finally perform the accurate measure-

Albert Einstein, 1921

ments that confirmed Einstein's law describing the energy of ejected photoelectrons—and he received the 1923 Nobel Prize for those efforts (as well as for his measurement of the electron's charge). When Einstein received his own 1921 Nobel Prize, the citation said simply, "To Albert Einstein for his services to theoretical physics and especially for his discovery of the law of the photoelectric effect."

The Nobel Committee had chosen to recognize Einstein's formula but ignore the intellectual revolution through which he had derived it. There was no mention in either prize of the light quanta, or of Einstein's contribution to quantum theory. Abraham Pais called this "an historic understatement but also an accurate reflection on the consensus in the physics community."

The doubts about the photon, and about quantum theory in general, would be laid to rest in the next decade with the creation of a formal theory of "quantum mechanics" that would unseat Newton's laws of motion as the fundamental principles governing how objects move and react to forces. When that theory finally came, Einstein would acknowledge its successes, but it would now be he who opposed the quantum.

Refusing to accept quantum theory as the final word, Einstein never stopped believing that it would eventually be replaced with an even more fundamental theory that would restore the traditional concept of cause and effect. In 1905 he had published his three papers, each of which changed the course of physics; for the rest of his life he tried in vain to do it again—to *reverse* what he had started. In 1951, in one of his last letters to his friend Michele Besso, Einstein admitted he had failed. "All these fifty years of pondering," he wrote, "have not brought me any closer to answering the question, What are light quanta?"

11

The Invisible Realm

Just after completing my Ph.D., I got a junior faculty position at Caltech and started fishing around for a topic to work on next so I wouldn't have to drop out of academia and take a more lucrative position waiting tables at the faculty club. After a seminar one afternoon, I got to talking to physicist Richard Feynman about a theory called string theory. Among his fellow physicists, Feynman, then in his sixties, was at the time probably the most revered scientist in the world. Today many (though far from all) view string theory as the leading candidate for a unified theory of all the forces of nature, the holy grail of theoretical physics. But back then, few people had heard of it, and most of those who had didn't care for it—including Feynman. He was grousing about it when a fellow who was visiting the department from a university in Montreal stepped over. "I don't think we should discourage young people from investigating new theories just because they are not accepted by the physics establishment," he said to Feynman.

Did Feynman reject string theory because it represented so great a break from his prior system of beliefs that he was unable to adjust his thinking? Or would he have come to the same conclusions about its shortcomings even if it hadn't represented such a break from prior theories? We can't know, but Feynman told the visitor that he wasn't advising me not to work on something new, just that I should be careful, because if it didn't work out I could end up wasting lots of time. The visitor said, "Well, I have been working on my own theory for twelve years," and then he proceeded to describe it in excruciating detail. When he was finished, Feynman turned to me and said, in front of the man who had just proudly described his work, "That's exactly what I mean about wasting your time."

The frontiers of research are enshrined in fog, and any active scientist is bound to waste effort following uninteresting or dead-end trails. But one of the traits that distinguishes the successful physicist is the knack (or luck) of choosing problems that prove both enlightening and solvable.

I've compared the passion of physicists to that of artists, but I have always felt that artists have a great advantage over physicists—in art, no matter how many of your colleagues and critics say your work stinks, no one can *prove* it. In physics they can. In physics, there is little consolation in thinking you had a "beautiful idea" if it was not a correct idea. And so in physics, as in any attempt at innovation, you have to maintain a difficult balance, being careful about the research problems you choose to pursue while not being so careful that you never do anything new. That's why the tenure system is so valuable to science—it makes it safe to fail, which is essential for fostering creativity.

Looking back, it appears that Einstein's exciting theory of photons, the light quanta, should have immediately stimulated a great deal of new research into the fledgling quantum theory. But to Einstein's contemporaries, who had yet to see a lot of evidence for the photon but had many good reasons for skepticism, to work on the photon would have required a great deal of intellectual adventurousness and courage.

Even young physicists, usually the most uninhibited when it comes to working on a problem that might not work or that might draw ridicule, and whose worldviews are still malleable, came and went, choosing for their doctoral and postdoctoral work anything but Einstein's crazy photon theory.

Almost ten years passed with virtually no progress. Einstein himself shot past thirty, old already for a pioneer theorist, and was spending a lot of his time on a different revolutionary idea: that of extending, or generalizing, his 1905 theory of special relativity to include gravity. (Special relativity was a modification of Newton's laws of motion; general relativity would replace Newton's law of gravity, but it required Einstein to alter special relativity.) Einstein's inattention to the photon theory led Robert Millikan to write, "Despite . . . the apparently complete success of the Einstein equation [for the photoelectric effect], the physical theory [of the photon] of which it is the expression is found so untenable that Einstein himself, I believe, no longer holds to it."

Millikan was wrong. Einstein hadn't given up on the photon, but since his attention was elsewhere at the time, it is easy to see why Millikan thought so. However, neither the photon nor the quantum concept that spawned it were dead. On the contrary, they would soon become stars, thanks, finally, to Niels Bohr (1885–1962), a twenty-something young man who was neither set in his ways nor experienced enough to know that he shouldn't risk wasting time by challenging our ideas about the laws that govern the world.

* * *

When Niels Bohr was of high school age, he would have been taught that the Greeks invented natural philosophy, and that Isaac Newton's equations, which described how objects react to the force of gravity, represented the first huge step toward the goal of understanding how the world works, because they allowed scientists to make precise quantitative predictions regarding the motion of falling and orbiting objects. Bohr would also have been taught that, shortly before his own birth, Maxwell had added to Newton's work a theory of how objects react to and generate electric and magnetic forces—thus advancing the Newtonian worldview to what we now know would turn out to be its zenith.

Physicists in Bohr's formative years seemed to have a theory of both forces and motion that included all the interactions of nature that were known at the time. What Bohr didn't know, as the century turned and he entered the University of Copenhagen to do his undergraduate work, was that after more than two hundred years of ever greater successes, the Newtonian worldview was about to collapse.

As we've seen, the challenge to Newton came because, though Maxwell's new theory initially made it seem possible to extend Newton's laws of motion to a whole new set of phenomena, in the end it revealed that phenomena such as blackbody radiation and the photoelectric effect violate the predictions of Newtonian (classical) physics. But the theoretical advances of Einstein and Planck were possible only because technical innovations had enabled experimentalists to explore physical processes involving the atom. And it was that turn of events that inspired Bohr, for he had a great appreciation—and considerable flair—for experimental research.

The years leading up to Bohr's dissertation were indeed exciting ones for anyone interested in experimental physics. In those years, technical advances like the development of evacuated glass tubes with an embedded electron source—a predecessor of the "cathode ray tubes" that were the screens in old-fashioned televisions—led to a number of important breakthroughs. For example: Wilhelm Röntgen's discovery of X-rays (1895); Thomson's discovery of the electron (1897); and New Zealand–born physicist Ernest Rutherford's realization that certain chemical elements such as uranium and thorium emit mysterious emanations (1899–1903). Rutherford (1871–1937) actually classified not one but three members of that zoo of mystery rays—the alpha, beta, and gamma rays. Those emanations, he speculated, were the debris produced when atoms of one element spontaneously disintegrated to form atoms of another.

Thomson's and Rutherford's discoveries in particular were a revelation because they related to the atom and its parts, which, it turned out, cannot be described employing Newton's laws, or even his conceptual framework. And so their observations, it would eventually be realized, demanded a whole new approach to physics.

But if both the theoretical and experimental developments of the day were dizzying, the initial reaction of the physics community to most of it was to take a chill pill and pretend none of it was happening. So not only were Planck's quantum and Einstein's photon dismissed, but so were these revolutionary experiments.

Prior to 1905, those who thought the atom was metaphysical nonsense treated talk of electrons—a supposed component of the atom—about as seriously as an atheist might treat a debate about whether God is a man or a woman. More surprising is the fact that those who *did* believe in atoms didn't like the electron either—because the electron was a supposed "part" of the atom, and the atom was supposed to be "indivisible." So outlandish did Thomson's electron seem that one distinguished physicist told him that upon hearing of his claim, he thought Thomson was "pulling their legs."

Similarly, Rutherford's idea that an atom of one element can decay into an atom of another was dismissed as if it had come from a man who had grown a long beard and donned an alchemist's robe. In 1941, scientists would learn how to convert mercury to gold—literally, the alchemists' dream—by bombarding that metal with neutrons in a nuclear reactor. But in 1903, Rutherford's colleagues were not adventurous enough to accept his bold claims regarding that transmutation of elements. (Ironically, they *were* adventurous enough to play with the

Ernest Rutherford

glowing radioactive trinkets Rutherford provided them, thus irradiating themselves with rays from the process they believed was not taking place.)

The spate of strange research papers from both theoretical and experimental physicists must have seemed to many like the physics equivalent of today's social psychology literature, in which researchers regularly claim to have made wild discoveries such as "People who eat grapes have more car accidents." But actually, though the physicists' conclusions sounded outlandish, they were correct. And eventually the accumulating experimental evidence, together with Einstein's theoretical arguments, forced physicists to accept the atom and its parts.

For his work leading to the discovery of the electron, Thomson won the 1906 Nobel Prize in Physics, while Rutherford received the Nobel Prize in 1908—but in Chemistry—for what amounted to the discovery that the robed alchemists had been on to something.

This, then, was the scene in 1909 when Niels Bohr entered the physics research picture. He was just five years younger than Einstein, but the gap was large enough to make him of a new generation, one that entered a field that had finally accepted both the atom and the electron—though still not the photon.

For his Ph.D. dissertation, Bohr chose to analyze and critique Thomson's theories. When he finished, he applied for and received a grant that allowed him to work in Cambridge so he could get the great man's reactions. The debate of ideas is a key feature of science, so for Bohr to approach Thomson with his criticism was not quite like an art student approaching Picasso to say his faces have too many angles—but it was close. And Thomson indeed did not prove anxious to grant his upstart critic an audience. Bohr would stay almost a year, but Thomson would not discuss the dissertation with him—or even read it.

Thomson's inattention would prove to be a blessing in disguise, for while Bohr was languishing in Cambridge, failing in his plan to engage Thomson, he met Rutherford, who had come visiting. Rutherford had himself worked under Thomson in his younger days, but by then he was the world's leading experimental physicist and director of a center for the study of radiation at the University of Manchester. Unlike Thomson, Rutherford appreciated Bohr's ideas, and he invited him to come work in his laboratory.

Rutherford and Bohr would make an odd pair. Rutherford was a huge, energetic man, both broad and tall, with a strong face and a thundering voice so loud that it sometimes disturbed sensitive equipment. Bohr had a delicate nature and was far gentler in both looks and manner, with droopy cheeks, a soft voice, and a slight speech impediment. Rutherford had a thick New Zealand accent; Bohr spoke poor, Danish-sounding English. Rutherford, when contradicted in a conversation, would listen with interest but then let the conversation end without making a reply. Bohr lived for debate, and had a hard time thinking creatively unless there was another person in the room to bounce ideas off and debate with.

The pairing with Rutherford was a lucky break for Bohr, for though Bohr went to Manchester thinking he might conduct experiments on the atom, after he arrived he instead became obsessed with a theoretical model of the atom that Rutherford was working out, based on Rutherford's own experimental studies. It was through the theoretical work he did on that "Rutherford atom" that Bohr would revive the dormant quantum idea and accomplish what Einstein's work on the photon hadn't: he would put the quantum idea on the map to stay.

* * *

When Bohr arrived in Manchester, Rutherford was doing experiments to investigate how the electric charge in the atom is distributed. He had decided to study that issue by analyzing the way charged particles are deflected when they are fired, like bullets, at an atom. For the charged projectiles he chose alpha particles—which he himself had discovered, and which we today know are simply positively charged helium nuclei.

Rutherford hadn't yet concocted his model of the atom, but he was assuming that the atom would conform pretty well with another model, one developed by Thomson. The proton and nucleus were not then known, and in Thomson's model the atom consisted of a diffuse fluid of positive charge in which enough tiny electrons circulate to offset the positive electricity. Since electrons weigh very little, Rutherford expected that, like marbles in the path of a cannonball, they'd have little effect on the course of the massive alpha particles. It was the much heavier fluid of positive charge—and the manner in which it was distributed—that Rutherford sought to study.

Rutherford's apparatus was simple. A beam of alpha particles was created from a radioactive substance such as radium and aimed at a thin piece of gold foil. Beyond the foil stood a small target screen. After the alpha particles passed through the foil, they would hit the screen, producing a tiny and very faint flash of light. Sitting in front of the screen with a magnifying glass, one could, with some effort, record its position and determine the degree to which the alpha particle had been deflected by the atoms in the foil.

Though Rutherford was world famous, the work and the work environment were the opposite of glamorous. The lab was in a damp, gloomy basement, with pipes running across both floor and ceiling. The ceiling was so low that you might bang your head on it, and the floor so uneven that you might then trip on one of the pipes in the floor before the pain in your head had even subsided. Rutherford himself lacked the patience required for making the measurements, and after trying it on one occasion for only two minutes, he swore and gave up. His assistant, German Hans Geiger, on the other hand, was a "demon" at the tedious task. Ironically, he would later negate the value of his own skill by inventing the Geiger counter.

Rutherford expected that most of the heavy, positively charged alpha particles would pass through the foil in the spaces between the gold atoms, too far from any of them to be noticeably deflected. But some, he theorized, would pass through one or more atoms and thus deviate a tiny bit from a straight path, repulsed by their diffuse positive charges.

The experiment would indeed elucidate the structure of the atom, but by luck rather than in the manner he had envisioned.

At first, all the data Geiger collected conformed to Rutherford's expectations, and appeared consistent with Thomson's model. Then, one day in 1909, Geiger suggested a "small research" project for a young undergraduate student named Ernest Marsden to get his feet wet. Rutherford, who had been sitting in on a class on probability theory in the math department, realized that there might be a small chance that a few of the alpha particles were being deflected by a somewhat larger angle than his apparatus was designed to detect. And so he suggested to Geiger that Marsden perform a variation on their experiment to look into that possibility.

Marsden got to work looking for particles that had experienced larger deviations than Geiger had previously been looking for—even very large deflections that, if they occurred, would violate everything Rutherford "knew" about the structure of the atom. That task was, in Rutherford's view, almost certain to be nothing more than a mammoth waste of time. In other words, it was a fine project for an undergrad.

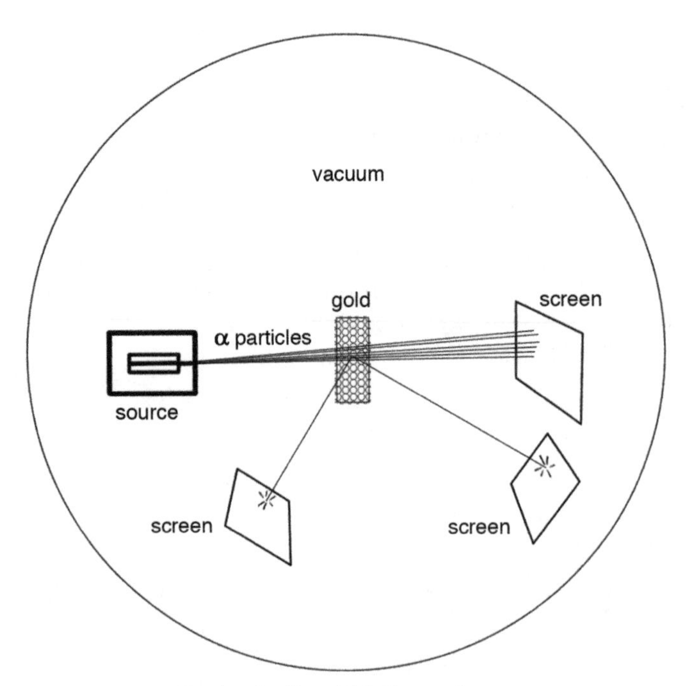

Rutherford's gold foil experiment

Marsden dutifully watched as alpha particle after alpha particle traveled through the foil as expected, with no dramatic deflections. And then that which was virtually unimaginable finally occurred: a scintillation appeared in the detection screen while it had been positioned far off center. In the end, of the many thousands of alpha particles Marsden observed, just a handful were deflected at wide angles, but one or two of those bounced back, almost boomerang-like. That was enough.

After hearing the news, Rutherford said it was "quite the most incredible event that had ever happened to me in my life. It was almost as incredible as if you had fired a 15-inch shell at a piece of tissue paper and it came back and hit you." He reacted this way because his mathematics told him that there had to be something inconceivably small and powerful in the gold foil to cause infrequent but large deflections like that. And so Rutherford hadn't elucidated the details of the Thomson model after all—he had discovered that the Thomson model was wrong.

In advance of carrying out Marsden's experiment, the project had seemed outlandish, the sort of activity Feynman was warning me not to engage in. In the century since its execution, however, the experiment has universally been praised as brilliant. And indeed, without it, there likely would have been no "Bohr atom," meaning that a coherent theory of the quantum would likely have come—assuming it still would have come—many years later. That, in turn, would have had a great effect on what we call technological progress. It would have delayed the development of the atom bomb, for one, meaning the bomb would not have been used on Japan, thus saving the lives of numerous innocent Japanese civilians but perhaps costing the lives of the numerous soldiers who'd have perished in an Allied invasion. It also would have put off many other inventions, such as the transistor, and thus the start of the computer age. It's hard to say exactly what all the effects of not performing that single, seemingly pointless undergraduate experiment would have been, but it is safe to say that the world today would look quite different. And so here again we see the fine line between an outlandish crackpot project and an innovative idea that changes everything.

In the end, Rutherford oversaw numerous further experiments in which Geiger and Marsden observed more than one million scintillations. From this data, he would fashion his theory about the structure of the atom, a theory that was different from Thomson's in that, while it still portrayed the electrons as moving in concentric orbits, the positive charge was no longer spread out, but rather concentrated at the atom's tiny center. Geiger and Marsden, though, would soon part ways. They would find themselves fighting on opposite sides in World War I, and then applying their science on opposite sides in World War II: while Marsden worked on the new technology of radar, Geiger, a Nazi supporter, worked to develop a German atom bomb.

The Rutherford atom is the model we all learn in grade school, in which electrons orbit the nucleus just as the planets orbit the sun. Like many concepts of science, it seems uncomplicated when boiled down to an everyday simile like that classroom model, but the idea's true brilliance lies in precisely those "technical" complexities that are lost in the distillation process that produces such simple pictures. To have an intuitive picture is helpful, but what brings an idea in physics to life is the mathematical consequences it implies. And so a physicist must be not just a dreamer but a technician.

To Rutherford the dreamer, the experiment said that the vast majority of the atom's mass, and all of its positive charge, must be concentrated at its center, in an incredibly tiny ball of charged matter so dense that a mere cupful of that material would weigh one hundred times as much as Mount Everest. He would later call that central core of the atom the "nucleus." (That you and I don't weigh anything near the weight of Mount Everest is a testament to the fact that the nucleus is only a minuscule dot at the center of the atom, which is otherwise made up mostly of empty space.)

Rutherford the technician ground through the complex, technical

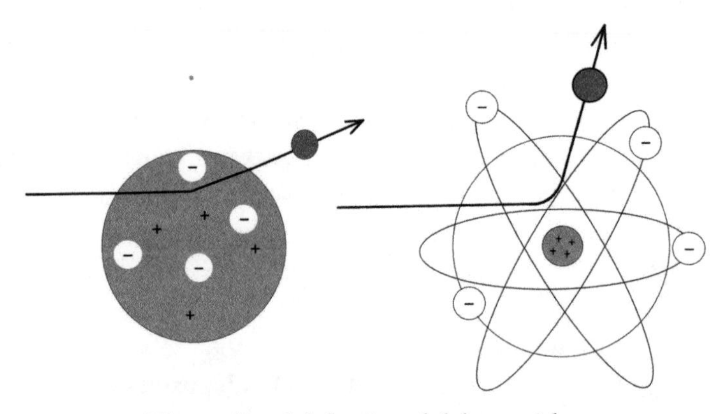

The predicted deflection of alpha particles, given the Thomson (left) and Rutherford atoms

mathematics to find that indeed, if the picture he'd imagined was true, his experiment would produce precisely the observations his team had made. Most of the fast, heavy alpha particles would pass through the gold foil, missing the minute atomic centers, and as a result they would be only slightly affected. Meanwhile, a few—those that passed close to a nucleus—would encounter an intense force field, causing a large deflection. The strength of that force field must have seemed as science-fiction-like to Rutherford as the force fields in the movies seem to us today. But even if we can't yet create such fields in our macro world, they do exist within the atom.

The important point that Rutherford had discovered is that the positive charge in the nucleus is concentrated at its center, not spread out. His portrayal of the electrons as orbiting the nucleus as the planets do the sun, on the other hand, was completely wrong—and he knew it.

First, the analogy of the solar system ignores the interaction *among* the planets in the solar system, and likewise *among* the different electrons in an atom. Those interactions are not at all alike. The planets, which have quite a bit of mass but no net electric charge, interact via gravity; electrons, which are charged but have little mass, interact through the electromagnetic force. Gravity is an extremely weak force, so the pull the planets exert on one another is so small that for many practical purposes it can be ignored; the electrons, however, exert on one another an enormously strong electromagnetic repulsion that would quickly disrupt those nice circular orbits.

The other glaring issue was that both planets and electrons that move in a circle will emit waves of energy—gravitational energy in the case of planets, and electromagnetic energy in the case of electrons. Again, because gravity is so weak, in the billions of years our solar system has existed, the planets wouldn't have lost more than a few percent of their energy. (In fact, this effect wasn't even known until Einstein's theory of gravity predicted it in 1916.) On the other hand, because the electromagnetic force is so strong, according to Maxwell's theory Rutherford's orbiting electrons would radiate away all their energy and plummet into the nucleus in about one one-hundred-millionth of a second. In other words, if Rutherford's model were true, the universe as we know it would not exist.

If ever there was a prediction you'd think was likely to sink a theory, it would be the prediction that the universe does not exist. So why was it taken seriously?

That illustrates another important point about progress in physics: most theories are not sweeping theories of grand scope, but rather specific models meant to account for a particular situation. And so, even if they have drawbacks, and one is aware that a model fails in some situations, it can be useful nonetheless.

In the case of the Rutherford atom, physicists who worked on the atom appreciated that his model made correct predictions regarding the nucleus, and assumed that some future experiment would reveal essential missing facts that would settle the issue of how the electrons entered the picture, and why the atom was stable. What was *not* apparent was that the atom was demanding not a more clever explanation, but a revolutionary one. The pale and modest Niels Bohr, however, had a different take on things. To young Bohr, the Rutherford atom and its contradictions were a haystack that held a golden needle. And he was determined to find it.

* * *

Bohr posed a question to himself: If the atom doesn't send off waves of energy as classical theory demands (at least according to Rutherford's model), could that be because the atom doesn't obey classical laws? To pursue that line of reasoning, Bohr turned to Einstein's work on the photoelectric effect. He asked what it would mean if the quantum idea applied to the atom as well. That is, what if the atom, like Einstein's light quanta, could have only certain energies? The idea led to him to revise Rutherford's model and create what would come to be called the Bohr atom.

Bohr explored his idea by focusing his attention on the simplest atom, hydrogen, which consists of a single electron orbiting a nucleus made of a single proton. The difficulty of Bohr's undertaking is illustrated by the fact that it wasn't even obvious at the time that hydrogen *did* have that simple structure—Bohr had to deduce that hydrogen has just one electron from a series of experiments performed by Thomson.

Newtonian physics predicts that an electron can circle a nucleus (which, in the case of hydrogen, is just a proton) at any distance, as long as its speed and energy have the proper values, which are determined by that distance. The smaller the electron's distance from the proton, the lower the atom's energy must be. But suppose, in the spirit of Einstein, we were to contradict Newtonian theory by adding a new law dictating that—for some as-yet-unknown reason—an atom is not free to have just any energy, but rather only a value belonging to some discrete set of possibilities. Since the orbital radius is determined by the energy, that restriction on the *energy* values allowed would translate into a restriction on the possible *radii* at which an electron can orbit. When we make this assumption, we say that the atom's energy and the radii of the electron orbits are "quantized."

Bohr postulated that if the properties of an atom are quantized, the atom cannot *continuously* spiral inward toward the nucleus and lose energy, as classical Newtonian theory predicts; instead the atom can lose energy only in "clumps" as it jumps from one allowed orbit to another. According to Bohr's model, when an atom is excited by an energy input—for instance, from a photon—the absorbed energy causes an electron to jump to one of the farther-out, higher-energy orbits. And each time a jump is made back to a smaller, lower-energy orbit, a quantum of light—a photon—is emitted with a frequency corresponding to the difference in energy between the two orbits.

Now suppose that, again for some yet-unknown reason, there is an *innermost* allowed orbit—an orbit of lowest energy, which Bohr dubbed the "ground state." In that case, when the electron reaches that state, it can no longer lose energy, so it doesn't plummet into the nucleus, as the Rutherford model predicts. Bohr expected that a similar, if more complicated, scheme would also work for the other elements, which have multiple electrons: he saw quantization as the key to the stability of the Rutherford atom—and, hence, of all the matter in the universe.

Like Planck's work on blackbody radiation and Einstein's explanation of the photoelectric effect, Bohr's ideas were not derived from a general theory of the quantum, but rather were ad hoc concepts concocted to explain only one thing—in this case, the stability of the Rutherford atom. It is a testament to human ingenuity that, despite the absence of any "mother theory" to birth his model, the Bohr picture, like Planck's and Einstein's, was essentially correct.

Bohr would later say that his ruminations on the atom had jelled only after a chance conversation with a friend in February 1913. That friend had reminded him of the laws in a field called spectroscopy—the study of the light given off by an element in gaseous form that has been "excited," say by an electric discharge or by intense heat. It had long been known that—for reasons that were not understood—in these situations each gaseous element emits a particular group of electromagnetic waves characterized by a limited set of frequencies. Those frequencies are called spectral lines, and form a kind of fingerprint from which elements can be identified. After talking to his friend, Bohr realized that he could employ his model of the atom to predict what the fingerprint of hydrogen ought to look like, thus connecting his theory to the test of experimental data. It is that step, of course, that in science elevates an idea from a promising or "beautiful" notion to a serious theory.

When Bohr finished the math, the result astonished even him: the differences in energy among his "allowed orbits" reproduced precisely all the frequencies of the many series of spectral lines that had been observed. It is hard to imagine the exhilaration the twenty-seven-year-old Bohr must have felt when he realized that, with his simple model, he had reproduced all the spectroscopists' puzzling formulas, and explained their origin.

Bohr published his masterpiece on the atom in July 1913. He had worked hard for that triumph. From the summer of 1912 to his moment of inspiration in February 1913, he had wrestled with his ideas day and night, putting in so many hours that even his diligent colleagues were in awe. In fact, they thought he might drop from exhaustion. One incident says it all: he had arranged to be married on August 1, 1912, and he was, but he canceled his honeymoon in scenic Norway and spent the time instead in a hotel room in Cambridge, dictating a paper on his work to his new wife.

Bohr's new theory, being such a hodgepodge, was clearly only the beginning. For example, he called the allowed orbits "stationary states," because the electrons, by not radiating, as classical theory required, were behaving as if they did not move. On the other hand, he spoke often of the electrons' "state of motion," picturing them circling the nucleus in their allowed orbits until they either jumped to a lower-energy orbit or became excited by incoming radiation and jumped to a higher-energy state. I mention this because it illustrates the fact that Bohr was employing two contradictory images. This is the approach of many pioneers in theoretical physics—in literature, one is told not to mix metaphors, but in physics, if we know that one metaphor is not totally appropriate, it is common to (carefully) mix in another.

In this case, Bohr was not terribly fond of the classical picture of a solar-system atom, but it was his starting point, and to create his new theory he employed the equations of classical physics that related the radius and energy of the electron orbits, while folding in new quantum ideas like the principle of stationary states, thereby creating a modified picture.

The Bohr atom initially met with mixed reactions. At the University of Munich, an influential physicist named Arnold Sommerfeld (1868–1951) not only immediately recognized the work as a milestone of science, but began working on the idea himself, in particular exploring its connection with relativity. Meanwhile, Einstein said that Bohr's was "one of the greatest discoveries [ever]." But perhaps most revealing with regard to just how shocking the Bohr atom appeared to a physicist of his era was another comment Einstein made. The man who had been brave enough to propose not just the existence of light quanta, but also the idea that space, time, and gravity are all intertwined, said of the Bohr atom that he had had similar ideas but, because of their "extreme novelty," had not had the courage to publish them.

That publication took courage is reflected by some of the other reactions Bohr received. For example, at the University of Göttingen, a leading German institution, there was, as Bohr later recalled, complete agreement "that the whole thing was some awful nonsense, bordering on fraud." One Göttingen scientist, an expert in spectroscopy, put the Göttingen attitude in writing: "It was regrettable in the highest degree that the literature should be contaminated by such wretched information, betraying so much ignorance." One of the grand old men of British physics, Lord Rayleigh, meanwhile, said he could not bring himself to believe that "Nature behaved in that way." Yet he added—presciently—that "men over seventy should not be hasty in expressing opinions on new theories." Another leading British physicist, Arthur Eddington, was also less than enthusiastic, having already dismissed the quantum ideas of Planck and Einstein as "a German invention."

Even Rutherford reacted negatively. For one, he had little taste for theoretical physics. But what bothered him about Bohr's work—which was, after all, a revision of his own model of the atom—was that his Danish colleague had provided no mechanism for the electron to carry out those jumps between energy levels that he had postulated. For example, if the electron, in moving to an energy level corresponding to a smaller orbit, is "jumping" to the new orbit and not "spiraling" continuously inward, then just what sort of path does this "jump" consist of, and what happens to cause it?

As it would turn out, Rutherford's objection went precisely to the heart of the matter. Not only would such a mechanism never be found, but when quantum theory matured into a general theory of nature, it would dictate that such questions have no answer, and hence no place in modern science.

What eventually convinced the physics world of the correctness of Bohr's ideas—and hence also of Planck's and Einstein's earlier work—came over a ten-year period, from 1913 to 1923. Applying Bohr's theory to atoms of elements that are heavier than hydrogen, Bohr and others realized that by ordering the elements by atomic number rather than by atomic weight, as Mendeleev had done, they could eliminate some errors in Mendeleev's periodic table.

The atomic weight is determined by the number of protons and neutrons in the atom's nucleus. The atomic number, by contrast, equals the number of protons, which, since atoms have no overall electric charge, is also the number of electrons that the atom possesses. Atoms with more protons in their nucleus generally have more neutrons as well, but not always, so the two measures can differ in their implications for ordering the elements. Bohr's theory showed that the atomic number is the proper parameter upon which to base the periodic table, because it is the protons and electrons that determine an element's chemical properties, not its neutrons. It had taken more than fifty years, but thanks to Bohr, science could now explain why Mendeleev's mysterious periodic table worked.

With the maturation of the quantum ideas into a general structure that would replace Newton's laws, physicists would eventually be able to write down equations from which, in principle, the behavior of all atoms could be derived—though in most cases it would take the technology of supercomputers to derive it. But no one had to wait for supercomputers to test Bohr's ideas regarding the significance of the atomic number: in the Mendeleev tradition, he predicted the properties of a yet-undiscovered element—ironically, one that Mendeleev, having based his system on atomic weight, had gotten wrong.

The element was discovered very shortly thereafter, in 1923, and named hafnium, after Hafnia, the Latin name for Bohr's hometown of Copenhagen. With that, no physicist (or chemist) could ever again doubt the truth of Bohr's theories. Some five decades later, Bohr's name would join Mendeleev's on the periodic table with the naming of element 107, bohrium, in 1997. That same year, his former mentor—and sometime critic—was also honored, with the naming of element 104, rutherfordium.*

*In addition to Mendeleev, Bohr, Rutherford, and Lise Meitner—whom I mentioned earlier—there are twelve other scientists with elements named for them: Vasili Samarsky-Bykhovets (samarium), Johan Gadolin (gadolinium), Marie Skłodowska-Curie and Pierre Curie (curium), Albert Einstein (einsteinium), Enrico Fermi (fermium), Alfred Nobel (nobelium), Ernest Lawrence (lawrencium), Glenn T. Seaborg (seaborgium), Wilhelm Röntgen (roentgenium), Nicolaus Copernicus (copernicium), and Georgy Flyorov (flerovium).

12

The Quantum Revolution

Despite all the brilliant and eager minds now focused on the quantum, and the isolated truths they had guessed at or discovered, by the early 1920s there was still no general theory of the quantum, or any hint that such a theory might be possible. Bohr had concocted certain principles that, if true, explained why atoms were stable and accounted for their line spectra, but why were those principles true, and how do you apply them to analyze other systems? No one knew.

Many quantum physicists were becoming discouraged. Max Born (1882–1970), a future Nobel Prize winner who would soon introduce the term "photon," wrote, "I am thinking hopelessly about quantum theory, trying to find a recipe for calculating helium and other atoms; but I am not succeeding in this. . . . The quanta are really in a hopeless mess." And Wolfgang Pauli (1900–1958), another future Nobel laureate who would propose and then work out the mathematical theory of the property called spin, put it this way: "Physics is very muddled at the moment; it is much too hard for me anyway, and I wish I were a movie comedian or something like that and had never heard anything about physics."

Nature offers us puzzles, and it is we who must make sense of them. One thing about physicists is that they invariably have deep faith that those puzzles harbor profound truths. We believe that nature is governed by general rules and is not just a hodgepodge of unrelated phenomena. The early quantum researchers did not know what the general theory of the quantum would be, but they trusted that such a theory would exist. The world they explored was obstinate in its resistance to explanation, but they imagined that it could be made sense of. Their dreams nourished them. They were vulnerable to moments of doubt and desperation, as we all are, yet they marched forward on difficult journeys that consumed years of their lives, motivated by their belief that there would be a reward of truth at the other end. As with any very difficult endeavor, we find that those who succeeded had very strong convictions, because those whose faith was weak dropped out before they could succeed.

It is easy to understand the despair of those like Born and Pauli, for not only was quantum theory challenging in and of itself, but it came of age in difficult times. Most of the quantum pioneers worked in Germany, or moved between Germany and the institute that Bohr had raised money to establish in 1921 at the University of Copenhagen, and so they were fated to conduct their search for a new scientific order at a time when the social and political order around them was dissolving into chaos. In 1922, the German foreign minister was assassinated. In 1923, the value of the German mark sank to a trillionth of its prewar value, and it took five hundred billion of those "German dollars" just to buy a kilogram of bread. Still, the new quantum physicists sought sustenance in understanding the atom and, more generally, the fundamental laws of nature that apply at that tiny scale.

The sustenance finally began to arrive in the middle of that decade. It came in fits and starts, beginning with a paper published in 1925 by a twenty-three-year-old named Werner Heisenberg (1901–1976).

* * *

Born in Würzburg, Germany, Heisenberg, the son of a classical languages professor, was recognized at an early age as brilliant—and competitive. His father encouraged that spirit of competition, and Heisenberg fought frequent battles with his year-older brother. The feud culminated in a bloody fight in which they beat each other with wooden chairs and then called a truce—one that would last mainly because they went their own ways, leaving home and not speaking for the remainder of their lives. In later years, Heisenberg would attack the challenges presented by his work with the same fierceness.

Werner would always take competition as a personal challenge. He had no particular talent for skiing, but he trained himself to become an excellent skier. He became a long-distance runner. He took up the cello and the piano. But most important of all, when in grade school he discovered that he had a talent for arithmetic, it sparked him to take a great interest in mathematics and its applications.

In the summer of 1920, Heisenberg decided to pursue a doctorate in math. To be accepted into a program required convincing a faculty member to sponsor you, and through a connection of his father's, Heisenberg managed to obtain an interview with a well-known mathematician named Ferdinand von Lindemann at the University of Munich. As it turned out, it wasn't the good kind of interview you sometimes get through connections, in which Heisenberg might have been offered tea and Black Forest cake and told they'd heard amazing stories about his brilliance. Rather, it was the bad kind, in which Lindemann—two years from retirement, partially deaf, and not very interested in first-year students—kept a poodle on his desk that barked so loudly he could hardly hear Heisenberg. In the end, though, what really seemed to have doomed Heisenberg's chances was that he mentioned having read a book on Einstein's theory of relativity, written by the mathematician Hermann Weyl. Upon learning of the young man's interest in a physics book, Lindemann, a number theorist, abruptly ended the interview, saying, "In that case you are completely lost to mathematics."

With his comment, Lindemann may have meant that showing interest in physics indicates bad taste, though, as a physicist I like to think he was really saying that, having been exposed to a far more interesting subject, Heisenberg would never have the patience for mathematics. In any case, Lindemann's arrogance and closed-mindedness changed the course of history, for had he embraced Heisenberg, physics would have the lost the man whose ideas would become the heart of quantum theory.*

Instead, after his rejection by Lindemann, Heisenberg did not see many options, and decided to try for the consolation prize of a physics doctorate under Arnold Sommerfeld, who had been a great supporter of the Bohr atom, and by then had made his own contributions to the theory. Sommerfeld, a slight, balding man with a large mustache and no poodles, was impressed that young Heisenberg had tackled Weyl's book. Not impressed enough to immediately embrace him, but enough to offer

*Ironically, Lindemann had once dabbled, without much success, in physics. He is best remembered as the man who proved that you cannot "square the circle"—that is, given a ruler and a compass alone, you cannot constuct a square with the same area as a given circle.

to sponsor him provisionally. "It may be that you know something; it may be that you know nothing," Sommerfeld said. "We shall see."

Heisenberg, of course, *did* know something. Enough that he completed his doctorate under Sommerfeld in 1923 and, in 1924, received an even more advanced degree called the "Habilitation," working under Born in Göttingen. But his path to immortality really started after that, on a visit to Niels Bohr in Copenhagen in the fall of 1924.

When Heisenberg arrived, Bohr was leading a misguided effort to amend his model of the atom, and Heisenberg joined it. I say "misguided" not just because the effort failed, but because of its aim: Bohr wanted to rid his model of the photon, Einstein's quantum of light. That might sound strange, since it was the idea of light quanta that had first inspired Bohr to think about the notion that an atom might be restricted to having only certain discrete energies. Still, Bohr, like most physicists, was reluctant to accept the reality of the photon, so he asked himself, Could one create a variant of the original Bohr atom that didn't incorporate it? Bohr believed he could. We've seen Bohr sweat over ideas and eventually succeed, but this was a case in which he would sweat over an idea and fail.

When I was a student, my friends and I idolized a number of physicists. Einstein for his airtight logic and radical ideas. Feynman and British physicist Paul Dirac (1902–1984) for inventing seemingly illegal mathematical concepts and applying them to get amazing results. (The mathematicians would later find a way to justify them after all.) And Bohr for his intuition. We thought of them as heroes, superhuman geniuses whose thinking was always clear, and whose ideas were always correct. That's not unusual, I suppose—artists, entrepreneurs, and sports fans can all name figures they consider larger than life.

In my student days, we were told that Bohr's intuition about quantum physics was so impressive that he seemed to have "a direct line to God." But while discussions of the early days of quantum theory often tell of Bohr's great insights, they rarely mention his many mistaken ideas. That's natural, because with time the good ideas survive and the bad ones are forgotten. Unfortunately, that leaves us with the mistaken impression that science is more straightforward and easier—at least for certain "geniuses"—than it really is.

Basketball great Michael Jordan once said, "I've missed more than nine thousand shots in my career. I've lost almost three hundred games.

Twenty-six times, I've been trusted to take the game-winning shot and missed. I failed over and over and over again in my life. And that is why I succeed." He said it in a Nike commercial, because to hear that even a legend failed and persisted through his failures is inspirational. But to anyone engaged in a field of discovery or innovation, it is of equal value to hear of Bohr's misguided notions, or Newton's fruitless efforts at alchemy, to recognize that our intellectual idols had wrong ideas and failures as mammoth as the ones we are quite aware of having had ourselves.

That Bohr should have considered his own Bohr atom to be too radical an idea is interesting, but not surprising, for science, like society, is built upon certain shared ideas and beliefs, and the Bohr atom did not conform. As a result, pioneers from Galileo and Newton to Bohr and Einstein—and beyond—had one foot in the past, even as their imaginations helped to create the future.

In that, the "revolutionaries" of science are no different from forward-thinking individuals in other fields. Consider, for example, Abraham Lincoln, champion of freeing the slaves in the American South, who was nevertheless unable to let go of his anachronistic belief that the races will never live together "in social and political equality." Lincoln himself realized that one could perceive his stand against slavery to be inconsistent with his tolerance of racial inequality. But he defended his acceptance of the supremacy of Caucasians by saying that whether it "accords with justice" is not the key issue, because white supremacy is a "universal feeling" that, whether "well or ill-founded, can not be safely disregarded." To abandon white supremacy, in other words, was too radical a step, even for him.

If you ask people why they believe this or that, they will not usually be as open or self-aware as Lincoln. Few will say, as he essentially did, that they believe something because everyone else believes it. Or "because I have always believed it," or "because I was indoctrinated to believe it at home and in school." But, as Lincoln remarked, that is often a good part of the reason. In society, shared beliefs create culture, and sometimes injustice. In science, art, and other areas where creativity and innovation are important, shared beliefs can create mental barriers to progress. That is why change often comes in fits and starts, and that is why Bohr got bogged down in trying to alter his theory.

If Bohr's new theory was doomed, it did have one very fortunate effect: it forced young Heisenberg to think deeply about the implications of Bohr's original theory of the atom. Gradually, his analysis started to move him toward a radical new view of physics: that it might be viable, and even desirable, to abandon the idea of a physical picture of the inner workings of the atom—the orbital motion of electrons, for example, which we imagine in our minds but in practice cannot observe.

Bohr's theory, like the theories of classical physics, rested on mathematical values assigned to features such as the position and orbital velocity of the electron. In the world of objects Newton studied—projectiles, pendulums, planets—position and velocity can be observed and measured. But experimenters in the lab *cannot* observe whether atomic electrons are here or there, or how fast they are moving, if indeed they are moving at all. If the classical concepts of position, speed, paths, orbits, and trajectories are not observable on the atomic level, Heisenberg reasoned, perhaps one should stop trying to create a science of the atom—or other systems—that is based on them. Why be wedded to those old ideas? They were mental ballast, Heisenberg decided—so *seventeenth century*.

Is it possible, Heisenberg asked himself, to develop a theory based only upon atomic data that can be directly measured, such as the frequencies and amplitudes of radiation that atoms emit?

Rutherford had objected to the Bohr atom because Bohr gave no mechanism for how the electron jumps among the atom's energy levels; Heisenberg would answer that criticism not by providing that mechanism, but by asserting that there is no mechanism, no path when one speaks of electrons, or at least that the very question is outside the realm of physics—because physicists measure the light absorbed or emitted in such processes but cannot witness the processes themselves. By the time Heisenberg returned to Göttingen in the spring of 1925 to work as a lecturer in Born's institute, it had become his dream—his goal—to invent a new approach to physics based purely on measurable data.

Creating a radically new science that abandoned Newton's intuitive description of reality and that disavowed concepts—such as position and velocity—that we can all picture and relate to was an audacious goal for anyone, let alone a twenty-three-year-old like Heisenberg. But like Alexander, who changed the political map of the world at age twenty-two, young Heisenberg would lead a march that reshaped the scientific map of the world.

* * *

The theory Heisenberg created from his inspiration would replace Newton's laws of motion as our fundamental theory of nature. Max Born would dub it "quantum mechanics," to distinguish it from Newton's laws, often called Newtonian mechanics or classical mechanics. But the theories of physics are validated by the accuracy of their predictions, not by common agreement or taste, so one might wonder how a theory based on exotic ideas like Heisenberg's could "replace" a well-established theory like Newton's, which had had so many successes.

The answer is that, though the conceptual framework behind quantum mechanics is far different from Newton's, the mathematical predictions of the theories usually differ only for systems on the atomic scale or smaller, where Newton's laws fail. And so, once it became fully developed, quantum mechanics would explain the strange behavior of the atom without contradicting the well-established description of everyday phenomena provided by Newtonian theory. Heisenberg and the others who worked to develop quantum theory knew that this must be the case, and they developed a mathematical expression of the idea that provided useful tests for their evolving theory. Bohr called it the "correspondence principle."

How did Heisenberg create a concrete theory from what was then little more than a philosophical preference? His challenge was to translate the notion that physics should be based on "observables"—quantities we measure—into a mathematical framework that, like Newton's, could be used to describe the physical world. The theory he invented would apply to any physical system, but he developed it in the context of the atomic world, with the initial aim of explaining, through a general mathematical theory, the reasons for the success of Bohr's ad hoc model of the atom.

Heisenberg's first step was to identify the observables that are appropriate for the atom. Since, in the atomic world, what we measure are the frequencies of light that atoms emit, and the amplitude, or intensity, of those spectral lines, it is those properties that he chose. Then he set out to employ the techniques of traditional mathematical physics to derive the relation between traditional Newtonian "observables," such as position and velocity, and that data on spectral lines. His aim was to use that connection to replace each observable in Newtonian physics with a quantum counterpart. That would prove to be a step that required both creativity and courage, because it would demand that Heisenberg turn position and momentum into mathematical entities that looked both new and bizarre.

The new type of variable was demanded by the fact that although position, say, is defined by specifying a single point, spectral data requires a different description. Each of the various properties of the light that atoms emit, such as color and intensity, forms not just a single number, but an entire array of numbers. The data form an array because there is a spectral line corresponding to a jump from any initial state of the atom to any final state—producing an entry for every possible *pair* of Bohr's energy levels. If that sounds complicated, don't worry—it is. In fact, when Heisenberg first came up with the scheme, he himself called it "very strange." But the gist of what he did was eliminate from his theory electron orbits one can visualize and replace them with purely mathematical quantities.

Those who worked on theories of the atom before Heisenberg had, like Rutherford, wanted to discover a mechanism behind atomic processes. They had thought of the inaccessible contents of the atom as being real, and tried to derive the nature of the observed spectral lines based on guesses about the behavior of what is inside—such as orbiting electrons. Their analyses had always assumed that the components of the atom would have the same basic character as the things we are used to in everyday life. Only Heisenberg thought differently, and had the

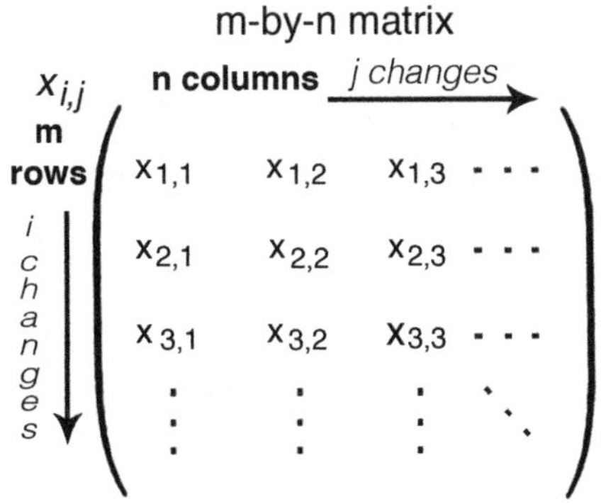

m-by-n matrix

$x_{i,j}$ **n columns** *j changes* →

m rows ↓ *i changes*

$$\begin{pmatrix} x_{1,1} & x_{1,2} & x_{1,3} & \cdots \\ x_{2,1} & x_{2,2} & x_{2,3} & \cdots \\ x_{3,1} & x_{3,2} & x_{3,3} & \cdots \\ \vdots & \vdots & \vdots & \ddots \end{pmatrix}$$

In Heisenberg's theory, position is represented by an infinite matrix, or array, of numbers rather than the familiar spatial coordinates courage to boldly declare that the orbits of the electrons are beyond the scope of observation, and hence are *not real* and have no place in the theory. This would be Heisenberg's approach not just to the atom, but to any physical system.

In insisting on such analyses, Heisenberg abandoned the Newtonian picture of the world as an arrangement of material objects that have an individual existence and definite properties such as speed and location. His theory, once perfected, would require us to accept, instead, a world based on a different conceptual scheme, one in which an object's path and even its past and future are not precisely determined.

Given that in today's world many people have trouble adjusting to new technologies like texting and social media, one can only imagine the mental openness it took to adjust one's thinking to a theory that said the electrons and nuclei you are made of don't have a concrete existence. But Heisenberg's approach demanded just that. It wasn't just a new kind of physics—it was an entirely new conception of reality. Such issues led Max Born to question the centuries-old split between physics and philosophy. "I am now convinced," he wrote, "that theoretical physics is actual philosophy."

As these ideas began to fall into place for Heisenberg, and his mathematical calculations progressed, he grew increasingly excited. But then he got sick with an attack of hay fever so severe he had to leave Göttingen and retreat to a rocky island in the North Sea on which little or nothing grew. His entire face was apparently horrendously swollen. And yet he kept working, day and night, and finished the research that would constitute his first paper on the ideas that would turn physics on its head.

When he returned home, Heisenberg wrote up his findings and gave one copy to his friend Pauli and another to Born. The paper outlined a methodology and applied it to a couple of simple problems, but Heisenberg hadn't been able to apply his ideas to calculate anything of practical interest. His work was very rough, horridly complicated, and extremely mysterious. To Born, confronting it must have been like talking to one of those people you meet at a cocktail party who go on and on but don't make any sense. Most people, faced with reading a paper that difficult, would go at it for a few minutes, then drop it and have a glass of wine. But Born kept at it. And in the end, he was so impressed by Heisenberg's work that he immediately wrote Einstein and told him that the young scientist's ideas were "surely correct and profound."

Like Bohr and Heisenberg, Born had been inspired by Einstein's relativity, and noted that Heisenberg's focus on what can be measured was analogous to Einstein's careful attention, in creating relativity, to the operational aspects of how time is gauged.

Einstein, though, didn't like Heisenberg's theory, and this would be the point in the evolution of quantum theory at which Einstein and the quantum began to part company: Einstein could not bring himself to endorse a theory that gave up on the existence of a well-defined objective reality in which objects have definite properties like position and velocity. That the properties of an atom might be explainable by a provisional theory that did not reference the atom's orbit, he could stomach. But a *fundamental* theory proclaiming that such orbits don't exist—this he would not subscribe to. As he would later write, "I incline to the belief that physicists will not be permanently satisfied with . . . an indirect description of Reality."

Heisenberg himself wasn't sure what he had created. He later recounted how giddy he was, working until 3 A.M. one night when he was on the threshold of discovery, and then being so excited by his new findings that he couldn't sleep. Yet while he was working on the manuscript for the first paper setting forth his ideas, he wrote to his father, "My own works are at the moment not going especially well. I don't produce very much and I don't know whether another [paper] will jump out of this."

Born, meanwhile, kept puzzling over Heisenberg's strange mathematics. Then one day he had an epiphany: he had seen a scheme like Heisenberg's elsewhere. The arrays, he recalled, were like something mathematicians called "matrices."

Matrix algebra was then an arcane and obscure subject, and Heisenberg had apparently reinvented it. Born asked Pauli to help him translate Heisenberg's paper into the mathematician's matrix language (and to extend that language to allow for the fact that Heisenberg's had an infinite number of rows and columns). The future Nobel laureate Pauli became agitated. He accused Born of trying to ruin his friend's beautiful "physical ideas" by introducing "futile mathematics" and "tedious and complicated formalism."

In reality, the language of matrices would prove a great simplification. Born found himself another matrix algebra helper, his student Pascual Jordan, and within months, in November 1925, Heisenberg, Born, and Jordan had submitted a paper on Heisenberg's quantum theory that is now a landmark in the history of science. Not long after that, Pauli digested their work and applied the new theory to derive the spectral lines of hydrogen, and to show how they are affected by electric and magnetic fields, something that had not previously been possible. It was the first practical application of the nascent theory that would soon unseat Newton's mechanics.

* * *

More than two thousand years had passed since the origin of the idea of the atom, more than two hundred since Newton invented mathematical mechanics, more than twenty since Planck and Einstein introduced the concept of the quantum. Heisenberg's theory was in a way the culmination of all those long threads of scientific thinking.

The problem was, once it was fully developed, Heisenberg's theory required thirty pages to explain the atom's energy levels, which Bohr's theory had explained in a few lines. To that, my ever-practical father the tailor would have said, "Oy, for *this* he had to study all those years?" And yet Heisenberg's theory *was* superior, for it produced its result on the basis of deep principles rather than the ad hoc assumptions of Bohr. For that reason, you'd think it would have been instantly embraced. But

Werner Heisenberg (left) with Niels Bohr

most physicists were not directly involved in searching for a theory of the quantum, and they seemed to think like my father. To them, requiring thirty pages rather than three lines did not seem like a step forward. They—Rutherford conspicuously among them—were unimpressed and uninterested, viewing Heisenberg the way you'd view an auto mechanic who tells you he could fix your problem with a new thermostat, but it would be better to replace the entire car.

The small group of quantum theory cognoscenti, however, had a different reaction. Almost without exception, they were floored. For Heisenberg's admittedly complicated theory explained, in a profound sense, why Bohr's provisional theory of the hydrogen atom had worked, and also provided a complete description of the observed data.

For Bohr, in particular, this was the culmination of a quest that he had helped to start. He knew his atom was only an ad hoc provisional model, destined to be explained, eventually, by a more general theory, and this, he was convinced, was it. "Due to the last work of Heisenberg," he wrote, "prospects have with a stroke been realized, which . . . have for a long time been the center of our wishes."

For a while, physics was in an odd state, like a World Cup stadium in which the winning goal was scored but only a handful of fans had noticed. Ironically, what in the end elevated quantum theory from a theory that held the interest of only the specialists to what is recognized as the fundamental theory behind all physics was the appearance a few months later, in January and February 1926, of two papers that together described *another* general theory of the quantum, one that employed far different concepts and methods—seemingly, a different view of reality.

The new competing theory described the electrons in an atom as a wave—a concept physicists were accustomed to visualizing, though certainly not in the context of electrons. Strangely, despite the differences, like Heisenberg's theory, it explained the Bohr atom. Ever since the Greeks, scientists had had to make do without any theory describing the atom. Now they seemed to have two. And they seemed incompatible, one viewing nature as consisting of waves of matter and energy, another that insisted it is pointless to view nature as consisting of anything, but prescribing instead that we consider only the mathematical relations among data.

The new quantum theory was the work of Austrian physicist Erwin Schrödinger (1887–1961) and it was about as different in style from

Heisenberg's as were the two men, and the venues in which they made their breakthroughs. While Heisenberg had done his work holed up alone on a rocky island with swollen sinuses, Schrödinger did his during a Christmas holiday he spent with a mistress in the Alpine resort town of Arosa. He "did his great work," a mathematician friend said, "during a late erotic outburst in his life." By "late," the mathematician meant Schrödinger's advanced age of thirty-eight.

Maybe the mathematician had a point about Schrödinger's advanced age. Time after time, we've seen young physicists accepting new ideas, and older ones longing for the traditional ways of doing things, as if the older one gets, the more difficult is it to embrace the shifts in a changing world. Schrödinger's work, it turns out, is actually another instance of that tendency—because, ironically, the motivation Schrödinger gave for constructing his new theory was a desire to have a quantum theory that, in contrast to Heisenberg's, looked like conventional physics: Schrödinger was fighting to maintain the familiar, not trying to overturn it.

Unlike the much younger Heisenberg, Schrödinger *did* envision the motion of the electrons in an atom. And though his exotic "matter waves" did not directly endow the electron with Newtonian properties the way Bohr's orbits had, his new "wave theory" of the quantum, which at first no one knew exactly how to interpret, held the promise of avoiding the distasteful view of reality that Heisenberg's theory required.

It was an alternative that physicists appreciated. Before Schrödinger, quantum mechanics had been slow to find acceptance. Heisenberg's unfamiliar mathematics, involving an infinite number of matrix equations, seemed horrendously complex, and physicists were uncomfortable with the abandonment of variables they could visualize, in favor of symbolic arrays. Schrödinger's theory, on the other hand, was easy to use and was based upon an equation resembling those physicists had already studied as undergraduates in the context of sound waves and water waves. That methodology was the bread-and-butter of classical physicists, making the transition to quantum physics relatively easy. Just as important, by providing a way to visualize the atom, even though still not employing Newtonian concepts such as orbits, Schrödinger had made quantum theory more palatable—and the antithesis of what Heisenberg had been striving for.

Even Einstein loved the Schrödinger theory—at first. He had considered the idea of matter waves himself and had worked with the Austrian in the past. "The idea of your work springs from true genius!" he wrote Schrödinger in April 1926. Ten days later, he wrote him again: "I am convinced that you have made a decisive advance with your formulation of the quantum condition, just as I am convinced that the Heisenberg-Born method is misleading." He wrote glowingly of Schrödinger's work again in early May.

However, that same month, May 1926, Schrödinger dropped another bombshell: he published a paper showing that, much to his *own* dismay, his theory and Heisenberg's were mathematically equivalent—they were both correct. That is, though the two theories employed different conceptual frameworks—different views of what is going on "under the hood" of nature (actually, Heisenberg refused to even look under the hood)—those proved to be mere differences in language: what the two theories said about what we *observe* was the same.

To complicate matters further (or to make them yet more interesting), two decades later, Richard Feynman would create yet a third formulation of quantum theory, quite different in its mathematics and conceptual framework from both Heisenberg's and Schrödinger's but also mathematically equivalent to the earlier theories—implying the same physical principles and making identical predictions.

Wallace Stevens wrote, "I was of three minds, / Like a tree / In which there are three blackbirds," but that situation might seem odd when translated to physics. If physics holds any "truth," can there be more than one "correct" theory? Yes, even in physics, there can be many ways of looking at things. That's especially true in modern physics, in which what we "look at," like atoms, electrons, or the Higgs particle, cannot literally be "seen," leading physicists to create their mental pictures from mathematics rather than from a palpable reality.

In physics, one person can phrase a theory in terms of one set of concepts while another phrases a theory of the same phenomenon in terms of a different set. What elevates the exercise above the right-wing-versus-left-wing struggles of politics is that, in physics, for a viewpoint to be considered valid, it must pass the test of experiment, and that means that the alternate theories must lead to the same conclusions—something political philosophies rarely do.

That brings us back to the issue of whether theories are discovered or invented. Without getting into the philosophical question of whether an external objective reality exists, one can say that the process of creating quantum theory was one of *discovery* in the sense that physicists stumbled upon so many of its principles as they explored nature, yet quantum theory was *invented* in that scientists designed and created several different conceptual frameworks that all do the same job. Just as matter can behave as a wave or particle, it seems, so, too, does the theory that describes it have two seemingly contradictory characters.

When Schrödinger published his paper showing the equivalence of his theory and Heisenberg's, no one yet understood the proper interpretation of his formulation. Still, his proof made it clear that future work would reveal that his approach raised the same philosophical issues that were already evident in Heisenberg's version of the theory. And so, after that paper, Einstein would never again write approvingly of quantum theory.

Even Schrödinger himself soon turned on quantum theory. He remarked that he might not have published his papers had he known "what consequences they would unleash." He had created his seemingly innocuous theory in an attempt to supersede Heisenberg's unpalatable alternative, but their equivalence meant that he had not understood the objectionable implications of his own work. In the end, he had only fed the fire and furthered the new quantum ideas that he preferred not to accept.

In an unusually emotional footnote in his equivalence paper, Schrödinger wrote that he "felt discouraged, not to say repelled," by Heisenberg's methods, "which appeared very difficult to me, and by the lack of visualizability." The revulsion was mutual. After reading the papers in which Schrödinger presented his theory, Heisenberg wrote to Pauli, saying, "The more I reflect on the physical portion of the Schrödinger theory, the more disgusting I find it . . . what Schrödinger wrote about the visualizability of his theory [is] crap."

The rivalry proved one-sided, for Schrödinger's method quickly won out as the formalism of choice for most physicists, and for solving most problems. The number of scientists working on quantum theory quickly rose, but the number of those employing Heisenberg's formulation fell.

Even Born, who had helped Heisenberg develop his theory, was won over to Schrödinger's method, and even Heisenberg's friend Pauli marveled at how much easier it was to derive the hydrogen spectrum by employing Schrödinger's equation. None of this pleased Heisenberg.

Bohr, meanwhile, focused on understanding more about the relationship between the theories. In the end, British physicist Paul Dirac gave the definitive explication of the deep connection between the theories, and even invented a hybrid formalism of his own—the formalism that is favored today—that allows one to move deftly between them, depending on the issues involved. By 1960 there had been more than 100,000 papers based on applications of quantum theory.

* * *

Despite all the advances in quantum theory, Heisenberg's approach will always lie at its heart, for he had been inspired by a drive to banish the classical picture of particles having trajectories or orbits through space, and in 1927 he finally published the paper that guaranteed victory in that battle. He showed once and for all that no matter what formalism you use, it is a matter of scientific principle—what we now know as the uncertainty principle—that to picture motion as Newton did is futile. Though Newton's concept of reality may *seem* to hold on macroscopic scales, on the more fundamental level of the atoms and molecules that make up macroscopic objects, the universe is governed by a far different set of laws.

The uncertainty principle restricts what we can know at any given time about certain pairs of observables, such as position and velocity.* That's not a restriction on measurement technology, or a limitation of human ingenuity; rather, it is a restriction imposed by nature itself. Quantum theory says that *objects don't have precise properties* such as position and velocity, and what's more, if you try to measure them, the more precisely you measure one, the less precisely you can measure the other.

In everyday life, we certainly *do* seem to be able to measure position and velocity as precisely as we wish. This seems to contradict the uncertainty principle, but when you run through the math of quantum theory, you find that the masses of everyday objects are so large that the uncertainty principle is irrelevant for the phenomena of daily life. That is why Newtonian physics worked well for such a long time—it was only when physicists began dealing with phenomena at the atomic scale that the limits of the Newtonian promise become apparent.

*Technically, the uncertainty principle constrains our knowledge of position and *momentum*, which is mass times velocity, but for our purposes the distinction is unimportant.

For example, suppose that electrons weighed as much as soccer balls. Then, if you pinpoint the position of an electron to within one millimeter in any direction, you could still measure its velocity to within a precision of better than one part in a billionth of a billionth of a billionth of a kilometer per hour. That's definitely sufficient for any use to which we might put such a calculation in daily life. But a real electron, because it is far lighter than a soccer ball, is a different story. If you measure the position of a real electron to a precision corresponding to roughly the size of an atom, the uncertainty principle says that the electron's velocity is not determined more precisely than to about plus or minus a thousand kilometers per hour—which is the difference between the electron sitting still and moving faster than a jumbo jet. And so here Heisenberg had his vindication: those unobservable atomic orbits that specify precise electron paths are banned by nature after all.

As quantum theory became better understood, it became clear that in the quantum world there is no certainty, only probabilities—there is no "Yes, this will happen," only "Sure, any of these things may happen." In the Newtonian worldview, the state of the universe at any given time in the future or past is seen as being imprinted upon the universe at the present, and, employing Newton's laws, it is there for anyone of sufficient intelligence to read. If we had sufficient data on the earth's interior, we could predict earthquakes; if we knew every physical detail relevant to the weather, we could, in principle, say with certainty whether it will rain tomorrow—or a century from tomorrow.

That Newtonian "determinism" lies at the heart of Newtonian science: the idea that one event causes the next, and so on, and that it all can be predicted using mathematics. It was part of Newton's revelation, a giddy kind of certainty that inspired everyone from economists to social scientists to "want to have what physics was having." But quantum theory tells us that at its heart—on the fundamental level of the atoms and particles that everything is made of—the world is not deterministic, that the present state of the universe doesn't determine future (or past) events, but only the probability that one of many alternative futures will occur (or pasts have occurred). The cosmos, quantum theory tells us, is like a giant bingo game. It is as a reaction to these ideas that Einstein made his famous pronouncement, in a letter to Born, that "[quantum] theory produces a good deal but hardly brings us closer to the secret of the Old One. I am at all events convinced that *He* does not play dice."

It is interesting that Einstein invoked the concept of God in that statement: the "Old One." Einstein did not believe in the traditional personal God of, say, the Bible. To Einstein "God" was not a player involved in the intimate details of our lives, but instead represented the beauty and logical simplicity of the laws of the cosmos. And so when Einstein said the Old One does not play dice, he meant that he couldn't accept a role for randomness in the grand scheme of nature.

My father was neither a physicist nor a dice player, and in his time living in Poland, he was ignorant of the grand developments in physics just hundreds of miles away. But when I explained quantum uncertainty to him, he had a much easier time with it than Einstein had. To my father, the quest to understand the universe centered not on the observations made by telescopes or microscopes, but rather on the human condition. And so, just as he understood, from his own life, Aristotle's distinction between natural and violent change, so too did his past make the randomness inherent in quantum theory an easy pill to swallow. He told me about the time he stood in a long line in the town marketplace, into which the Nazis had funneled thousands of Jews. When the roundup started, he had hidden in a latrine, with a fugitive underground leader he had been assigned to protect. But neither he nor the fugitive could take the stench, so they eventually emerged. The fugitive bolted, and no one ever saw him again. My father was herded into the line, and joined it near its end.

The line moved slowly, and my father could see that everyone was being loaded onto trucks. As he neared the front, the SS officer who was in charge cut the line off at the last four, one of whom was my father. They required three thousand Jews, the man said, and the line had apparently held 3,004. Wherever they were going, they would be going without him. He later discovered that the destination was the local cemetery, where everyone was ordered to dig a mass grave and then was shot dead and buried in it. My father had drawn number 3,004 in a death lottery in which German precision trumped Nazi brutality. To my father, that was an instance of randomness that was difficult for his mind to grasp. The randomness of quantum theory, by contrast, was easy.

Like our lives, a scientific theory can rest upon bedrock or it can be built upon sand. Einstein's unlimited hope for the physical world was that quantum theory would prove to have been erected upon the latter, a weak foundation that would, in the long run, lead to its collapse. When the uncertainty principle came, he suggested that it was not a fundamental principle of nature, but rather a limitation of quantum mechanics—an indication that the theory did not stand on solid ground.

Objects *do* have definite values for quantities like position and velocity, he believed, but quantum theory just cannot handle them. Quantum mechanics, Einstein said, though undeniably successful must be an incomplete embodiment of a deeper theory that restores objective reality. Though few other than Einstein shared that belief, for many years it was a possibility that no one could rule out, and Einstein went to his grave thinking he would someday be vindicated. In recent decades, however, sophisticated experiments based on the very clever work of Irish theoretical physicist John Bell (1928–1990) *have* ruled out that possibility. Quantum uncertainty is here to stay.

"Einstein's verdict," Born confided, "came as a hard blow." Born, along with Heisenberg, had made important contributions to the probabilistic interpretation of quantum theory, and he'd hoped for a more positive reaction. He adored Einstein and felt a sense of loss, as if he'd been forsaken by a revered leader. Others felt similarly, and were even moved to tears at having to reject Einstein's ideas. But Einstein soon found himself virtually alone in his opposition to quantum theory, singing, as he put it, "my solitary little song" and looking "quite bizarre as seen from the outside." In 1949, some twenty years after his initial letter rejecting Born's work, and just six years before his death, he wrote Born again to say, "I am generally regarded as a sort of petrified object, rendered blind and deaf by the years. I find this role not too distasteful, as it corresponds very well with my temperament."

* * *

Quantum theory was created by a concentration of scientific brainpower in Central Europe that surpassed or at least rivaled that of any of the intellectual constellations we've encountered in our journey through the ages. Innovation starts with the right physical and social environment, so the fact that those in distant lands contributed little is no accident: spurred by technical advances that revealed a torrent of new phenomena related to the atom, theoretical physicists lucky enough to be part of the community at that time and place traded insights and observations regarding aspects of the universe that were being revealed for the first time in human history. It was a magical time in Europe, with burst after burst of imagination lighting the sky, until the outline of a new realm of nature began to appear.

Quantum mechanics arose from the sweat and genius of many scientists working in a small cluster of countries, exchanging ideas and arguing, but all allied in their passion and dedication to the same goal. Both the alliances and the conflicts of those great minds, though, would soon be eclipsed by the chaos and savagery that was about to overtake their continent. The stars of quantum physics would be scattered like playing cards flying off a bad shuffle.

The beginning of the end came in January 1933, when Field Marshal Paul von Hindenburg, Germany's president, appointed Adolf Hitler chancellor of Germany. The very next night, in the great university town of Göttingen—where Heisenberg, Born, and Jordan had collaborated on Heisenberg's mechanics—uniformed Nazis marched through the streets, waving torches and swastikas, singing patriotic songs, and taunting the Jews. Within months, the Nazis held book-burning ceremonies all over the country and proclaimed a purge of non-Aryan academics from universities. Suddenly, many of the most esteemed German intellectuals were forced either to abandon their homes, or, like my father the tailor in Poland with no such option, to stay and face the growing Nazi threat. It is estimated that within five years, nearly two thousand top scientists had fled, due either to their ancestry or their political beliefs.

Of Hitler's rise, though, Heisenberg reportedly remarked, with great glee, "Now we at least have order, an end is put to the unrest, and we have a strong hand governing Germany which will be to the good of Europe." Ever since his teens, Heisenberg had been unhappy with the direction of German society. He'd even been active in a nationalistic youth group that mixed long hikes into the wilderness with campfire discussions decrying the moral decadence of Germany and the loss of a common purpose and tradition. As a scientist, he aimed to be aloof from politics, but he seemed to have seen in Hitler a strong hand that could restore Germany's pre–World War I greatness.

Yet the new physics that Heisenberg championed, and indeed helped invent, was bound to irk Hitler. In the nineteenth century, German physics had established its primacy and prestige primarily through the gathering and analysis of data. Certainly mathematical hypotheses were made and analyzed, but this was not generally the physicist's focus. In the early decades of the twentieth century, though, theoretical physics blossomed as a field, and, as we've seen, it achieved startling successes. The Nazis, however, dismissed it as overly speculative and mathematically abstruse. Like the "degenerate" art they hated so much, it was seen as disgustingly surreal and abstract. Worst of all, much of it was the work of scientists of Jewish heritage (Einstein, Born, Bohr, Pauli).

The Nazis came to call the new theories—relativity and quantum theory—"Jewish physics." As a result, they were not just wrong; they, too, were degenerate, and the Nazis banned them from being taught at the universities. Even Heisenberg got grief, for he had worked on "Jewish physics" and with Jewish physicists. The attacks angered Heisenberg, who, despite numerous offers of prestigious foreign positions, had remained in Germany, loyal to his government, and had done everything the Third Reich had asked.

Heisenberg tried to quash his problems by appealing directly to Heinrich Himmler, head of the Schutzstaffel (the SS) and the man who would be in charge of building the concentration camps. His mother and Himmler's had known each other for years, and Heisenberg used that connection to get a letter to Himmler. Himmler responded with an intense eight-month-long investigation that gave Heisenberg nightmares for years to come but ended in Himmler's declaration that "I believe that Heisenberg is decent, and we could not afford to lose or to silence this man, who is relatively young and can educate a new generation." Heisenberg agreed, in exchange, to disavow the Jewish creators of Jewish physics, and to avoid uttering their names in public.

Of the other leading pioneers of the quantum, Rutherford was then at Cambridge. There he helped found, and was president of, an organization to aid academic refugees. He died in 1937 at age sixty-six, the result of delaying an operation on a strangulated hernia. Dirac, who had become Lucasian professor at Cambridge (the post Newton and Babbage had held, and Hawking would later hold), worked for a while on issues relevant to the British A-bomb project, and was then invited to work on the Manhattan Project but declined for ethical reasons. He spent his last years at Florida State University, in Tallahassee, where he died in 1984, at age eighty-two. Pauli, who was a professor in Zurich at the time, was, like Rutherford, the head of an international refugee project, but when the war broke out, he was denied Swiss citizenship and fled to the United States, where he was living when awarded the Nobel Prize shortly after the war's end. In his later years, he became increasingly interested in mysticism and psychology, especially dreams, and was a founding member of the C. G. Jung Institute, in Zurich. He died in a Zurich hospital in 1958, at age fifty-eight, of pancreatic cancer.

Schrödinger, like Pauli, was Austrian, but was living in Berlin when Hitler took power. With regard to Hitler, as in so many other ways, Schrödinger proved to be the antithesis of Heisenberg: he was an outspoken anti-Nazi and soon left Germany to take up a position in Oxford. Shortly thereafter, he received the Nobel Prize along with Dirac. Heisenberg, who was attempting to hold German physics together, resented Schrödinger's departure, "since he was neither Jewish nor otherwise endangered."

As it turned out, Schrödinger didn't last long at Oxford. The difficulty arose because he lived with both his wife and his mistress—whom he regarded as more of a second wife. As his biographer Walter Moore wrote, in Oxford, "wives were regarded as unfortunate female appendages . . . It was deplorable to have one wife at Oxford—to have two was unspeakable."

Schrödinger would eventually settle in Dublin. He died of tuberculosis in 1961, at age seventy-three. He had first contracted the disease in 1918, while fighting in World War I, and the respiratory issues he'd suffered from ever since were the reason for his fondness for the Alpine resort Arosa, where he had developed his version of quantum theory.

Einstein and Born were living in Germany when Hitler came to power, and timely emigration was a matter of survival, due to their Jewish ancestry. Einstein, who was then a professor in Berlin, was by chance visiting Caltech in the United States the day Hitler was appointed. He decided not to return to Germany, and he never set foot there again. The Nazis confiscated his personal property, burned his works on relativity, and put a five-thousand-dollar bounty on his head. But he had not been caught off guard: as they left for California, Einstein had told his wife to take a good look at their house. "You will never see it again," he told her. She thought he was being foolish.

Einstein became an American citizen in 1940, but also retained his Swiss citizenship. He died in 1955 and was brought to a crematorium where twelve close friends had quietly gathered. After a brief remembrance, the body was cremated and his ashes were scattered at an undisclosed location, but a pathologist at Princeton Hospital had removed his brain, and it has been studied on and off in the decades since. What's left of it resides in the U.S. Army's National Museum of Health and Medicine, in Silver Spring, Maryland.

Born, barred from teaching and worried about the ongoing harassment of his children, also immediately sought to leave Germany. Heisenberg worked hard to have Born exempted from the non-Aryan work prohibition, but instead, with help from Pauli's refugee organization, Born left for an appointment at Cambridge in July 1933, and later moved to Edinburgh. Having been passed over for the Nobel Prize when Heisenberg won it in 1932—for work they had done together—he was awarded his Nobel in 1954. He died in 1970. On his tombstone is inscribed the epitaph "$pq - qp = h/2\pi$," one of the most famous equations of quantum theory, a mathematical statement that would become the foundation of Heisenberg's uncertainty principle—and one that he and Dirac discovered independently of each other.*

Bohr, living in Denmark, where he ran what is now called the Niels Bohr Institute, was for a while a bit more insulated from Hitler's actions, and he helped Jewish refugee scientists find posts in the United States, Britain, and Sweden. But in 1940 Hitler invaded Denmark, and in the fall of 1943 Bohr was tipped off by the Swedish ambassador in Copenhagen that he faced immediate arrest as part of the plan to deport all of Denmark's Jews. As it happens, he was originally meant to be arrested the month prior, but the Nazis felt it would raise less of a furor if they waited until the height of the mass arrests. The delay saved Bohr, and he fled with his wife to Sweden. The next day, Bohr met with King Gustav V and convinced him to publicly offer asylum to Jewish refugees.

Bohr himself, however, was in danger of being abducted. Sweden was crawling with German agents, and though he was being housed at a secret location, they knew he was in Stockholm. Soon Winston Churchill got word to Bohr that the British would evacuate him, and he was packed onto a mattress in the bomb compartment of a de Havilland

*I am proud to trace my own Ph.D. lineage back to Max Born. The sequence is: Born/J. Robert Oppenheimer (became head of the Manhattan Project)/Willis Lamb (Nobel prize winner and one of the founders of the laser)/Norman Kroll (made fundamental contributions to the theory of light and atoms)/Eyvind Wichmann (my Ph.D. advisor, an important figure in mathematical physics).

Pioneers of quantum theory at the Fifth Solvay International Conference on Electrons and Photons, Brussels, 1927. Back row: Schrödinger (sixth from left), Pauli (eighth), Heisenberg (ninth). Middle row: Dirac (fifth), Born (eighth), Bohr (ninth). Front row: Planck (second), Einstein (fifth).

Mosquito, an unarmed and fast high-flying bomber that could avoid German fighters. Along the way, Bohr passed out for lack of oxygen, but he made it alive, still dressed in the clothes he'd been wearing when he left Denmark. His family followed. From England, Bohr fled to the United States, where he became an adviser to the Manhattan Project. After the war he returned to Copenhagen, where he died in 1962, aged seventy-seven.

Of the great quantum theorists, only Planck, Heisenberg, and Jordan remained in Germany. Jordan, like the great experimentalist Geiger, was an enthusiastic Nazi. He became one of the German army's three million storm troopers and proudly wore his brown uniform with jackboots and swastika armband. He tried to interest the Nazi party in various schemes for advanced weapons, but ironically, due to his involvement in "Jewish physics," he was ignored. After the war, he entered German politics and won a seat in the Bundestag, the German parliament. He died in 1980, aged seventy-seven, the only one of those early pioneers who did not receive the Nobel Prize.

Planck did not have sympathy for the Nazis, but neither did he do much to resist them, even quietly. Instead, like Heisenberg, his priority seemed to be to preserve as much of German science as possible, while complying with all Nazi laws and regulations. He had a meeting with Hitler in May 1933, aimed at dissuading him from the policies that had chased Jews from German academia, but of course the meeting changed nothing. Years later, Planck's youngest son, with whom he was close, tried to change the Nazi party in a bolder manner—he was part of the plot to assassinate Hitler on July 20, 1944. Arrested with the others, he was tortured and executed by the Gestapo. For Planck it was the tragic culmination of a life full of tragedies. Of his five children, three others had died young—his oldest son was killed in action in World War I, and two daughters died in childbirth. It is said, though, that his son's execution finally extinguished Planck's desire to live. He died two years later, aged eighty-nine.

Heisenberg, despite his initial enthusiasm, soured on the Nazis. Still, he held high scientific positions throughout the Third Reich and performed his duties without complaint. When Jews were dismissed from the universities, he did his best to preserve German physics by attracting the best possible replacements. He never joined the Nazi party, but he stayed at his post and never broke with the regime.

When the German atom bomb project got started in 1939, Heisenberg jumped in and immersed himself with enormous energy. He soon completed calculations showing that a nuclear fission chain reaction might be possible, and that pure uranium 235, a rare isotope, would make a good explosive. It is one of the many ironies of history that the early wartime successes of Germany may have led to their ultimate defeat: the regime didn't deploy many resources toward the bomb project at first because the war was going so well, and by the time the tide had turned it was too late—the Nazis were defeated before they could build one.

After the war, Heisenberg was held briefly by the Allies, along with nine other leading German scientists. After his release, he returned to working on fundamental questions in physics, on rebuilding German science, and on rehabilitating his reputation among scientists outside his home country. Heisenberg died at his home in Munich on February 1, 1976, having never regained the stature he had once enjoyed.

The mixed reaction of the physics community to Heisenberg after the war is perhaps reflected in my own behavior. When, as a student in 1973, I had the chance to attend a talk he gave at Harvard on the development of quantum theory, I couldn't bring myself to go. But years later, when I was an Alexander von Humboldt fellow at the institute where he had been director, I often stood outside the office he had occupied and pondered the spirit that had helped invent quantum mechanics.

* * *

Though the quantum theory developed by the great quantum pioneers doesn't alter our description of the gross physics of the macroscopic world, it has revolutionized the way we live, creating a change in human society as mammoth as that of the industrial revolution. The laws of quantum theory underlie all of the information and communication technologies that have remade modern society: the computer, the Internet, satellites, cell phones, and all electronics. But just as important as its practical applications is what quantum theory tells us about nature, and about science.

The triumphalism of the Newtonian worldview had promised that, with the right mathematical calculations, mankind could predict and explain all natural phenomena, and had therefore inspired scientists in every field of endeavor to want to "Newtonize" their subjects. The quantum physicists of the first half of the twentieth century extinguished those aspirations, and uncovered a truth that is ultimately both empowering and profoundly humbling. Empowering because quantum theory demonstrates that we can understand and manipulate an unseen world beyond our experience. And humbling because for millennia, the progress made by scientists and philosophers suggested that our capacity for understanding was infinite, but now nature, speaking through the great discoveries of the quantum physicists, is telling us that there are limits to what we can know, and to what we can control. What's more, the quantum reminds us that other unseen worlds may exist, that the universe is a place of extraordinary mystery, and that fluttering just beyond the horizon may be further inexplicable phenomena demanding new revolutions in thought and theory.

In these pages we've traveled on a journey of millions of years, beginning with the first human species, which were far different from ourselves, both physically and mentally. In this four-million-year journey it is only in the last blink of an eye that we entered the present era, in which we have learned that laws govern nature, but that there is more to those laws than that which we experience in our everyday existence—more things in heaven and earth, as Hamlet put it to Horatio, than are dreamt of in our philosophy.

Our knowledge will continue to increase for the foreseeable future, and given the exponential growth in the number of people practicing science, it seems reasonable to believe that the next hundred years will bring advances as great as the last thousand. But if you are reading this book, you know that there is more to the questions people ask about our surroundings than the technical side—we humans see beauty in nature, and we seek meaning. We don't just want to know how the universe operates; we want to understand how we fit in. We want to give a context to our lives and our finite existence, and to feel connected to other humans, to their joys and sorrows, and to the vast cosmos in which those joys and sorrows play but a tiny role.

Understanding and accepting our place in the universe can be difficult, but it has, from the very start, been one of the aims of those who study nature—from the early Greeks, who considered science alongside metaphysics, ethics, and aesthetics as a branch of philosophy, to pioneers like Boyle and Newton, who undertook the study of nature as a means to understand the nature of God. For me, the connection between insights into the physical world and the human world arose most starkly one day while I was in Vancouver, on the set of the television show *MacGyver.* I had written the episode they were filming, and was instructing the props people and set designers regarding what a low-temperature physics laboratory looks like. Suddenly, in the midst of those mundane technical discussions, I was for the first time made to face the fact that we humans are not above nature, but rather come and go like the flowers, or Darwin's finches.

It all started when a phone call was relayed from the production office to the set to me. In those days, before every twelve-year-old had a cell phone, to take a call on the set was unusual, and I normally received phone messages hours after the fact, scrawled on irregular scraps of paper. Messages like *Leonard: <illegible> wants you to <illegible>. He said it is urgent! Call him at <illegible>.* This time was different. This time a production assistant brought me a phone.

On the other end of the connection was a doctor at the University of Chicago hospital. He informed me that my father had had a stroke and was in a coma—a delayed result of that surgery my dad had had, months earlier, to repair his aorta. By nightfall I was at the hospital, gazing at my father, who was lying on his back, eyes closed, looking peaceful. I sat beside him and stroked his hair. He felt warm and alive, as if he were asleep, as if he might wake up at any moment, smile at my presence, reach out to touch me, and ask if I wanted to share some rye bread and pickled herring for breakfast.

I spoke to my dad. I told him I loved him—much as I would, on occasion, many years later, say the same thing to my sleeping children. But the doctor stressed that my father was not sleeping. He could not hear my voice, this doctor said. He told me that my father's brain readings indicated that he was all but dead. My father's warm body, apparently, was like the *MacGyver* physics lab—a facade, in fine shape on the exterior, but just a shell, incapable of any meaningful function. The doctor told me that my father's blood pressure would gradually fall, and his breathing gradually slow, until he died.

I hated science just then. I wanted it to be proved wrong. Who are scientists and doctors to tell you the fate of a human being? I would have given anything, or everything, to have my father back, or even to have him back for just another day, for an hour or even a minute, to say I loved him, and to say good-bye. But the end came exactly as the doctor had said it would.

The year was 1988, and my father was seventy-six years old. After his death, our family "sat shiva," meaning we engaged in a traditional seven-day period of mourning during which you pray three times a day and do not leave the home. All my life I had sat in our living room and talked with him, but now as I sat there he was just a memory, and I knew I would never talk to him again. Thanks to our human intellectual journey, I knew that his atoms still existed, and always would; but I also knew that, though his atoms hadn't died with him, they would now disperse. Their organization into the being I knew as my father was gone and would never exist again, except as a shadow in my mind and in the minds of the others who loved him. And I knew that, in a few decades, the same would happen to me.

To my surprise, I now felt that what I had learned thanks to my human struggle to understand the physical world didn't make me callous—it gave me strength. It helped me to rise above my own heartbreak, to feel less alone because I was part of something greater. It opened my eyes to the awesome beauty of our existence, whatever years we are each granted. My father, despite never having had the chance to even enter high school, had also had a great appreciation for, and curiosity about, the nature of the physical world. I had told him once, in one of those living room conversations of my youth, that I would write a book about it someday. Finally, decades later, this has been that book.

My father, the night he proposed to my mother,
New York, 1951

Epilogue

There is an old brainteaser about a monk who one day leaves his monastery at sunrise to hike up to a temple at the summit of a tall mountain. The mountain has just one path, quite narrow and winding, and he takes it slowly at times, as sections of it are rather steep, but he reaches the temple shortly before sunset. The next morning, he descends along the path, again beginning at sunrise, and reaches his monastery again at sunset. The question is: Is there a spot along the path that he will come to at exactly the same time on both days? The point is not to identify the spot, only to say whether there is one—or not.

This is not one of those riddles that depends on a trick, on camouflaged information, or on a novel interpretation of some word. There is no altar along the path that the monk prays at each day at noon, nothing you need to know about the speed of his ascent or descent, no other missing detail you'd have to guess at to solve the riddle. Nor is this like the riddle that tells you a butcher stands six feet tall and then asks what he weighs, to which the answer is "meat." No, the situation in this puzzle is quite straightforward, and chances are you understood from just one reading everything you need to know to determine the answer.

Think about it for a while, because your success at solving this riddle, like many questions scientists have tried to answer through the ages, may depend on your capacity for patience and persistence. But even more to the point, as all good scientists know, it will depend on your ability to ask the question in the right way, to step back and see the problem from a slightly different angle of vision. Once you do that, the answer is easy. It's finding that angle of vision that can be hard. That's why Newton's physics and Mendeleev's periodic table and Einstein's relativity required people of towering intellect and originality to create them—and yet they can be understood, when properly explained, by any college student majoring in physics and chemistry today. And that's why what boggles the mind of one generation becomes common knowledge to those who follow, allowing scientists to scale ever greater heights.

To arrive at the solution to the monk puzzle, rather than replaying in your mind the picture of the monk climbing up the mountain one day and down the next day, let's do a thought experiment and visualize the problem differently. Imagine that there are two monks—one who walks up and another who walks down, both departing at sunrise on the *same day*. Obviously, they will pass each other along the way. The point at which they pass is the spot that the monk of the riddle will come to at the same time on both days. So the answer to the riddle is "Yes."

That the monk will reach a particular point along the path at the same time on both his ascent and his descent can seem like an unlikely coincidence. But once you free your mind to entertain the fantasy of two monks ascending and descending on the same day, you see that it's not a coincidence—it's an inevitability.

In a way, the advance of human understanding was made possible by a succession of such fantasies, each created by someone capable of looking at the world just a little bit differently. Galileo imagining objects falling in a theoretical world devoid of air resistance. Dalton imagining how elements might react to form compounds if they were made of unseeable atoms. Heisenberg imagining that the realm of the atom is governed by bizarre laws that are nothing like those we experience in everyday life. One end of the spectrum of fantastical thinking is labeled "crackpot," and the other "visionary." It is due to the earnest efforts of a long parade of thinkers whose ideas originate at various points in between that our understanding of the cosmos has progressed to where it is today.

If I achieved my goal, the preceding pages have imparted an appreciation of the roots of human thought about the physical world, the kinds of questions those who study it concern themselves with, the nature of theories and research, and the ways in which culture and belief systems affect human inquiry. That's important for understanding many of the social, professional, and moral issues of our time. But much of this book has also been about the way scientists and other innovators think.

Twenty-five hundred years ago, Socrates likened a person going through life without thinking critically and systematically to an artisan such as a potter who practices his craft without following proper procedures. Making pottery may appear simple, but it isn't. In Socrates's time, it involved procuring clay from a pit south of Athens, placing the clay on a specially made wheel, spinning it at just the proper speed for the diameter of the part being made, and then sponging, scraping, brushing, glazing, drying, and firing twice in a kiln, each time at the right temperature and humidity. Departing from any of these procedures will result in pottery that is misshapen, cracked, discolored, or just plain ugly. Powerful thinking, Socrates pointed out, is also a craft, and it is one worth doing well. After all, we all know people who, applying it poorly, have created lives that are misshapen or otherwise sadly flawed.

Few of us study atoms or the nature of space and time, but we all form theories about the world we live in, and use those theories to guide us at work and at play, and as we decide how to invest, what's healthy to eat, and even what makes us happy. Also, like scientists, in life we all have to innovate. That might mean figuring out what to make for dinner when you have little time or energy, improvising a presentation when your notes have gone missing and the computers are all down—or something as life-changing as knowing when to let go of the mental baggage of the past, and when to hold on to the traditions that sustain you.

Life itself, especially modern life, presents us with intellectual challenges analogous to those scientists face, even if we don't think of ourselves as such. And so, of all the lessons that might have been gleaned from this adventure, perhaps the most important are those that have exposed the character of successful scientists, the flexible and unconventional thinking, the patient approach, the lack of allegiance to what others believe, the value of changing one's perspective, and the faith that there are answers and that we can find them.

* * *

Where is our understanding of the universe today? The twentieth century saw huge advances on all fronts. Once physicists solved the riddle of the atom and invented quantum theory, these advances in turn made others possible, so that the pace of scientific discovery grew ever more frenzied.

Aided by new quantum technologies such as the electron microscope, the laser, and the computer, chemists came to understand the nature of the chemical bond, and the role of the shape of molecules in chemical reactions. Meanwhile, the technology to create and harness those reactions had also exploded. By the middle of the century, the world had been remade. No longer dependent on substances from nature, we learned how to create new artificial materials from scratch, and to alter old materials for new uses. Plastics, nylon, polyester, hardened steel, vulcanized rubber, refined petroleum, chemical fertilizers, disinfectants, antiseptics, chlorinated water—the list goes on and on, and as a result, food production grew, mortality plunged, and our life spans rocketed upward.

At the same time, biologists made great progress in detailing how the cell operates as a molecular machine, deciphering how genetic information is passed among generations, and describing the blueprint for our own species. Today we can analyze DNA fragments drawn from bodily fluids to identify mysterious infectious agents. We can splice sections of DNA into existing organisms to create new ones. We can place optical fibers into rats' brains and control them as if they were robots. And we can sit before a computer and watch people's brains form thoughts, or experience feelings. In some cases we can even read their thoughts.

But though we have come far, it is almost certainly wrong to believe that we are near any final answers. To think so is a mistake that has been made throughout history. In ancient times, the Babylonians felt sure that the earth was created from the corpse of the sea goddess Tiamat. Thousands of years later, after the Greeks made incredible advances in our understanding of nature, most were equally positive that all objects in the terrestrial world were made from some combination of earth, air, fire, and water. And after another two millennia had passed, the Newtonians believed that everything that has occurred or will occur, from the motion of atoms to the orbits of planets, could in principle be explained and predicted by employing Newton's laws of motion. All these were fervently held convictions, and all were wrong.

At whatever time we live in, we humans tend to believe that we ourselves stand at the apex of knowledge—that although the beliefs of those before us were flawed, our *own* answers are correct, and will never be superseded as theirs were. Scientists—even great ones—are no less prone to this kind of hubris than anyone else. Witness Stephen Hawking's pronouncement in the 1980s that physicists would have their "theory of everything" by the end of the century.

Are we today, as Hawking suggested decades ago, on the verge of having answered all our fundamental questions about nature? Or are we in a situation like that at the turn of the nineteenth century, in which the theories we think are true will soon be replaced by something completely different?

There are more than a few clouds on the horizon of science that would indicate that we may be in the latter scenario. Biologists still don't know how and when life first originated on earth, or how likely it would be to originate on other earthlike planets. They don't know the selective advantages that drove the evolutionary development of sexual reproduction. Perhaps most important of all, they don't know how the brain produces the experiences of the mind.

Chemistry, too, has great unanswered questions, from the mystery of how water molecules form hydrogen bonds with their neighbors to create the magical properties of that vital liquid, to how long chains of amino acids fold to form the precise spaghetti-like proteins that are vital for life. It is in physics, though, that the most potentially explosive issues lie. In physics, the open questions have the potential to make us revise everything we think we now know about the most fundamental aspects of nature.

For example, although we have built a very successful "standard model" of forces and matter that unites electromagnetism and the two nuclear forces, almost nobody believes that model is acceptable as the final word. One major drawback is that the model excludes gravity. Another is that it has many adjustable parameters—"fudge factors"—which are fixed on the basis of experimental measurement but cannot be accounted for by any overarching theory. And progress on string theory/M-theory, which once seemed to hold the promise of meeting both of those challenges, seems stalled, calling into question the high hopes that many physicists had for it.

At the same time, we now suspect that the universe we can see with even the most powerful of our instruments is but a tiny fraction of what is out there, as if most of creation is a ghostlike netherworld destined to remain, at least for a while, a mystery. More precisely stated, the ordinary matter and light energy we detect with our senses and in our laboratories seem to make up just 5 percent of the matter and energy in the universe, while an unseen, never detected type of matter called "dark matter" and an unseen, never detected form of energy called "dark energy" are thought to make up the rest.

Physicists postulate the existence of dark matter because the matter we *can* see in the heavens seems to be pulled on by gravity of unknown origin. Dark energy is equally mysterious. The popularity of the idea stems from a 1998 discovery that the universe is expanding at an ever accelerating rate. That phenomenon could be explained by Einstein's theory of gravity—general relativity—which allows for the possibility that the universe is infused throughout with an exotic form of energy that exerts an "antigravity" effect. But the origin and nature of that "dark energy" has yet to be discovered.

Will dark matter and dark energy prove to be explanations that fit into our existing theories—the standard model and Einstein's relativity? Or, like Planck's constant, will they eventually lead us to a completely different view of the universe? Will string theory prove true, or, if not, will we ever discover a unified theory of all the forces in nature, and one that is free of "fudge factors"? No one knows. Of all the reasons I wish I could live forever, living to know the answers to these questions is near the top of my list. I guess that's what makes me a scientist.

Acknowledgments

Over the years in which I was finally putting these ideas to paper, I was privileged to benefit from the insight of many friends who are scholars in various aspects of science and its history, and others who read parts of various drafts of the manuscript and offered their constructive criticism. I am particularly grateful to Ralph Adolphs, Todd Brun, Jed Buchwald, Peter Graham, Cynthia Harrington, Stephen Hawking, Mark Hillery, Michelle Jaffe, Tom Lyon, Stanley Oropesa, Alexei Mlodinow, Nicolai Mlodinow, Olivia Mlodinow, Sandy Perliss, Markus Pössel, Beth Rashbaum, Randy Rogel, Fred Rose, Pilar Ryan, Erhard Seiler, Michael Shermer, and Cynthia Taylor. I am also indebted to my agent and friend, Susan Ginsburg, for her guidance regarding the content of the book and all aspects of publishing, and, just as important, for the fabulous wine-laden dinners during which much of that guidance was given. Another person who helped me enormously was my patient editor, Edward Kastenmeier, who provided valuable criticism and suggestions throughout the book's evolution. I am also grateful to Dan Frank, Emily Giglierano, and Annie Nichol at Penguin Random House, and Stacy Testa at Writer's House, for their help and advice. And, finally, a great debt of thanks to my other editor, the one who was on call twenty-four hours each day, my wife, Donna Scott. She tirelessly read draft after draft, going through every paragraph and offering profound and valuable suggestions and ideas, and much encouragement, also often accompanied by wine, but (almost) never by impatience. This book has brewed in my mind ever since, as a child, I started speaking about science to my father. He was always interested in what I had to say, and provided his own streetwise wisdom in response. I like to think that, had he lived to see this book, he would have treasured it.

Notes

1. Our Drive to Know

5 the chariot was invented: Alvin Toffler, *Future Shock* (New York: Random House, 1970), 26.

6 still using carrier pigeons: "Chronology: Reuters, from Pigeons to Multimedia Merger," *Reuters*, February 19, 2008, accessed October 27, 2014, http://www.reuters.com/article/2008/02/19/us-reuters-thomson-chronology-idUSL1849100620080219.

6 "The world of today": Toffler, *Future Shock*, 13.

9 "as subjective and psychologically conditioned": Albert Einstein, *Einstein's Essays in Science* (New York: Wisdom Library, 1934), 112.

2. Curiosity

11 ratlike creatures: Maureen A. O'Leary et al., "The Placental Mammal Ancestor and the Post–K-Pg Radiation of Placentals," *Science* 339 (February 8, 2013): 662–67.

12 "all life evolved to a certain point": Julian Jaynes, *The Origin of Consciousness in the Breakdown of the Bicameral Mind* (Boston: Houghton Mifflin, 1976), 9.

13 One of the first nearly human creatures: For the story of Lucy and her significance, see Donald C. Johanson, *Lucy's Legacy* (New York: Three Rivers Press, 2009). See also Douglas S. Massey, "A Brief History of Human Society: The Origin and Role of Emotion in Social Life," *American Sociological Review* 67 (2002): 1–29.

13 Lucy's large teeth: B. A. Wood, "Evolution of Australopithecines," in *The Cambridge Encyclopedia of Human Evolution*, ed. Stephen Jones, Robert D. Martin, and David R. Pilbeam (Cambridge, U.K.: Cambridge University Press, 1994), 239.

14 the structure of her spine and knees indicates: Carol. V. Ward et al., "Complete Fourth Metatarsal and Arches in the Foot of Australopithecus afarensis," *Science* 331 (February 11, 2011): 750–53.

14 If you did that . . .: 4×10^6 years ago = 2×10^5 generations; 2×10^5 houses × 100-foot-wide lot for each house ÷ 5,000 feet per mile = 4,000 miles.

14 About halfway there: James E. McClellan III and Harold Dorn, *Science and Technology in World History*, 2nd ed. (Baltimore: Johns Hopkins University Press, 2006), 6–7.

15 Brain size varies considerably *among:* Javier DeFelipe, "The Evolution of the Brain, the Human Nature of Cortical Circuits, and Intellectual Creativity," *Frontiers in Neuroanatomy* 5 (May 2011): 1–17.

15 Handy Man stood upright: Stanley H. Ambrose, "Paleolothic Technology and Human Evolution," *Science* 291 (March 2, 2001): 1748–53.

16 Handy Men had used their stone cutters: "What Does It Mean to Be Human?" Smithsonian Museum of Natural History, accessed October 27, 2014, www.humanorigins.si.edu.

17 In their struggle to procure food: Johann De Smedt et al., "Why the Human Brain Is Not an Enlarged Chimpanzee Brain," in *Human Characteristics: Evolutionary Perspectives on Human Mind and Kind,* ed. H. Høgh-Olesen, J. Tønnesvang, and P. Bertelsen (Newcastle upon Tyne: Cambridge Scholars, 2009), 168–81.

17 bonobos fail to become proficient: Ambrose, "Paleolothic Technology and Human Evolution," 1748–53.

17 Recent neuroimaging studies: R. Peeters et al., "The Representation of Tool Use in Humans and Monkeys: Common and Uniquely Human Features," *Journal of Neuroscience* 29 (September 16, 2009): 11523–39; Scott H. Johnson-Frey, "The Neural Bases of Complex Tool Use in Humans," *TRENDS in Cognitive Sciences* 8 (February 2004): 71–78.

17 Sadly, there are rare cases: Richard P. Cooper, "Tool Use and Related Errors in Ideational Apraxia: The Quantitative Simulation of Patient Error Profiles," *Cortex* 43 (2007): 319; Johnson-Frey, "The Neural Bases," 71–78.

18 *Homo erectus,* or "Erect Man": Johanson, *Lucy's Legacy,* 192–93.

20 all descendants of those mere hundreds: Ibid., 267.

20 There is some controversy: András Takács-Sánta, "The Major Transitions in the History of Human Transformation of the Biosphere," *Human Ecology Review* 11 (2004): 51–77. Some researchers believe that modern human behavior emerged first earlier, in Africa, and then was brought to Europe in a "second out-of-Africa" migration. See, for example, David Lewis-Williams and David Pearce, *Inside the Neolithic Mind* (London: Thames and Hudson, 2005), 18; Johanson, *Lucy's Legacy,* 257–62.

20 the second-most-expensive organ: Robin I. M. Dunbar and Suzanne Shultz, "Evolution in the Social Brain," *Science* 317 (September 7, 2007): 1344–47.

20 Yet nature chose: Christopher Boesch and Michael Tomasello, "Chimpanzee and Human Cultures," *Current Anthropology* 39 (1998): 591–614.

21 For example, they would: Lewis Wolpert, "Causal Belief and the Origins of Technology," *Philosophical Transactions of the Royal Society A* 361 (2003): 1709–19.

21 They examined the blocks: Daniel J. Povinelli and Sarah Dunphy-Lelii, "Do Chimpanzees Seek Explanations? Preliminary Comparative Investigations," *Canadian Journal of Experimental Psychology* 55 (2001): 185–93.

21 scribbled down all the *whys*: Frank Lorimer, *The Growth of Reason* (London: K. Paul, 1929); quoted in Arthur Koestler, *The Act of Creation* (London: Penguin, 1964), 616.

22 similar rising intonation: Dwight L. Bolinger, ed., *Intonation: Selected Readings.* (Harmondsworth, U.K.: Penguin, 1972), 314; Alan Cruttenden, *Intonation* (Cambridge, U.K.: Cambridge University Press, 1986), 169–17.

22 six-month-olds sat in front of a horizontal track: Laura Kotovsky and Renee

Baillargeon, "The Development of Calibration-Based Reasoning About Collision Events in Young Infants," *Cognition* 67 (1998): 313–51.

3. Culture

25 the abundance of the land: James E. McClellan III and Harold Dorn, *Science and Technology in World History*, 2nd ed. (Baltimore: Johns Hopkins University Press, 2006), 9–12.

25 For example, they built homes: Many of these developments had had precursors among older nomadic groups, but the technologies did not flourish, for the products did not fit the wandering lifestyle. See McClellan and Dorn, *Science and Technology*, 20–21.

26 early farmers had more spinal issues: Jacob L. Weisdorf, "From Foraging to Farming: Explaining the Neolithic Revolution," *Journal of Economic Surveys* 19 (2005): 562–86; Elif Batuman, "The Sanctuary," *New Yorker*, December 19, 2011, 72–83.

26 a "material plenty": Marshall Sahlins, *Stone Age Economics* (New York: Aldine Atherton, 1972), 1–39.

27 studies of the San people: Ibid., 21–22.

27 Göbekli Tepe: Andrew Curry, "Seeking the Roots of Ritual," *Science* 319 (January 18, 2008): 278–80; Andrew Curry, "Gobekli Tepe: The World's First Temple?," *Smithsonian Magazine*, November 2008, accessed November 7, 2014, http://www.smithsonianmag.com/history-archaeology/gobekli-tepe.html; Charles C. Mann, "The Birth of Religion," *National Geographic*, June 2011, 34–59; Batuman, "The Sanctuary."

28 "Within a minute of first seeing it": Batuman, "The Sanctuary."

29 The emergence of group-based religious ritual: Michael Balter, "Why Settle Down? The Mystery of Communities," *Science* 20 (November 1998): 1442–46.

29 "You can make a good case": Curry, "Gobekli Tepe."

30 Neolithic peoples began to settle: McClellan and Dorn, *Science and Technology*, 17–22.

30 Çatalhöyük: Balter, "Why Settle Down?," 1442–46.

31 The difference between a village and a city: Marc Van De Mieroop, *A History of the Ancient Near East* (Malden, Mass.: Blackwell, 2007), 21. See also Balter, "Why Settle Down?," 1442–46.

32 Their extended family units often remained: Balter, "Why Settle Down?," 1442–46; David Lewis-Williams and David Pearce, *Inside the Neolithic Mind* (London: Thames and Hudson, 2005), 77–78.

32 inhabitants would open a grave: Ian Hodder, "Women and Men at Çatalhöyük," *Scientific American*, January 2004, 81.

32 no longer viewed as partnering with animals: Ian Hodder, "Çatalhöyük in the Context of the Middle Eastern Neolithic," *Annual Review of Anthropology* 36 (2007): 105–20.

32 The new attitude would eventually lead: Anil K. Gupta, "Origin of Agriculture and Domestication of Plants and Animals Linked to Early Holocene Climate Amelioration," *Current Science* 87 (July 10, 2004); Van De Mieroop, *History of the Ancient Near East*, 11.

34 turned out a series of papers: L. D. Mlodinow and N. Papanicolaou, "SO (2, 1) Algebra and the Large N Expansion in Quantum Mechanics," *Annals of Physics* 128 (1980): 314–34; L. D. Mlodinow and N. Papanicolaou, "Pseudo-Spin Structure and Large N Expansion for a Class of Generalized Helium Hamiltonians," *Annals of Physics* 131 (1981): 1–35; Carl Bender, L. D. Mlodinow, and N. Papanicolaou, "Semiclassical Perturbation Theory for the Hydrogen Atom in a Uniform Magnetic Field," *Physical Review A* 25 (1982): 1305–14.

34 "dimensional scaling": Jean Durup, "On the 1986 Nobel Prize in Chemistry," *Laser Chemistry* 7 (1987): 239–59. See also D. J. Doren and D. R. Herschbach, "Accurate Semiclassical Electronic Structure from Dimensional Singularities," *Chemical Physics Letters* 118 (1985): 115–19; J. G. Loeser and D. R. Herschbach, "Dimensional Interpolation of Correlation Energy for Two-Electron Atoms," *Journal of Physical Chemistry* 89 (1985): 3444–47.

34 In reality, Watt concocted: Andrew Carnegie, *James Watt* (New York: Doubleday, 1933), 45–64.

35 "Immature poets imitate": T. S. Eliot, *The Sacred Wood and Major Early Essays* (New York: Dover Publications, 1997), 72. First published in 1920.

35 evolutionarily adapted to teach other humans: Gergely Csibra and György Gergely, "Social Learning and Cognition: The Case for Pedagogy," in *Processes in Brain and Cognitive Development*, ed. Y. Munakata and M. H. Johnson (Oxford: Oxford University Press, 2006): 249–74.

35 researchers studying distinct groups of chimps: Christophe Boesch, "From Material to Symbolic Cultures: Culture in Primates," in *The Oxford Handbook of Culture and Psychology*, ed. Juan Valsiner (Oxford: Oxford University Press, 2012), 677–92. See also Sharon Begley, "Culture Club," *Newsweek*, March 26, 2001, 48–50.

35 The best-documented example: Boesch, "From Material to Symbolic Cultures." See also Begley, "Culture Club"; Bennett G. Galef Jr., "Tradition in Animals: Field Observations and Laboratory Analyses," in *Interpretation and Explanation in the Study of Animal Behavior*, ed. Marc Bekoff and Dale Jamieson (Oxford: Westview Press, 1990).

36 evidence of culture in many species: Boesch, "From Material to Symbolic Cultures." See also Begley, "Culture Club."

36 "Chimps can show other chimps": Heather Pringle, "The Origins of Creativity," *Scientific American*, March 2013, 37–43.

36 "cultural ratcheting": Michael Tomasello, *The Cultural Origins of Human Cognition* (Cambridge, Mass.: Harvard University Press, 2001), 5–6, 36–41.

36 Archaeologists sometimes compare: Fiona Coward and Matt Grove, "Beyond the Tools: Social Innovation and Hominin Evolution," *PaleoAnthropology* (special issue, 2011): 111–29.

37 At Bell Labs: Jon Gertner, *The Idea Factory: Bell Labs and the Great Age of American Knowledge* (New York: Penguin, 2012), 41–42.

37 "It's not how smart you are": Pringle, "Origins of Creativity," 37–43.

4. Civilization

39 dwarfs and pygmies: Robert Burton, in *The Anatomy of Melancholy* (1621); George Herbert, in *Jacula Prudentum* (1651); William Hicks, in *Revelation*

Revealed (1659); Shnayer Z. Leiman, "Dwarfs on the Shoulders of Giants," *Tradition*, Spring 1993. Use of the phrase actually seems to go all the way back to the twelfth century.

40 Despite that wiggle room: Marc Van De Mieroop, *A History of the Ancient Near East* (Malden, Mass.: Blackwell, 2007), 21–23.

40 the great walled city of Uruk: Ibid., 12–13, 23.

41 a tenfold increase in size: Some scholars estimate the population as high as 200,000. For example, see James E. McClellan III and Harold Dorn, *Science and Technology in World History*, 2nd ed. (Baltimore: Johns Hopkins University Press, 2006), 33.

42 specialized professions required a new understanding: Van De Mieroop, *History of the Ancient Near East*, 24–29.

42 Bread became the product: McClellan and Dorn, *Science and Technology in World History*, 41–42.

42 Also, the bones reveal: David W. Anthony, *The Horse, the Wheel, and Language: How Bronze-Age Riders from the Eurasian Steppes Shaped the Modern World* (Princeton, N.J.: Princeton University Press, 2010), 61.

44 the need for police: Van De Mieroop, *History of the Ancient Near East*, 26.

44 The inhabitants in each city believed: Marc Van De Mieroop, *The Ancient Mesopotamian City* (Oxford: Oxford University Press, 1997), 46–48.

44 "Goods were received by the god of the city": Van De Mieroop, *History of the Ancient Near East*, 24, 27.

45 But their ability to go beyond simple requests: Elizabeth Hess, *Nim Chimpsky* (New York: Bantam Books, 2008), 240–41.

45 "It's about as likely": Susana Duncan, "Nim Chimpsky and How He Grew," *New York*, December 3, 1979, 84. See also Hess, *Nim Chimpsky*, 22.

46 five hundred tribes of indigenous peoples: T. K. Derry and Trevor I. Williams, *A Short History of Technology* (Oxford: Oxford University Press: 1961), 214–15.

46 "No mute tribe": Steven Pinker, *The Language Instinct: How the Mind Creates Language* (New York: Harper Perennial, 1995), 26.

46 The magnitude of the task: Georges Jean, *Writing: The Story of Alphabets and Scripts* (New York: Henry N. Abrams, 1992), 69.

47 We are *certain* of only one other: Jared Diamond, *Guns, Germs and Steel* (New York: W. W. Norton, 1997), 60, 218. Regarding the New World, see María del Carmen Rodríguez Martinez et al., "Oldest Writing in the New World," *Science* 313 (September 15, 2006): 1610–14; John Noble Wilford, "Writing May Be Oldest in Western Hemisphere," *New York Times*, September 15, 2006. These describe a block with a hitherto unknown system of writing that has recently been found in the Olmec heartland of Veracruz, Mexico. Stylistic and other dating of the block places it in the early first millennium B.C., the oldest writing in the New World, with features that firmly assign this pivotal development to the Olmec civilization of Mesoamerica.

48 When Thomas Edison invented: Patrick Feaster, "Speech Acoustics and the Keyboard Telephone: Rethinking Edison's Discovery of the Phonograph Principle," *ARSC Journal* 38, no. 1 (Spring 2007): 10–43; Diamond, *Guns, Germs and Steel*, 243.

48 the religious community at one temple employed: Jean, *Writing: The Story of Alphabets*, 12–13.

49 Most of the remaining 15 percent: Van De Mieroop, *History of the Ancient Near East*, 30–31.

49 For example, humans: Ibid., 30; McClellan and Dorn, *Science and Technology in World History*, 49.

49 There were also compound signs: Jean, *Writing: The Story of Alphabets*, 14.

50 Pictograms for a hand and a mouth: Derry and Williams, *A Short History of Technology*, 215.

50 "tablet houses": Stephen Bertman, *Handbook to Life in Ancient Mesopotamia* (New York: Facts on File, 2003), 148, 301.

51 the tablet houses were able to broaden their scope: McClellan and Dorn, *Science and Technology in World History*, 47; Albertine Gaur, *A History of Writing* (New York: Charles Scribner's Sons, 1984), 150.

51 By around 2000 B.C.: Sebnem Arsu, "The Oldest Line in the World," *New York Times*, February 14, 2006, 1.

52 the Phoenician script: Andrew Robinson, *The Story of Writing* (London: Thames and Hudson, 1995), 162–67.

52 And from Greece it spread: Derry and Williams, *A Short History of Technology*, 216.

52 "a covenant with the devil": Saint Augustine, *De Genesi ad Litteram* (*The Literal Meaning of Genesis*), completed in A.D. 415.

53 In the first cities of Mesopotamia: Morris Kline, *Mathematics in Western Culture* (Oxford: Oxford University Press, 1952), 11.

53 Young infants: Ann Wakeley et al., "Can Young Infants Add and Subtract?," *Child Development* 71 (November–December 2000): 1525–34.

54 The first arithmetic abbreviations: Morris Kline, *Mathematical Thought from the Ancient to Modern Times*, vol. 1 (Oxford: Oxford University Press, 1972), 184–86, 259–60.

55 Each year, after the fields were flooded: Kline, *Mathematical Thought*, 19–21.

55 able to level a fifty-foot beam: Roger Newton, *From Clockwork to Crapshoot* (Cambridge, Mass.: Belknap Press of the Harvard University Press, 2007), 6.

56 The idea of law originated: Edgar Zilsel, "The Genesis of the Concept of Physical Law," *The Philosophical Review* 3, no. 51 (May 1942): 247.

56 the peoples of early Mesopotamia: Robert Wright, *The Evolution of God* (New York: Little, Brown, 2009), 71–89.

57 first instance of a higher power: Joseph Needham, "Human Laws and the Laws of Nature in China and the West, Part I," *Journal of the History of Ideas* 12 (January 1951): 18.

57 a transcendent god laid down laws: Wright, *Evolution of God*, 87–88.

57 It decreed, for example: "Code of Hammurabi, c. 1780 BCE," Internet Ancient History Sourcebook, Fordham University, March 1998, accessed October 27, 2014, http://www.fordham.edu/halsall/ancient/hamcode.asp; "Law Code of Hammurabi, King of Babylon," Department of Near Eastern Antiquities: Mesopotamia, the Louvre, accessed October 27, 2014, http://www.louvre.fr/en/oeuvre-notices/law-code-hammurabi-king-babylon; Mary Warner Marien and William Fleming, *Fleming's Arts and Ideas* (Belmont, Calif.: Thomson Wadsworth, 2005), 8.

58 laws governing . . . the inanimate world: Needham, "Human Laws and the Laws of Nature," 3–30.

58 "pay fine and penalty to each other": Zilsel, "The Genesis of the Concept of Physical Law," 249.

58 "The sun will not transgress": Ibid.
58 though Kepler wrote occasionally: Ibid., 265–67.
58 "Man seems to be inclined to interpret nature": Ibid., 279.
60 "Out yonder was this huge world": Albert Einstein, *Autobiographical Notes* (Chicago: Open Court Publishing, 1979), 3–5.

5. Reason

62 With the conquest of Mesopotamia: Daniel C. Snell, *Life in the Ancient Near East* (New Haven, Conn.: Yale University Press, 1997), 140–41.
63 a staple of Greek education: A. A. Long, "The Scope of Early Greek Philosophy," in *The Cambridge Companion to Early Greek Philosophy*, ed. A. A. Long (Cambridge, U.K.: Cambridge University Press, 1999).
63 "one should expect a chaotic world": Albert Einstein to Maurice Solovine, March 30, 1952, *Letters to Solovine* (New York: Philosophical Library, 1987), 117.
63 "the most incomprehensible thing": Albert Einstein, "Physics and Reality" in *Ideas and Opinions*, trans. Sonja Bargmann (New York: Bonanza, 1954), 292.
64 a magical land of grapevines: Will Durant, *The Life of Greece* (New York: Simon and Schuster, 1939), 134–40; James E. McClellan III and Harold Dorn, *Science and Technology in World History*, 2nd ed. (Baltimore: Johns Hopkins University Press, 2006), 56–59.
64 the vanguard of the Greek enlightenment: Adelaide Glynn Dunham, *The History of Miletus: Down to the Anabasis of Alexander* (London: University of London Press, 1915).
67 the first to prove geometric truths: Durant, *The Life of Greece*, 136–37.
67 "Be patient toward": Rainer Maria Rilke, *Letters to a Young Poet* (1929; New York: Dover, 2002), 21.
69 The name Pythagoras: Durant, *The Life of Greece*, 161–66; Peter Gorman, *Pythagoras: A Life* (London: Routledge and Kegan Paul, 1979).
70 "for the honor he gives to number": Carl Huffman, "Pythagoras," Stanford Encyclopedia of Philosophy, Fall 2011, accessed October 28, 2014, http://plato.stanford.edu/entries/pythagoras.
73 In Aristotle's theory of the world: McClellan and Dorn, *Science and Technology*, 73–76.
75 "the first to write like a professor": Daniel Boorstin, *The Seekers* (New York: Vintage, 1998), 54.
76 "the tyranny of common sense": Ibid., 316.
76 "What everyone believes is true": Ibid., 55.
76 "it originates in the constitution of the universe": Ibid.
77 he complained loudly: Ibid., 48.
80 "Mine is the first step": See George J. Romanes, "Aristotle as a Naturalist," *Science* 17 (March 6, 1891): 128–33.
81 "the same things are not proper": Boorstin, *The Seekers*, 47.
81 He supplied a lofty reason: "Aristotle," The Internet Encyclopedia of Philosophy, accessed November 7, 2014, http://www.iep.utm.edu.

6. A New Way to Reason

86 A remark by Cicero: Morris Kline, *Mathematical Thought from Ancient to Modern Times*, vol. 1 (Oxford: Oxford University Press, 1972), 179.

87 Cities shrank, the feudal system arose: Kline, *Mathematical Thought*, 204; J. D. Bernal, *Science in History*, vol. 1 (Cambridge, Mass.: MIT Press, 1971), 254.

87 a period of hundreds of years: Kline, *Mathematical Thought*, 211.

87 But by the thirteenth and fourteenth centuries: David C. Lindberg, *The Beginnings of Western Science: The European Scientific Tradition in Philosophical, Religious, and Institutional Context, 600 B.C. to A.D. 1450* (Chicago: University of Chicago Press, 1992), 180–81.

88 all instruction had to center on religion: Toby E. Huff, *The Rise of Early Modern Science: Islam, China, and the West* (Cambridge, U.K.: Cambridge University Press, 1993), 74.

88 Without the salutary benefits: Ibid., 77, 89. Huff and George Saliba disagree on the origin and nature of Islamic science, especially the role of astronomy, which has led to a productive and stimulating discussion in the field. For more on Saliba's argument, see his *Islamic Science and the Making of the European Renaissance* (Cambridge, Mass.: MIT Press, 2007).

88 China, another grand civilization: For more on the situation, see Huff, *Rise of Early Modern Science*, 276–78.

89 In India, too: Bernal, *Science in History*, 334.

89 The revival of science in Europe: Lindberg, *Beginnings of Western Science*, 203–5.

89 the new value placed on knowledge: J. H. Parry, *Age of Reconnaissance: Discovery, Exploration, and Settlement, 1450–1650* (Berkeley: University of California Press, 1982). See especially Part 1.

89 the development of a new institution: Huff, *Rise of Early Modern Science*, 187.

90 The revolution in education: Lindberg, *Beginnings of Western Science*, 206–8.

90 scholars would come together: Huff, *Rise of Early Modern Science*, 92.

91 "Even if the statue": John Searle, *Mind, Language, and Society: Philosophy in the Real World* (New York: Basic Books, 1999), 35.

92 don't have to hunt cats: For more on fourteenth-century conditions, see Robert S. Gottfried, *The Black Death* (New York: Free Press, 1985), 29.

94 no one had any idea how to measure time: For a sweeping and readable examination of the history of the concept of time, see David Landes, *Revolution in Time: Clocks and the Making of the Modern World* (Cambridge, Mass.: Belknap Press of the Harvard University Press, 1983).

94 the "Merton rule": Lindberg, *Beginnings of Western Science*, 303–4.

95 Still, the rule made quite a splash: Clifford Truesdell, *Essays in the History of Mechanics* (New York: Springer-Verlag, 1968).

95 "Do not worry": Albert Einstein, in a letter dated January 7, 1943, quoted in Helen Dukas and Banesh Hoffman, *Albert Einstein: The Human Side; New Glimpses from His Archives* (Princeton, N.J.: Princeton University Press, 1979), 8.

95 "the book [of nature] cannot be understood": Galileo Galilei, *Discoveries and Opinions of Galileo* (New York: Doubleday, 1957), 237–38.

96 Why didn't they rush: Henry Petroski, *The Evolution of Useful Things* (New York: Knopf, 1992), 84–86.

97 Their technological innovations: James E. McClellan III and Harold Dorn, *Science and Technology in World History*, 2nd ed. (Baltimore: Johns Hopkins University Press, 2006), 180–82.

98 But with their setup: Elizabeth Eisenstein, *The Printing Press as an Agent of Change* (Cambridge, U.K.: Cambridge University Press, 1980), 46.

98 Within a few years: Louis Karpinski, *The History of Arithmetic* (New York: Russell and Russell, 1965), 68–71; Philip Gaskell, *A New Introduction to Bibliography* (Oxford, U.K.: Clarendon Press, 1972), 251–65.

98 as European society evolved: Bernal, *Science in History*, 334–35.

99 Vincenzo came from a noble family: My discussion of Galileo's life draws heavily from J. L. Heilbron, *Galileo* (Oxford: Oxford University Press, 2010), and from Stillman Drake, *Galileo at Work* (Chicago: University of Chicago Press, 1978).

101 "See if he is too small": Heilbron, *Galileo*, 61.

101 By the time Galileo got to Padua: Galileo may have been suffering from multiple disenchantments. William A. Wallace argues in his *Galileo, the Jesuits, and the Medieval Aristotle* (Burlington, Vt.: Variorum, 1991) that Galileo, in preparation for his tenure at Pisa, actually appropriated much of his material from lectures given by Jesuits at the Colegio Romano between 1588 and 1590. Wallace also has a chapter called "Galileo's Jesuit Connections and Their Influence on His Science" in Mordechai Feingold's collection *Jesuit Science and the Republic of Letters* (Cambridge, Mass.: MIT Press, 2002).

102 when he got a result that surprised him: Bernal, *Science in History*, 429.

103 more damage if dropped from higher up: G. B. Riccioli, *Almagestum novum astronomiam* (1652), vol. 2, 384; Christopher Graney, "Anatomy of a Fall: Giovanni Battista Riccioli and the Story of G," *Physics Today* (September 2012): 36.

103 We must "cut loose from these difficulties": Laura Fermi and Gilberto Bernardini, *Galileo and the Scientific Revolution* (New York: Basic Books, 1961), 125.

105 a rare instance of Newton's giving credit: Richard Westfall, *Force in Newton's Physics* (New York: MacDonald, 1971), 1–4. In reality, Jean Buridan, who had been Oresme's teacher in Paris, had stated a similar law within the framework of the Merton scholars, though not nearly as clearly as Galileo. See John Freely, *Before Galileo: The Birth of Modern Science in Medieval Europe* (New York: Overlook Duckworth, 2012), 162–63.

106 "an impossible amalgam": Westfall, *Force in Newton's Physics*, 41–42.

106 The idea of a heliocentric universe: Bernal, *Science in History*, 406–10; McClellan and Dorn, *Science and Technology*, 208–14.

106 "I think it easier to believe this": Bernal, *Science in History*, 408.

107 "one convex and the other concave": Daniel Boorstin, *The Discoverers* (New York: Vintage, 1983), 314.

108 "uneven, rough, and full of cavities": Freely, *Before Galileo*, 272.

110 The visit seemed to end in a draw: Heilbron, *Galileo*, 217–20; Drake, *Galileo at Work*, 252–56.

111 "dismissed the fixed-earth philosophers": Heilbron, *Galileo*, 311.

111 what they objected to was a renegade: William A. Wallace, "Gallieo's Jesuit Connections and Their Influence on His Science," in Mordechai Feingold, ed., *Jesuit Science and the Republic of Letters* (Cambridge, Mass.: MIT Press, 2002), 99–112.

111 "I, Galileo, son of the late Vincenzo": Károly Simonyi, *A Cultural History of Physics* (Boca Raton, Fla.: CRC Press, 2012), 198–99.

113 "I find everything disgusting": Heilbron, *Galileo*, 356.

113 "it is not good to build mausoleums": Ibid.

113 "touches not just Florence": Drake, *Galileo at Work*, 436.

7. The Mechanical Universe

114 "the Laws of Nature": Pierre Simon Laplace, *Théorie Analytique des Probabilities* (Paris: Ve. Courcier, 1812).

115 But had we not been raised in a Newtonian culture: To understand Sir Isaac Newton in the context of upheaval in seventeenth-century England, see Christopher Hill, *The World Turned Upside Down: Radical Ideas During the English Revolution* (New York: Penguin History, 1984), 290–97.

115 "I don't know what I may seem": Richard S. Westfall, *Never at Rest* (Cambridge, U.K.: Cambridge University Press, 1980), 863. This is *the* authoritative biography of Newton, and I have relied on it accordingly.

117 greater tendency to go into a scientific career: Ming-Te Wang et al., "Not Lack of Ability but More Choice: Individual and Gender Differences in Choice of Careers in Science, Technology, Engineering, and Mathematics," *Psychological Science* 24 (May 2013): 770–75.

117 "One of the strongest motives": Albert Einstein, "Principles of Research," address to the Physical Society, Berlin, in Albert Einstein, *Essays in Science* (New York: Philosophical Library, 1934), 2.

118 "not finally reducible": Westfall, *Never at Rest*, ix.

118 For Newton believed that God: W. H. Newton-Smith, "Science, Rationality, and Newton," in Marcia Sweet Stayer, ed., *Newton's Dream* (Montreal: McGill University Press, 1988), 31.

119 "threatening my father and mother": Westfall, *Never at Rest*, 53.

120 "herding sheep and shoveling dung": Ibid., 65.

122 the basis of a whole new scientific tradition: Ibid., 155.

126 "your theory is *not* crazy enough!": William H. Cropper, *Great Physicists: The Life and Times of Leading Physicists from Galileo to Hawking* (New York: Oxford University Press, 2004), 252.

126 The Trinity College election: Westfall, *Never at Rest*, 70–71, 176–79.

127 Barrow effectively arranged: Richard Westfall, *The Life of Isaac Newton* (Cambridge, U.K.: Cambridge University Press, 1993), 71, 77–81.

127 Newton attacked the study of light: See the chapter "A Private Scholar & Public Servant," in "Footprints of the Lion: Isaac Newton at Work," Cambridge University Library—Newton Exhibition, accessed October 28, 2014, www.lib.cam.ac.uk/exhibitions/Footprints_of_the_Lion/private_scholar.html.

129 After Newton derived his law of gravity: W. H. Newton-Smith, "Science, Rationality, and Newton," in *Newton's Dream*, ed. Marcia Sweet Stayer (Montreal: McGill University Press, 1988), 31–33.

129 a Bible-based prediction for the end: Richard S. Westfall, *Never at Rest*, 321–24, 816–17.

130 "One nature rejoices": Paul Strathern, *Mendeleev's Dream* (New York: Berkley Books, 2000), 32.

130 "Dissolve volatile green lion": Westfall, *Never at Rest*, 368.
130 I felt that pressure: I wrote a memoir about that period in my life; see Leonard Mlodinow, *Feynman's Rainbow: A Search for Beauty in Physics and in Life* (New York: Vintage, 2011).
131 a "good Newton" and a "bad Newton": Newton-Smith, "Science, Rationality, and Newton," 32–33.
132 "Had Newton died in 1684": Westfall, *Never at Rest*, 407.
134 "Now I am upon this subject": Ibid., 405.
137 the momentous idea: Richard Westfall, *Force in Newton's Physics* (New York: MacDonald, 1971), 463.
137 For example, employing: As measured in "Parisian feet," which are 1.0568 of the usual feet.
137 better than one part in three thousand: Robert S. Westfall, "Newton and the Fudge Factor," *Science* 179 (February 23, 1973): 751–58.
140 If you sneeze: Murray Allen et al., "The Accelerations of Daily Living," *Spine* (November 1994): 1285–90.
142 "The study of nature": Francis Bacon, *The New Organon: The First Book*, in *The Works of Francis Bacon*, ed. James Spedding and Robert Leslie Ellis (London: Longman, 1857–70), accessed November 7, 2014, http://www.bartleby.com/242/.
142 "if the law of forces were known": R. J. Boscovich, *Theiria Philosophiae Naturalis* (Venice, 1763), reprinted as *A Theory of Natural Philosophy* (Chicago: Open Court Publishing, 1922), 281.
143 "the most perfect mechanics": Westfall, *Life of Isaac Newton*, 193.
143 the prickly Newton believed: Michael White, *Rivals: Conflict as the Fuel of Science* (London: Vintage, 2002), 40–45.
144 Newton's reaction to the criticism: Ibid.
144 "These principles I consider": Westfall, *Never at Rest*, 645.
145 He ruled the Society with an iron fist: Daniel Boorstin, *The Discoverers* (New York: Vintage, 1983), 411.
145 "The Chair being Vacant": Westfall, *Never at Rest*, 870.
146 levels of lead, arsenic, and antimony: John Emsley, *The Elements of Murder: A History of Poison* (Oxford: Oxford University Press, 2006), 14.
146 even the Catholic astronomers in Italy: J. L. Heilbron, *Galileo* (Oxford: Oxford University Press, 2010), 360.
146 "Here is buried Isaac Newton": "Sir Isaac Newton," Westminster Abbey, accessed October 28, 2014, www.westminster-abbey.org/our-history/people/sir-isaac-newton.

8. What Things Are Made Of

150 in the Jewish underground: Joseph Tenenbaum, *The Story of a People* (New York: Philosophical Library, 1952), 195.
150 But in 1786: Paul Strathern, *Mendeleev's Dream* (New York: Berkley Books, 2000), 195–98.
151 "all kinds of screwdrivers": From an interview I taped with my father, c. 1980. I have many hours of those interviews and have used them as a source for the stories I tell here.

151 "one of those bodies into which other bodies": J. R. Partington, *A Short History of Chemistry*, 3rd. ed. (London: Macmillan, 1957), 14.

152 What is really being emitted when wood burns: Rick Curkeet, "Wood Combustion Basics," EPA Workshop, March 2, 2011, accessed October 28, 2014, www.epa.gov/burnwise/workshop2011/WoodCombustion-Curkeet.pdf.

154 It was the world's first: Robert Barnes, "Cloistered Bookworms in the Chicken-Coop of the Muses: The Ancient Library of Alexandria," in Roy MacLeod, ed., *The Library at Alexandria: Centre of Learning in the Ancient World* (New York: I. B. Tauris, 2005), 73.

157 "less talent than my ass": Henry M. Pachter, *Magic into Science: The Story of Paracelsus* (New York: Henry Schuman, 1951), 167.

159 I speak of Robert Boyle: The definitive biography of Boyle is Louis Trenchard More, *The Life and Works of the Honorable Robert Boyle* (London: Oxford University Press, 1944). See also William H. Brock, *The Norton History of Chemistry* (New York: W. W. Norton, 1992), 54–74.

160 fell in love with science: More, *Life and Works*, 45, 48.

161 When you burn a log, Boyle observed: Brock, *Norton History of Chemistry*, 56–58.

161 Poor Hooke: J. D. Bernal, *Science in History*, vol. 2 (Cambridge, Mass.: MIT Press, 1971), 462.

162 the invention of new laboratory equipment: T. V. Venkateswaran, "Discovery of Oxygen: Birth of Modern Chemistry," *Science Reporter* 48 (April 2011): 34–39.

163 Due to the controversy: Isabel Rivers and David L. Wykes, eds., *Joseph Priestley, Scientist, Philosopher, and Theologian* (Oxford: Oxford University Press, 2008), 33.

163 The term itself: Charles W. J. Withers, *Placing the Enlightenment: Thinking Geographically About the Age of Reason* (Chicago: University of Chicago Press, 2007), 2–6.

165 "This air is of exalted nature": J. Priestley, "Observations on Different Kinds of Air," *Philosophical Transactions of the Royal Society* 62 (1772): 147–264.

165 It was left to a Frenchman: For the life of Lavoisier, see Arthur Donovan, *Antoine Lavoisier* (Oxford: Blackwell, 1993).

166 "agents in nature able to make the particles": Isaac Newton, *Opticks*, ed. Bernard Cohen (London, 1730; New York: Dover, 1952), 394. Newton first published *Opticks* in 1704, but his final thoughts on the matter are represented by the fourth edition, the last revised by Newton himself, which came out in 1730.

166 "founded on only a few facts": Donovan, *Antoine Lavoisier*, 47–49.

167 "a fabric woven of experiments": Ibid., 139. See also Strathern, *Mendeleev's Dream*, 225–41.

169 "The republic has no need of scientists": Douglas McKie, *Antoine Lavoisier* (Philadelphia: J. J. Lippincott, 1935), 297–98.

170 "merited the esteem of men": J. E. Gilpin, "Lavoisier Statue in Paris," *American Chemical Journal* 25 (1901): 435.

170 The sculptor, Louis-Ernest Barrias: William D. Williams, "Gustavus Hinrichs and the Lavoisier Monument," *Bulletin of the History of Chemistry* 23 (1999): 47–49; R. Oesper, "Once the Reputed Statue of Lavoisier," *Journal of Chemistry Education* 22 (1945): October frontispiece; Brock, *Norton History of Chemistry*, 123–24.

170 scrapped during the Nazi occupation: Joe Jackson, *A World on Fire* (New York: Viking, 2007), 335; "Lavoisier Statue in Paris," *Nature* 153 (March 1944): 311.

170 Ironically, in 1913: "Error in Famous Bust Undiscovered for 100 Years," *Bulletin of Photography* 13 (1913): 759; and Marco Beretta, *Imaging a Career in Science: The Iconography of Antoine Laurent Lavoisier* (Sagamore Beach, Mass.: Science Histories Publications, 2001), 18–24.

172 John Dalton: Frank Greenaway, *John Dalton and the Atom* (Ithaca, N.Y.: Cornell University Press, 1966); Brock, *Norton History of Chemistry*, 128–60.

175 Defined as "the disposition to pursue": A. L. Duckworth et al., "Grit: Perseverance and Passion for Long-Term Goals," *Journal of Personality and Social Psychology* 92 (2007): 1087–101; Lauren Eskreis-Winkler et al., "The Grit Effect: Predicting Retention in the Military, the Workplace, School and Marriage," *Frontiers in Psychology* 5 (February 2014): 1–12.

175 Dmitri Mendeleev: See Strathern, *Mendeleev's Dream;* Brock, *Norton History of Chemistry*, 311–54.

176 the Bankograph: Kenneth N. Gilpin, "Luther Simjian Is Dead; Held More Than 92 Patents," *New York Times*, November 2, 1997; "Machine Accepts Bank Deposits," *New York Times*, April 12, 1961, 57.

181 Mendeleev published his table: Dmitri Mendeleev, "Ueber die beziehungen der eigenschaften zu den atom gewichten der elemente," *Zeitschrift für Chemie* 12 (1869): 405–6.

9. The Animate World

183 Robert Hooke sharpened his penknife: Anthony Serafini, *The Epic History of Biology* (Cambridge, Mass.: Perseus, 1993), 126.

183 about thirty trillion cells: E. Bianconi et al., "An Estimation of the Number of Cells in the Human Body," *Annals of Human Biology* 40 (November–December 2013): 463–71.

184 ten million species on our planet: Lee Sweetlove, "Number of Species on Earth Tagged at 8.7 Million," *Nature*, August 23, 2011.

184 up to ten insect fragments: "The Food Defect Action Levels," Defect Levels Handbook, U.S. Food and Drug Administration, accessed October 28, 2014, http://www.fda.gov/food/guidanceregulation/guidancedocumentsregulatoryinformation/ucm056174.htm.

184 a serving of broccoli: Ibid.

184 microscopic organisms that live on your forearm: "Microbiome: Your Body Houses 10x More Bacteria Than Cells," *Discover*, n.d., accessed October 28, 2014, http://discovermagazine.com/galleries/zen-photo/m/microbiome.

185 Aristotle was an enthusiastic biologist: For Aristotle's work on biology, see Joseph Singer, *A History of Biology to About the Year 1900* (New York: Abelard-Schuman, 1959); Lois Magner, *A History of the Life Sciences*, 3rd. ed. (New York: Marcel Dekker, 2002).

186 people believed that some sort of life force: Paulin J. Hountondji, *African Philosophy*, 2nd ed. (Bloomington: Indiana University Press, 1996), 16.

188 "With this tube": Daniel Boorstin, *The Discoverers* (New York: Vintage, 1983), 327.

188 One of the greatest champions: Magner, *History of the Life Sciences*, 144.
189 "the most ingenious book": Ruth Moore, *The Coil of Life* (New York: Knopf, 1961), 77.
190 experiments being mocked on the stage: Tita Chico, "Gimcrack's Legacy: Sex, Wealth, and the Theater of Experimental Philosophy," *Comparative Drama* 42 (Spring 2008): 29–49.
190 One man who didn't doubt: For Leeuwenhoek's work on the microscope, see Moore, *Coil of Life*.
190 "has devised microscopes which far surpass": Boorstin, *The Discoverers*, 329–30.
191 "the outcome of my own unaided impulse": Moore, *Coil of Life*, 79.
191 "little eels, or worms": Boorstin, *The Discoverers*, 330–31.
192 "stuck out two little horns": Moore, *Coil of Life*, 81.
192 a handful of his microscopes remain intact: Adriana Stuijt, "World's First Microscope Auctioned Off for 312,000 Pounds," *Digital Journal*, April 8, 2009, accessed November 7, 2014, http://www.digitaljournal.com/article/270683; Gary J. Laughlin, "Editorial: Rare Leeuwenhoek Bids for History," *The Microscope* 57 (2009): ii.
193 "most important series": Moore, *Coil of Life*, 87.
193 "Anton van Leeuwenhoek considered": "Antony van Leeuwenhoek (1632–1723)," University of California Museum of Paleontology, accessed October 28, 2014, http://www.ucmp.berkeley.edu/history/leeuwenhoek.html.
193 its Newton was Charles Darwin: For Darwin's life, I relied largely on Ronald W. Clark, *The Survival of Charles Darwin: A Biography of a Man and an Idea* (New York: Random House, 1984); Adrian Desmond, James Moore, and Janet Browne, *Charles Darwin* (Oxford: Oxford University Press, 2007); and Peter J. Bowler, *Charles Darwin: The Man and His Influence* (Cambridge, U.K.: Cambridge University Press, 1990).
193 "it would have been unfortunate": "Charles Darwin," Westminster Abbey, accessed October 28, 2014, http://www.westminster-abbey.org/our-history/people/charles-darwin.
193 "Everybody is interested in pigeons": Clark, *Survival of Charles Darwin*, 115.
194 "God knows what the public will think": Ibid., 119.
195 "great curiosity about facts": Ibid., 15.
195 "I saw two rare beetles": Ibid., 8.
195 "I hate a barnacle": Charles Darwin to W. D. Fox, October 1852, Darwin Correspondence Project, letter 1489, accessed October 28, 2014, http://www.darwinproject.ac.uk/letter/entry-1489.
195 "sufficient to check any strenuous effort": Clark, *Survival of Charles Darwin*, 10.
196 "What a fellow that Darwin is": Ibid., 15.
197 "I have just room to turn round": Ibid., 27.
197 It was *not* during the voyage: Bowler, *Charles Darwin: The Man*, 50, 53–55.
197 "Your situation is above envy": Charles Darwin to W. D. Fox, August 9–12, 1835, Darwin Correspondence Project, letter 282, accessed October 28, 2014, http://www.darwinproject.ac.uk/letter/entry-282.
198 the beaks of the finches: Desmond, Moore, and Browne, *Charles Darwin*, 25, 32–34.
200 "It at once struck me": Ibid., 42.
201 "When I am dead": Bowler, *Charles Darwin*, 73.

202 the death . . . of the Darwins' second child: Adrian J. Desmond, *Darwin* (New York: W. W. Norton, 1994), 375–85.

202 "We have lost the joy of the Household": Charles Darwin's memorial of Anne Elizabeth Darwin, "The Death of Anne Elizabeth Darwin," accessed October 28, 2014, http://www.darwinproject.ac.uk/death-of-anne-darwin.

203 as alike "as two peas": Desmond, Moore, and Browne, *Charles Darwin*, 44.

203 "the extreme verge of the world": Ibid., 47.

203 "most solemn and last request": Ibid., 48.

203 "If it be accepted even by one competent judge": Ibid., 49.

204 "poisoning the foundations of science": Anonymous [David Brewster], "Review of *Vestiges of the Natural History of Creation*," *North British Review* 3 (May–August 1845): 471.

205 a "foul book": Evelleen Richards, "'Metaphorical Mystifications': The Romantic Gestation of Nature in British Biology," in *Romanticism and the Sciences*, eds. Andrew Cunningham and Nicholas Ardine (Cambridge, U.K.: Cambridge University Press, 1990), 137.

207 "[Wallace] has to-day sent me": "Darwin to Lyell, June 18, 1858," in *The Life and Letters of Charles Darwin, Including an Autobiographical Chapter*, ed. Francis Darwin (London: John Murray, 1887), available at http://darwin-online.org.uk/converted/published/1887_Letters_F1452/1887_Letters_F1452.2.html, accessed October 28, 2014.

208 Thomas Bell, who on his way out lamented: Desmond, *Darwin*, 470.

208 "weak as a child": Desmond, Moore, and Browne, *Charles Darwin*, 65.

209 "There is grandeur in this view": Bowler, *Charles Darwin*, 124–25.

209 "I have read your book with more pain than pleasure": Clark, *Survival of Charles Darwin*, 138–39.

211 "Everywhere Darwinism became": Desmond, Moore, and Browne, *Charles Darwin*, 107.

211 Gregor Mendel: See Magner, *History of the Life Sciences*, 376–95.

212 "I have everything to make me happy": Darwin to Alfred Russel Wallace, July 1881, quoted in Bowler, *Charles Darwin*, 207.

10. The Limits of Human Experience

216 atoms were far too small to be seen: In 2013, scientists were finally able to go a step further and "see" individual molecules reacting. See Dimas G. de Oteyza et al., "Direct Imaging of Covalent Bond Structure in Single-Molecule Chemical Reactions," *Science* 340 (June 21, 2013): 1434–37.

217 they are "hypothetical conjectures": Niels Blaedel, *Harmony and Unity: The Life of Niels Bohr* (New York: Springer Verlag, 1988), 37.

219 "the willingness to endure a condition": John Dewey, "What Is Thought?," in *How We Think* (Lexington, Mass.: Heath, 1910), 13.

220 "physics is a branch": Barbara Lovett Cline, *The Men Who Made a New Physics* (Chicago: University of Chicago Press, 1965), 34. See also J. L. Heilbron, *The Dilemmas of an Upright Man* (Cambridge, Mass.: Harvard University Press, 1996), 10.

220 Max Planck: Much of the material on Planck comes from Heilbron, *Dilemmas of an Upright Man*. See also Cline, *The Men Who Made a New Physics*, 31–64.

220 "disinclined to questionable adventures": Heilbron, *Dilemmas of an Upright Man*, 3.

220 "investigate the harmony": Ibid., 10.

220 "My maxim is always this": Ibid., 5.

221 But the phenomenon can be explained: Leonard Mlodinow and Todd A. Brun, "Relation Between the Psychological and Thermodynamic Arrows of Time," *Physical Review E* 89 (2014): 052102–10.

221 "Despite the great success": Heilbron, *Dilemmas of an Upright Man*, 14.

223 "nobody . . . had any interest whatever": Ibid., 12; Cline, *The Men Who Made a New Physics*, 36.

226 there are "certain forces by which": Richard S. Westfall, *Never at Rest* (Cambridge, U.K.: Cambridge University Press, 1980), 462.

226 "local motions which . . . cannot be detected": Ibid.

229 "A new scientific truth": The original quote, often misquoted, is "*Eine neue wissenschaftliche Wahrheit pflegt sich nicht in der Weise durchzusetzen, daß ihre Gegner überzeugt werden und sich als belehrt erklären, sondern vielmehr dadurch, daß ihre Gegner allmählich aussterben und daß die heranwachsende Generation von vornherein mit der Wahrheit vertraut gemacht ist.*" It appeared in *Wissenschaftliche Selbstbiographie: Mit einem Bildnis und der von Max von Laue gehaltenen Traueransprache* (Leipzig: Johann Ambrosius Barth Verlag, 1948), 22. The translation comes from *Max Planck, Scientific Autobiography and Other Papers*, trans. F. Gaynor (New York: Philosophical Library, 1949), 33–34.

231 "Of course, I am aware that Planck's law": John D. McGervey, *Introduction to Modern Physics* (New York: Academic Press, 1971), 70.

232 "Why abandon a belief": Robert Frost, "The Black Cottage," in *North of Boston* (New York: Henry Holt, 1914), 54.

232 "It was as if the ground": Albert Einstein, *Autobiographical Notes* (1949; New York: Open Court, 1999), 43.

235 "the irreproachable student": Carl Sagan, *Broca's Brain* (New York: Random House, 1974), 25.

236 "My son is deeply unhappy": Abraham Pais, *Subtle Is the Lord: The Science and Life of Albert Einstein* (Oxford: Oxford University Press, 1982), 45.

236 His first two papers: Ibid., 17–18.

240 "no revolution at all": Ibid., 31.

241 a letter he wrote to a friend in 1905: Ibid., 30–31.

242 "to find facts which would guarantee": Ronald Clark, *Einstein: The Life and Times* (New York: World Publishing, 1971), 52.

244 the 1913 recommendation: Pais, *Subtle Is the Lord*, 382–86.

245 "To Albert Einstein for his services": Ibid., 386.

245 "an historic understatement": Ibid.

246 "All these fifty years of pondering": Jeremy Bernstein, *Albert Einstein and the Frontiers of Physics* (Oxford: Oxford University Press, 1996), 83.

11. The Invisible Realm

247 "I don't think we should discourage": Leonard Mlodinow, *Feynman's Rainbow: A Search for Beauty in Physics and in Life* (New York: Vintage, 2011), 94–95.

248 "Despite . . . the apparently complete": Abraham Pais, *Subtle Is the Lord: The Science and Life of Albert Einstein* (Oxford: Oxford University Press, 1982), 383.

249 When Niels Bohr was of high school age: For more on Bohr's life and science and his relationship with Ernest Rutherford, see Niels Blaedel, *Harmony and Unity: The Life of Niels Bohr* (New York: Springer Verlag, 1988), and Barbara Lovett Cline, *The Men Who Made a New Physics* (Chicago: University of Chicago Press, 1965), 1–30, 88–126.

250 So outlandish did Thomson's electron seem: "Corpuscles to Electrons," American Institute of Physics, accessed October 28, 2014, http://www.aip.org/history/electron/jjelectr.htm.

250 In 1941, scientists: R. Sherr, K. T. Bainbridge, and H. H. Anderson, "Transmutation of Mercury by Fast Neutrons," *Physical Review* 60 (1941): 473–79.

253 The proton and nucleus were not then known: John L. Heilbron and Thomas A. Kuhn, "The Genesis of the Bohr Atom," in *Historical Studies in the Physical Sciences*, vol. 1, ed. Russell McCormmach (Philadelphia: University of Pennsylvania Press, 1969), 226.

255 "quite the most incredible event": William H. Cropper, *Great Physicists: The Life and Times of Leading Physicists from Galileo to Hawking* (Oxford: Oxford University Press, 2001), 317.

256 They would find themselves: For more on Geiger, see Jeremy Bernstein, *Nuclear Weapons: What You Need to Know* (Cambridge, U.K.: Cambridge University Press, 2008), 19–20; and Diana Preston, *Before the Fallout: From Marie Curie to Hiroshima* (New York: Bloomsbury, 2009), 157–58.

256 one hundred times as much as Mount Everest: Actually, it would be a hundred billion tons, for Everest weighs about a billion tons. See "Neutron Stars," *NASA Mission News*, August 23, 2007, accessed October 27, 2014, http://www.nasa.gov/mission_pages/GLAST/science/neutron_stars_prt.htm.

258 The difficulty of Bohr's undertaking: John D. McGervey, *Introduction to Modern Physics* (New York: Academic Press, 1971), 76.

261 "one of the greatest discoveries": Stanley Jaki, *The Relevance of Physics* (Chicago: University of Chicago Press, 1966), 95.

261 "It was regrettable in the highest degree": Blaedel, *Harmony and Unity*, 60.

261 Lord Rayleigh, meanwhile: Jaki, *Relevance of Physics*, 95.

261 "men over seventy": Ibid.

261 Arthur Eddington: Ibid., 96.

262 What eventually convinced: Blaedel, *Harmony and Unity*, 78–80; Jagdish Mehra and Helmut Rechenberg, *The Historical Development of Quantum Theory*, vol. 1 (New York: Springer Verlag, 1982), 196, 355.

262 With that, no physicist: Blaedel, *Harmony and Unity*, 79–80.

12. The Quantum Revolution

264 "I am thinking hopelessly about quantum theory": William H. Cropper, *Great Physicists: The Life and Times of Leading Physicists from Galileo to Hawking* (Oxford: Oxford University Press, 2001), 252.

264 "I wish I were a movie comedian": Ibid.

265 recognized at an early age as brilliant: The definitive biography of Heisenberg

is David C. Cassidy, *Uncertainty: The Life and Times of Werner Heisenberg* (New York: W. H. Freeman, 1992).

266 "In that case you are completely lost": Ibid., 99–100.

267 "It may be that you know something": Ibid., 100.

267 Could one create a variant: Olivier Darrigol, *From c-Numbers to q-Numbers: The Classical Analogy in the History of Quantum Theory* (Berkeley: University of California Press, 1992), 218–24, 257, 259; Cassidy, *Uncertainty*, 184–90.

267 "I've missed more than nine thousand shots": "Failure," television commercial, 1997, accessed October 27, 2014, https://www.youtube.com/watch?v=45mMioJ5szc.

268 Consider, for example, Abraham Lincoln: Lincoln-Douglas Debate at Charleston, Illinois, September 18, 1858, accessed November 7, 2014, http://www.nps.gov/liho/historyculture/debate4.htm.

268 white supremacy is a "universal feeling": Abraham Lincoln, address at Peoria, Illinois, October 16, 1854; see Roy P. Basler, ed., *The Collected Works of Abraham Lincoln*, vol. 2 (New Brunswick, N.J.: Rutgers University Press, 1953–55), 256, 266.

270 Max Born would dub it "quantum mechanics": William A. Fedak and Jeffrey J. Prentis, "The 1925 Born and Jordan Paper 'On Quantum Mechanics,'" *American Journal of Physics* 77 (February 2009): 128–39.

271 he himself called it "very strange": Niels Blaedel, *Harmony and Unity: The Life of Niels Bohr* (New York: Springer Verlag, 1988), 111.

272 "I am now convinced . . . actual philosophy": Max Born, *My Life and Views* (New York: Charles Scribner's Sons, 1968), 48.

272 he immediately wrote Einstein: Mara Beller, *Quantum Dialogue: The Making of a Revolution* (Chicago: University of Chicago Press, 1999), 22.

273 Like Bohr and Heisenberg, Born: Cassidy, *Uncertainty*, 198.

273 "I incline to the belief that physicists": Abraham Pais, *Subtle Is the Lord: The Science and Life of Albert Einstein* (Oxford: Oxford University Press, 1982), 463.

273 "My own works are at the moment": Cassidy, *Uncertainty*, 203.

273 Pauli became agitated: Charles P. Enz, *No Time to Be Brief* (Oxford: Oxford University Press, 2010), 134.

275 "Due to the last work of Heisenberg": Blaedel, *Harmony and Unity*, 111–12.

276 He "did his great work": Walter Moore, *A Life of Erwin Schrödinger* (Cambridge, U.K.: Cambridge University Press, 1994), 138.

277 "The idea of your work springs": Ibid., 149.

277 "I am convinced that you have made a decisive advance": Ibid.

277 "I was of three minds": Wallace Stevens, "Thirteen Ways of Looking at a Blackbird," *Collected Poems* (1954; New York: Vintage, 1982), 92.

278 "what consequences they would unleash": Pais, *Subtle Is the Lord*, 442.

278 "felt discouraged, not to say repelled": Cassidy, *Uncertainty*, 215.

278 "The more I reflect on the physical portion": Ibid.

279 more than 100,000 papers: Moore, *Life of Erwin Schrödinger*, 145.

280 "closer to the secret of the Old One": Albert Einstein to Max Born, December 4, 1926, in *The Born-Einstein Letters*, ed. M. Born (New York: Walker, 1971), 90.

282 "Einstein's verdict . . . came as a hard blow": Pais, *Subtle Is the Lord*, 443.

282 "my solitary little song": Ibid., 31.

282 "I am generally regarded as a sort of petrified object": Ibid., 462.

283 "Now we at least have order": Graham Farmelo, *The Strangest Man: The Hidden Life of Paul Dirac, Mystic of the Atom* (New York: Basic Books, 2009), 219–20.
284 "I believe that Heisenberg is decent": Cassidy, *Uncertainty*, 393.
285 Heisenberg . . . resented Schrödinger's departure: Ibid., 310.
285 As his biographer Walter Moore: Moore, *Life of Erwin Schrödinger*, 213–14.
285 "You will never see it again": Philipp Frank, *Einstein: His Life and Times* (Cambridge, Mass.: Da Capo Press, 2002), 226.
286 What's left of it resides: Michael Balter, "Einstein's Brain Was Unusual in Several Respects, Rarely Seen Photos Show," *Washington Post*, November 26, 2012.
287 proudly wore his brown uniform: Farmelo, *The Strangest Man*, 219.
288 Instead, like Heisenberg, his priority: Cassidy, *Uncertainty*, 306.
288 When the German atom bomb: Cassidy, *Uncertainty*, 421–29.

Epilogue

293 There is an old brainteaser: Martin Gardner, "Mathematical Games," *Scientific American*, June 1961, 168–70.
294 Socrates likened a person: Alain de Botton, *The Consolations of Philosophy* (New York: Vintage, 2000), 20–23.

Index

Page numbers in *italics* refer to illustrations.

A-bomb, 284
acceleration, 97
  law of free fall, 104
  Merton rule, 94–5, *96*
  Newton's laws of, 138–40
  rolling-ball experiment, 103–4
action, law of, 140–2
adaptation, 26, 199, 201
agriculture/farming, 26–7, 32, 41–3
air:
  as classical element, 73, 154, 155, 159, 161–2, 296
  composition of, 152, 154, 155, 161–2, 164, 175
  resistance, 103, 104, 125, 138, 139, 141, 142, 151, 294
alchemy, 7, 118, 155, 166, 168
  chemistry's origins in, 153
  converting mercury to gold, 250
  earliest practitioners, 130
  Newton's study of, 129–32, 135, 144, 149, 268
  origins of, 154–5
  Paracelsus's work in, 153, 156–8
Alexander the Great, 61–2, 71, 153
Alexandria, 153–4
algebra, 54, 273
alpha particles, 253–4, *256*
Anaximander, 58
animals:
  bonobos, 12, 20, 22
  Boyle's breathing experiments on, 161–2
  chimpanzees (*see* chimpanzees)
  communication among, 45–8
  cultural transmission by, 35–6
  human dominance over, 32
  macaque monkeys, 35–6
  tool use by, 29, 35
anthropology:
  Çatalhöyük village, 30–3
  cultural ratcheting, 36–7
  DNA analysis, 19
  Göbekli Tepe monument, 28
  group dynamics, 44
  Lucy (representative of *Australopithecus afarensis*), 13–17
  modern human behavior development, 20
  nomadic cultures, 27
antigravity effect, 298
Aquinas, St. Thomas, 110
Arabs (ancient), 87–9
archaeological discoveries:
  Çatalhöyük village, 30–3
  Code of Hammurabi block, 57–8
  cultural innovations compared to viruses, 36
  Göbekli Tepe monument, 27–30
  Lucy *(Australopithecus afarensis)*, 13–17
  mathematical calculations, 54
  writing tablets, 48–9, 51
Aristotelian physics:
  abandonment of, 91, 99, 102, 104, 105, 107, 108, 109, 114, 121, 137, 143, 160
  and common sense, 75
  free fall, 102–4
  Galileo's break from, 75, 79–80, 102–5
  Newton's view of, 143
  theory of motion, 93
Aristotle, 62, 66, 68, 71–81, *72*
  on biological organisms, 185
  on creation of new life, 186–7
  doubts cast on theories of, 79–80
  influence of, 72, 79, 80, 81, 99, 106, 121, 146, 151, 154, 186, 216

Aristotle *(continued):*
on Ionian thinkers, 68
life of, 71, 75, 80–1
natural versus violent change, 281
Newton's study of writings of, 121
and Plato, 71, 72, 75, 81
realm of the heavens, 106, 108, 137
role of purpose in theories of, 78, 79, 80
Roman Catholic Church's adoption of philosophy of, 81
students of, 61–2
on superiority of humans, 185
theory of biology, 185, 186–7, 188
theory of elements, 72, 151, 152, 161, 162
theory of force, 74, 78, 79–80, 105, 106
theory of matter, 154
arithmetic, 53–5; *see also* mathematics
arrow of time, 221
art/artists, 28, 99, 219
Asimov, Isaac, 8
astrology, 99–100
astronomy, 58
ancient Greek, 106
changing theories of, 158
Copernicus's model of, 106–7
Galileo's contribution to, 106–10
moon's orbit, 137–8
Newton's laws of gravity, 147
orbital motion, 132–3, 135
Ptolemaic, 107
telescope, 107–9
atom bomb, 255, 288
atomic number, 262
atomic orbits, 280
atomic weight, 172, 179, 262
atoms/atomic theory, 171–4
acceptance of, 251
advances due to, 241
alpha particle experiments, 253–5
Bohr atom, 255, 258–64
Boltzman's research on, 228–30
chemical reactions, 171–4
decay of, 266, 268, 274, 275, 276
electric charges in, 252–4
excitation of, 259
Heisenberg's theory, *271*, 271–2
Higgs particle, 217–19
motion of electrons in, 276
new equations for describing, 241
nucleus of atoms, 256
Planck's theory of blackbody radiation and, 228
quantum laws of, 216, 258
reality of, 217, 240
Rutherford's discoveries (*see* Rutherford, Ernest)
Rutherford's model, 256–69, *261*
structure of atoms, 257
study and use of, 215–16
theory of light in, 242
Thomson's model, 253, 254, 255
see also electrons
attraction, law of, 138, 166
Augustine, St., 52
Australopithecus afarensis, 13–17

Babbage, Charles, 199
Babylon, *53*, 54, 296
Babylonian Empire, 57
Bacon, Francis, 142
Barberini, Card. Maffeo, 110
Barrias, Ernest, 170
Barrow, Isaac, 127
base ten system, 54
Beagle, 196–8
behavior:
evolution of human, 19–21
governance of, 57
passing on of, by animals, 35–6
beliefs/belief systems:
abandonment of, 231–2
ancient, 186–8
Darwin's, 201–2
Einstein's, 281
in God, 38, 79
Newton's, 129–30
people's reasons for, 268
see also worldviews
Bell, John, 282
Bell Labs, 37
Berlin Physical Society, 228
Bernstein, Jeremy, 126
Berzelius, Jöns Jakob, 177
Besso, Michele, 246
Bible:
Darwin on moral authority of, 197
Hebrew, 28
Newton's analysis of, 129
see also religion/religious issues

biological organisms:
 assumptions about creation of, 186–7
 classification of, 177
 in food/on skin, 185
 microscopic, 188–93
 spontaneous generation theory, 187
 see also Darwin, Charles
biology, 183–212
blackbody radiation, 216, 225, 227–8, 230–1, 242–3, 249, 259
blood, 165, 192
Bohr, Niels:
 correspondence principle, 270
 critique of Thomson's theories by, 251–2
 derivation of ideas/concepts of, 259–60
 early years and education, 249
 evacuation from Germany, 286–7
 Heisenberg's visit to, 267
 on his own work, 268–9
 mentioned, 8, 126, 239–40, 249–50, *263n*, *274*, 284, 287
 Niels Bohr Institute, 285
 partnership with Rutherford, 252–3
 principles explaining atom stability, 264
 provisional theory of hydrogen atom, 275
 on Schrödinger's theory, 279
 theory of energy levels, 271
Bohr atom, 255, 258–63, 268, 270, 275
bohrium, 263
Boisbaudran, Paul-Émile Lecoq de, 179–80
Bolos of Mendes, 130
Boltzmann, Ludwig, 228–30, *231*, 240
bonobos, 12, 20, 22; *see also* chimpanzees
Boorstin, Daniel, 76
Born, Max:
 academic descendants of, *286n*
 Einstein on quantum theory to, 280
 epitaph of, 285
 mentioned, 264, 270, 272, 284, *287*
 paper on quantum theory, 273–4
 probabilistic interpretation of quantum theory by, 282
 on Schrödinger's theory, 278
Boscovich, Roger, 142
Boulding, Kenneth, 6
Boyle, Robert, 127, 159–62
 The Sceptical Chymist, 160–1
Brahe, Tycho, 132
brain, 12–15, *14n*, 15, 18, 20–1, 297
breathing experiments, 161–2
Brown, Robert, 240
Brownian motion, 240
Bruno, Giordano, 112
builders, of Göbekli Tepe monument, 29
burial practices, 32
Burton, Robert, 39
Bush, George W., 235–6
Bushmen (San people of Africa), 27

calculus:
 development of, 123
 differential and integral, *124n*
 invention of, 122, 143
 modern uses of, 125
calx (oxide of mercury), 164–5, 168–9
carbon dioxide, 164
carbonated beverages, 164
Caritat, Nicholas de (Marquis de Condorcet), 170, *171*
Çatalhöyük village, 30–3
Catholic Church:
 adoption of Aristotle's philosophy, 81
 edict against Copernicanism, 111
 endorsement of Aristotelianism, 110
 Galileo's conflicts with doctrine of, 111
 heliocentric view of universe, 106–7
 ownership of truth, 89
 see also religion/religious issues
causality:
 Aristotle's research on, 72–4
 human demand for, 63
 innate understanding of, 22
 Newton's understanding of, 135, 139, 141, 147
 in quantum universe, 216–17
 seeking reasons/purpose, 79
 traditional notions of, 233
cause and effect, 63, 216, 246
cells, Hooke's study of, 183
Chambers, Robert, *203n*, 205
chance, role of, 216
change:
 analysis of, 123
 Aristotle's concept of, 72–4
 causes of, 147–8
 effects and results of, 8–9
 mathematics of, 125

change *(continued):*
  natural and violent, 281
  Newton's conception of, 123–4
  of state by force, 138–9
  study of, 72–3
  to substances, 151–2
  see also physics
change of motion, 139–40; *see also* motion (change of position)
change of position, *see* motion (change of position)
chaos, 49, 56, 62, 64, 265, 283
charitable trusts (madrassas), 88
chemical elements, *see* elements
chemical equations, 169
chemical reactions, 168, 171–4
chemistry, 148–82, *218*
  birth of, 151
  Boyle's work in, 160–2
  calx (oxide of mercury), 164–5, 168–9
  Dalton's work in, 172–4
  Elementary Treatise on Chemistry (Lavoisier), 170
  International Union of Pure and Applied Chemistry, 237*n*
  language of, 174
  Lavoisier's work in, 166–71
  light versus matter principle, 238–9
  Mendeleev's work in, 177–82
  Nobel Prize in, 182
  physics versus, 149–51
  Priestley's study of gases, 162–4
  technological advances for studying, 152–3
  theoretical and experimental aspects of, 168
  see also alchemy
children, behavior studies of, 21
chimpanzees:
  common ancestry with humans, 12, 14, 20–2
  communication by sign language, 22, 45–6
  cultural transmission of behaviors, 35–6
  Köhler's experiments on, 21
  see also bonobos
China (ancient), 88–9, 98
Chomsky, Noam, 45–6
Chopra, Deepak, 85
Christianity, 87, 129; *see also* religion/religious issues
Church, *see* Catholic Church
Church of England, 144
Churchill, Winston, 286
circles, 136
circular motion, *136*
cities, 30–3, 39–43
Clark, William, 119
classical physics, *see* Newtonian (classical) physics
Clausius, Rudolf, 223
Clement, William, 124
Code of Hammurabi, 57–8
color, Newton's study of, 127–8
common sense, 75–7
communication:
  language ability, 45–8
  signing by chimpanzees/bonobos, 22, 45–6
  technology for, 6
  by theoretical ideas, 68
  see also language(s); writing systems/written word
Company of General Farmers, 167
compound elements, 173–4
Condorcet, Marquis de (Nicholas de Caritat), 170, *171*
conflict resolution, creation of police forces, 44
connectedness, 37, 43, 49
consciousness, existential, 30
conservation of mass, law of, 168, 170
Constantinus Africanus, 89
conventional thinking, 76–7
copernicium, 263*n*
Copernicus, Nicolaus, 91, 106–9, 111*n*, 112, 158, 263, 263*n*
  Revolutionibus Orbium Coelsestium (On the Revolutions of the Heavenly Spheres), 158
corpuscles theory, 128, 172
correspondence principle, 270
cosmos, 63, 280
creativity, 8, 9, 88, 235, 248, 268
critical thinking, 219
cultural ratcheting, 36–7
cultural revolution, 27
culture(s):
  ancient Greek, 62–8
  death/dying, 31–2

definition of, 35
development of, 12, 20, 23
formation of villages/cities, 31
of medieval period, 87
Mesopotamia's written, 51
in nonhuman species, 35
see also European society/culture
cuneiform pictograms, 50
Curie, Pierre, *263n*
curium, *263n*

Dalton, John, 172–4, 217, 241, 294
A New System of Chemical Philosophy, 172
dark energy, 298
dark matter, 297–8
Darwin, Charles:
Beagle voyage, 196–8
belief systems of, 201–2
bird identification, 198–9
books published by, 193–4
on Chambers's book on creation, 203–4
death of, 212
deaths of children of, 202, 208
early life and education, 195–6
experiments and investigations by, 205–6
and finches, 198
illness of, 202–3, 205
influence of, 209, 211
law of succession, 198
mentioned, *210*, 212, 240
natural selection theory, 186, 194, 200–1
On the Origin of Species by Means of Natural Selection; or, the Preservation of Favoured Races in the Struggle for Life, 194, 208–9
on population growth, 200
publication of views of, 203–4
study of species differences, 198–9
theory of diversity, 205–6
theory of evolution, 186, 194, 199–200
da Vinci, Leonardo, 99
death/dying, 24
ancient Egyptian concerns for, 153
cultural aspects, 31–2
embalming industry, 153
humans' awareness of, 37–8
De Motu Corporum in Gyrum (On the Motion of Bodies in Orbit) (Newton), 134, 138
Descartes, René, 39
determinism, 280
Dewey, John, 219
Dialogue Concerning the Two Chief World Systems (Galileo), 160
Dicke, Robert, 237 and *n*
diet/nutrition, 16–17, 166
differential calculus, 124*n*
dimensional scaling, 34
Dirac, Paul, 267, 279, 284, *287*
discoveries:
archaeological (*see* archaeological discoveries)
elements, 151–2
Galileo's, 107–9
inventions versus, 222
of new approximation technique, 33–4
oxygen, 152, 162–3, 165, 168, 170, 173–4, 178
Rutherford's, 250–1, *251*
of thinkers, 8
diseases, causes of, 156
diversity, 205–6
divine intelligence, 143, 185
DNA analysis, 19–20, 46, 211, 296
drugs/medicines, 89, 156–7
duality, 76
Dyson, Freeman, 237 and *n*

earthquakes, 67
Eddington, Arthur, 261
Edison, Thomas, 48, 131
education, 51–5, 88–90
Egypt (ancient):
biological organisms theory, 186–7
class system in, 50
concern for fate of the dead, 153
embalming practices in, 153
engineering capabilities in, 55
export of commodities from, 66
frogs' appearance theory, 154
intellectual inventions in, 50
knowledge of chemistry, 154
mathematical rules developed in, 66–7
use of geometry, 67
writing systems in, 47

Einstein, Albert:
on Bohr atom, 261
on chaos, 63
concept of God to, 281
death of, 285–6
education and employment, *234*, 235–6
general relativity theory, 95, 143, 248, 298
mentioned, 34, 176, 229, *245*, 263*n*, 272–3, 284, *287*, 288
on objective reality, 282
"On a Heuristic Viewpoint Concerning the Production and Transformation of Light," 241
papers on physics, 236
photoelectric effect, 243
on physical laws, 60
on Planck's quantum theory, 232
preservation of brain of, 286
Prussian Academy of Sciences, 244
and quantum theory, 128, 216, 232–3, 240, 241, 242, 245, 248, 258, 261, 273, 276–7, 278, 280–2
relativity theory, 143–4, 235, 238–40
on Schrödinger's theory, 276–7
on science, 9
special relativity theory, 143, 248
theory of gravity, 257
theory of light quanta, 128, 242–6, 248–50, 258, 261, 267
work on thermodynamics, 242
einsteinium, 263*n*
electricity, 163, 243
electromagnetic repulsion, 257
electromagnetism theory, 225, 237*n*, 238, 260; *see also* spectral lines
electrons, 243, 250, 256
energy emissions of, 257
measuring velocity of, 280
motion of, 276
positive/negative charge, 257
Rutherford's model, 258
state of motion of, 260
elementary particles of matter, *218*
Elementary Treatise on Chemistry (Lavoisier), 170
elements:
application of Bohr's theory to, 262
Aristotle's theory of, 161
atomic properties of, 173–4
Bohr and, 259–60, 262–3
Boyle and, 159, 161–2
combined, 173
Dalton and, 173–4
emanations from chemical, 250
four fundamental, 73, 152, 159, 170
incompatible, 106
Lavoisier and, 168, 169, 170, 171
Mendeleev and, 177–82
metallic, 152, 177
naming of, 182, 263 and *n*
periodic table of, 177–82, *181*, 262
principles of chemical, 241
transmutation of, 250–1
see also elements by name; periodic table of elements
Eliot, T. S., 35
energy:
in allowed orbits, 260
Bohr's theory of, 271
dark, 298
gravitational, 257
loss of atomic, 259
quantum of light, 230
energy output, 259
engineering, 53–5
Enlightenment era, 163
environment, innate understanding of physical world, 22–3
equations, chemical, 169
Erect Man *(Homo erectus)*, lifestyle of, 19
Essay on the Principle of Population (Malthus), 200
ethical codes, 56–7
Euclid, 55, 67
European society/culture:
Christianity in, 87
development of educational system, 88, 90–1
the elite, 98
influence of Merton's rule, 95
introduction of arithmetic, 54
medieval period, 87–8, 90–3, 97–9
Pythagoras's influence on, 71
Renaissance period, 87, 89–93, 97, 99
science revival in, 89–90
written alphabet use, 52
Eustace, Aland, 142
evolution:
of brain, 12–15, 18
of cities, 30–3, 39–43
of farming/agriculture, 26–7

human (*see* human evolution)
of language, 46
Malthus's essay on, 200
of search for knowledge, 4
of specialized professions, 42–3
of sports, 19
of technology, 5–6
see also Darwin, Charles; natural selection theory
existential consciousness, 30
experiments (in general), 174, 190, 227, 250; *see also specific researcher or object of experiment*

facial expression, 23
failures:
effects of, 223
Einstein's, 246
fear of, 267
Newtonian physics, 132, 216, 226, 270
persistence after, 268
Priestley's experiments, 165
of revolutionary thinkers, 165, 267–8
safety for allowing, 248, 258
farming/agriculture, 26–7, 32, 41–3
Fermi, Enrico, 263*n*
fermium, 263*n*
Feynman, Richard, 116–17, 247, 267, 277
fire, 18–19
fission, 288
Fitzroy, Robert, 196
fixed-earth philosophers, 111
flerovium, 263*n*
Flyorov, Georgy, 263*n*
foraging lifestyle, 26
force fields, 133, 257
force(s), 226
Aristotle's concept of, 74
attraction, 166 (*see also* gravity)
Kepler's law of, 133
in motion, 105–6
Newton's laws of, 125, 138–40
radial motion, 135
fossils, 198
Foundation trilogy (Asimov), 8
four-element theory, 73, 152, 159, 170
free fall, 102–4, 137, 141–2
French Revolution, 169
frequencies, high, 227–8
Frost, Robert, 231–2
Fry, Art, 96
fudge factors, 297

Gadolin, Johan, 263*n*
gadolinium, 263*n*
Galen, 155–6
Galileo Galilei:
books by, 110–13
conflicts with Catholic Church doctrine, 111
death of, 146
Dialogue Concerning the Two Chief World Systems, 111, 160
Discourses and Mathematical Demonstrations Relating to Two New Sciences, 113, 114, 121
early life and education, 99–102
experiments conducted by, 102–6
free fall, 102–4
on heliocentric universe, 106–7
ideas on motion, 104–6, 112–13
law of inertia, 105, 135, 138
mentioned, 95, 97, 99–104, *101*, 104–13, 137, 294
teaching of Copernicanism by, 111*n*
use of telescope, 107–8
view of Aristotle's theories, 102–6, 107–9, 110–11
Gamow, George, 237 and *n*
gases, 225
air, composition of, 161–2, 164–5, 168
Boyle's study of, 161–2
carbon dioxide, 164
Dalton's study of, 172–3
halogens, 177
invisible, 162
Priestley's study of, 163–4
spectroscopy, 259–60
see also oxygen
Geiger, Hans, 253–6, 287
general relativity, 95, 143, 248, 298
genetics, 46, 211
Geological Society of London, 198
geometry, 55, 67
Germany/German physicists, 283–8
Göbekli Tepe monument, 27–30
God:
belief in, 38, 79
Einstein's concept of, 281
laws of reproduction, 199
role in creating life, 187
see also religion/religious issues

gods, 44–5, 58, 62–3, 67
gold, 154–5
Google, 36–7
graphs, 123–4
gravitational energy, 257
gravitational phenomena, 226, 257
gravity, 106
  antigravity effect, 298
  Einstein's theory of, 143–4, 257, 298
  Newton's theory/laws of, 122, 129, 137–8, 140, 143–4, 147
  planetary, 137, 147
  relationship to time and space, 261
Greece (ancient), 289, 296
  alchemy in, 154–5
  Alexandria, 153–4
  culture and mythology of, 62–8
  education in, 63
  Ionia, 64, 65, *65*
  mathematics, 55, 91
  physics, 154
  Roman conquest of, 86–7
  thinkers of, 60
  trade in Miletus, 65–6
  translations of science of, 87
grit, 174–5
Grossmann, Marcel, 34, 235–6
ground state, 259
group dynamics, 44

hafnium, 262–3
Hahn, Otto, 237
Halley, Edmond, 34, 132–4, 142–3
halogens, 177
Hawking, Stephen, 85, 117, 174, 296
Heilbron, J. L., 108, 111
Heisenberg, Werner:
  approach to quantum theory, 279
  approach to theory of object properties, 271–2
  on Bohr's theory of the atom, 269
  death of, 288
  early life and education, 265–7
  Einstein on theory of objective reality, 273
  equivalence of theory to Schrödinger's, 277
  importance of work of, 274
  mentioned, 8, 239–40, *274*, 276, 287, *287*, 288
  and Nazis, 283–5, 288
  paper on quantum theory, 273–4
  on Schrödinger's theory, 278
  theory of energy levels of atoms, 274–5
  visit to Bohr, 267
heliocentric universe, 106–7, 110
Helmholtz, Hermann von, 223
Helmont, Jan Baptist van, 186–7
Hendenburg, Paul von, 283
Henshilwood, Christopher, 36
Herbert, George, 39
heredity, 211
Herodotus, 64
Herschbach, Dudley, 34
Hertz, Heinrich, 243
Hesiod, 63
Hicks, William, 39
Higgs particle, 217–19
high frequencies, 227–8
High Middle Ages, 88
Himmler, Heinrich, 284
Hindu caste structure, 89
History of Fishes, The, 142–3
Hitler, Adolf, 283–5, 288
HMS *Beagle*, 196–8
Hodder, Ian, 29
Holocaust, 59
Homer, 63
Homo genus, 11
  diet/nutrition of, 16–17, 26
  Homo erectus (Erect Man), 18–19
  Homo floresiensis, 29
  Homo habilis (Handy Man), 14–17, *16*, 17–18
  Homo sapiens sapiens (Wise, Wise Man), 19–20
  near extinction of, 19–20
  see also human beings
Hooke, Robert:
  conflict with Newton, 128, 135, 143–5
  experiments with Boyle, 161
  letter to Newton on orbital motion, 135
  mentioned, 39, 127–8, 133, 143
  Micrographia, 189, *189*
  microscopic studies by, 188–91
  and Royal Society, 160, 188, 190, 192
  study and naming of cells, 183
Hooker, Joseph Dalton, 207–8
Huffman, Carl, 70–1

human beings:
Aristotle on "deformities" of, 185
art/artists, 99, 219
capacity for language, 46–8
connection of physical world to, 290
curiosity of, 3, 5, 7, 21–2
desire for knowledge and meaning, 4
family life, 31–2
intellectual journey of, 290–1
migration patterns, 25–6
misery as natural state of, 200
music, 70, 219
near extinction of, 19–20
poems/poetry, 51, 101
self-preservation instinct, 10–11
social skills, 117
understanding through experience, 59, 78, 294
weight of brain/body size, 15
see also Homo genus
human culture, *see* culture(s)
human evolution:
behavior, 19–21
development of self-awareness, 24
diet of earliest *Homo* genuses, 16–17
discovery of fire, 18–19
evolutionary basis of sports, 19
first use of tools, 15–17
intellectual development, 7–8, 12–15, 21–2
Lucy, 13
Protungulatum donnae, 11–12
thought processes, 24–5
see also Australopithecus afarensis; Homo genus
hunter-gatherer lifestyle, 19, 26–30, 32–3, 42
Huxley, T. H., 206
hydrogen, 255, 260, 274–5

iatrochemia, 156–7
ideas:
Aristotle's devotion to, 75
building on others', 39
generation of, 33–5
openness to contradictory, 229
requirements for thriving of, 36
revolutionary, 76–7
theoretical, 68
see also innovation(s); inventions
illness, causes of, 156
imagination, 24, 105
India (ancient), 89, 98
Indian mathematics, 54
industrial revolution, 164
inertia, 105
infants, cognition study, 22–3
infinitesimal lengths, 124, 136
influence, through language, 45–8
infrared radiation, 227
innovation(s), 6, 8, 36–7, 48, 54, 66, 88
instincts, 10–11
integral calculus, 124*n*
intellectual development, 7–8, 12–15, 21–2
intellectual inventions, 50
intellectual professions, 43
interconnectedness, 37, 43, 49
International Union of Pure and Applied Chemistry, 237*n*
intuition, development of, 22–3
inventions:
agriculture, 32
base ten system, 54
calculus, 95
discoveries versus, 222
intellectual, 50
lightbulb, 227
microscopes, 188
modern technology, 8, 16
myths about, 34
by Paleolithic humans, 25–6
patents for, 176
Post-it notes, 96–7
printing press, 98
quantum theory, 222
sound recording, 48
telescopes/spyglasses, 107–8
transportation machines, 5
water clock, 102–3
writing systems, 48
inverse square law of attraction, 133–4, 136
invisible/intangible objects:
dark matter, 297–8
gases, 162
observation of, 219
reality of, 226, 240–1, 267
unobservable orbital motion, 280
see also atoms/atomic theory
Ionia, 64, *65*
Ionian thinkers, 68

Ionian understanding of world, 114
Islam (ancient), 87, 98

James II, King of England, 144
Jaynes, Julian, 22–3
Jeans, James, 231
"Jewish physics," 284
Jewish scientists, 283–9
Jobs, Steve, 176
Johanson, Donald, 13
Jordan, Michael, 267–8, 287
Jordan, Pascual, 273–4

Kant, Immanuel, 163
Kepler, Johannes, 58, 107, 132
Kepler's laws, 132–4, 136–7
Keynes, John Maynard, 118
Kirchhoff, Gustave, 222, 225
knowledge:
  acquisition of in early civilizations, 42–3
  development of mental tools, 38
  drive to search for, 3–4
  evolution of search for, 4
  hunting and gathering, 33–7
  requirements for thriving of, 36
  written language and accumulation of, 46–7
Köhler, Wolfgang, *The Mentality of Apes*, 21
Kroll, Norman, 286*n*

Lamb, Willis, 286*n*
language(s):
  animal communication, 22, 45–6
  of chemistry, 174
  learning foreign, 48
  mathematics as, 54
  spoken, 46–8, 69–70
  written, 46–7
  see also writing systems/written word
Lavoisier, Antoine, 165–71, *171*
Lawrence, Ernest, 263*n*
lawrencium, 263*n*
laws, mathematical, *see* mathematical laws
laws, quantum theory, *see* quantum physics
laws, scientific, 56–7, 59–60
laws of nature, 38, 241
  abstract, 72
  cause and effect, 63, 216, 246
  Code of Hammurabi, 57–8
  conservation of mass, 168, 170
  of force, 226
  inertia, 105, 135
  invention of concept of, 55
  mathematical (*see* mathematical laws)
  Newton's (*see* Newton's laws)
  personal nature of, 59
  physical versus human, 59–60
  Planck's constant, 230–2, 298
  quantitative, 78
  of quantum physics, 148
  scientific, 56–7
  testable, 217
  types and meaning of, 56
  see also specific law or theorist
Leakey, Louis, 17
Leeuwenhoek, Anton van, 190–3
Leibniz, Gottfried Wilhelm, 143
leisure, 66
Leonardo da Vinci, 99
life, creation/origin of, 186–7, 297
light:
  Lavoisier's study of lack of, 166–7
  Newton's study of, 127–8
  photoelectric effect, 243
  photon theory of (*see* photons)
  properties of, 271
  quantum of light theory, 128, 230, 242–6, 248–50, 261, 267
  spectroscopy, 259–60
  speed of, 238–9
  see also photons
limiting cases, 124
Lincoln, Abraham, 268
Lindemann, Ferdinand von, 266–7, 266*n*
Linnaeus, Carl, 177
Linnean Society of London, 207–8
Lippershey, Hans, 107–8
living beings, design of, 185
Locke, John, 143
logical thinking, 52
Lorimer, Frank, 21–2
Lucy *(Australopithecus afarensis)*, 13–17
Lyell, Charles, 206

macaque monkeys, 35–6
madrassas (charitable trusts), 88
maggots/flies experiments, 187
Malthus, T. R., 200
Manhattan Project, 286*n*, 287

Marduk, 58
Marsden, Ernest, 254–6
mass, law of conservation of, 168, 170
mathematical formulas, Planck's, 227–8
mathematical laws, 58, 70
  finding instantaneous speed, 124–5
  inverse square law, 133, 135–6, *136*
  motion (*see* motion [change of position])
  music, 70
  scientific agreement on, 86
Mathematical Principles of Natural Philosophy (Philosophiæ Naturalis Principia Mathematica), see Newton, Isaac; *Principia Mathematica* (Newton)
mathematical symbols, 54–5
mathematics:
  algebra, 54, 273
  calculus, 122–3, 124*n*, 125, 143
  of change, 125
  development of, 52–5
  evidence of Higgs particle, 217–19
  founder of Greek, 69
  geometry, 55, 67
  Heisenberg's, 273
  language of, 68–9
  as language of physics, 95
  matrix algebra, 273
  Merton's rule, 94–5
  optics, 127–8
  philosophy as, 77
  role in theory of physics of, 219
matrix algebra, 273
matter, science of, *see* chemistry
matter versus light principle, 238–9
Maxwell, James Clerk, 225–6, 238–9, 241
meaning, searching for underlying, 4
measurement, 53, 78, 93–4, 169
medical practices, Galen's, 155–6
Medici, Cosimo II de', 109
medicinal plants, 35
medicines/drugs, 89, 156–7
medieval period, 87–8, 87*n*, 90–3, 97
Meitner, Lise, 237 and *n*, 263*n*
Mendel, Gregor, 211
Mendeleev, Dmitri, 175–82, *181*, 237*n*, 241, 262–3, 263*n*
  "On the Relationship of the Properties of the Elements to their Atomic Weights," 181
Mentality of Apes, The (Köhler), 21
mercuric oxide (calx), 164–5, 168–9
Merton College scholars, 91, 93–5, 158
Merton rule, 94–5, *96*, 97
Mesopotamia (ancient), 41
  authority in, 44–5
  conquest of, 62
  first use of written word in, 47
  idea of laws in, 56
  intellectual inventions in, 50
  roles of professors in, 45
  sources of authority in, 44
  use of mathematics in, 53
  warfare, 43
  written culture of, 51
metallic elements, 152, 177
metaphysics, 86, 222, 233, 250
microbiology/microscopes, 188–93
Micrographia (Hooke), 189, *189*
Middle Ages, 88
Mierop, Marc Van De, 44–5
Miletus (ancient), 64–7
Millikan, Robert, 244–5, 248
Milton, John, 112
Ming Dynasty, 88
Mlodinow, Simon (author's father), 3–4, 37–8, 59, 60, 72–3, 79, 105, 115–16, 118, 150–1, 233, 274, 281, 283, 290–2, *292*
momentum, 139, 271
monk on the mountain riddle, 293–4
moon's orbital motion, 137–8
mortality, *see* death/dying
motion (change of position), 93
  Aristotle's concept of, 73–4
  arrow of time, 221
  circular, *136*
  forces causing, 79–80
  free fall, 102–4, 137
  Galileo's ideas on, 105, 112–13
  Merton rule, 94–5, *96*
  Newton's laws of, 123, 125, 138–42, 147, 225–6, 248
  orbital, 137
  radial, tangential, orbital, 135
  total quantity of motion, 140
M-theory, 297
Murray, John, 193–4, 208
Museum of Alexandria, 154
music, 70, 219
Muslims (medieval), 87
mythology, Greek, 62–3, 130

National Geographic, 28
natural change, 73–4
natural laws, *see* laws of nature
natural phenomena, 226
natural selection theory, 186, 194, 200–1, 207; *see also* Darwin, Charles; evolution
nature:
  abhorrence of vacuum in, 80
  Bacon on study of, 142
  fundamental principles of, 232
  general theory of, 262
  human approach to, 27
  interpretation of, 58–9
  laws of (*see* laws of nature)
  mathematical rules followed by, 71
  Newtonian laws of, 114–15
  obedience to quantitative laws by, 78
  power over, 32
  purpose, 58, 59, 78–9, 93, 114, 201, 233
  puzzles in, 264–5
  rational approach to, 64
  structure and order in, 63
  Thales on operation of, 67
  theories/understanding of, 114
  unified theory of, 57
nature's plan, 79
Nazi party, 73, 79, 170, 281, 283–8
Near East, map of ancient, *40*
Neolithic era (New Stone Age), 25, 30–3; *see also* Paleolithic era (Old Stone Age)
Neolithic revolution, 27
New Stone Age (Neolithic era), 25
New System of Chemical Philosophy, A (Dalton), 172
Newton, Isaac:
  contribution to science of, 132
  death of, 145–7
  early life and education, 115, 118–22
  God's plan for universe, 199
  ideas about fundamental particles (atoms), 147, 166, 226
  life in London, 144–5
  mentioned, 34, 39, 68, 73, 91, *145*
  move away from academics, 144–5
  Paracelsus's influence on, 157
  predictions by, 129
  pursuit of wrong/crazy ideas, 126–32
  questioning of conventional thinking by, 137
  reasons for off-course ideas, 131
  social life, 116–17, 118–19, 128, 131, 134, 144–5
  studies at Cambridge, 120
  view of Aristotle's theories, 121, 137, 143
  work on theology and alchemy, 129–30, 144
Newtonian (classical) physics:
  challenges to, 249
  De Motu Corporum in Gyrum (On the Motion of Bodies in Orbit) (Newton), 134, 138
  determinism in, 280
  electron behavior, 258–9
  energy loss from atoms, 259
  failures of, 216
  forces of attraction, 166
  Kepler's laws, 134
  and Maxwell's theory, 225–6, 244, 249
  of motion, 123, 131–2
  Principia Mathematica, 122, 130, 134–5, 138, 142–4
  relativity theory as modification of, 239
  unseating of, 274
  zenith of, 249
Newtonian worldview, 289
Newton's laws, 151
  First Law (change in state of being), 138–9
  Second Law (force equals mass times acceleration), 139–40
  Third Law (action/equal and opposite reaction), 140–2
  acceleration, 138–9
  applications of, 140–2
  attraction, 138
  criticism of, 144
  failure of, 270
  gravity, 122, 129, 137–8, 140, 143–4, 147
  nature, 59–60, 114
  replacement of, 262
Newton-Smith, W. H., 127
Niels Bohr Institute, 285
Nobel, Alfred, *263n*
nobelium, *263n*
Nobel Prize:
  Born, 264, 285
  chemistry, 182
  Einstein, 236–7, 242, 245

Hahn, 237
Lamb, 286*n*
mentioned, 15, 34, 263*n*, 288
Pauli, 273
Planck, 230
reasons for awarding of, 237*n*
Rutherford, 251, 285
scientists passed by, 237
Thomson, 251
nomadic lifestyle, 26–7, 29, 31–2
notation, mathematical, 54
nuclear fission, 288
nucleus of atoms, 256–7

objective reality, 273, 282
objects:
collision of, 125
in free fall, 103–4, 141
invisible/intangible (*see* invisible/intangible objects)
properties of, 279
values for quantities of, 282
observables, 270–1
observation(s):
Aristotle's, 78
Boyle's, 161–2
of invisible matter, 219 (*see also* atoms/atomic theory)
unobservable orbital motion, 280
ways of describing, 68–9
Oldenburg, Henry, 190
Old Stone Age (Paleolithic era), 25
"On a Heuristic Viewpoint Concerning the Production and Transformation of Light" (Einstein), 241
On the Motion of Bodies in Orbit (De Motu Corporum in Gyrum) (Newton), 134, 138
On the Origin of Species by Means of Natural Selection; or, the Preservation of Favoured Races in the Struggle for Life (Darwin), 194, 208–9
"On the Relationship of the Properties of the Elements to their Atomic Weights" (Mendeleev), 181
Oppenheimer, J. Robert, 286*n*
optics, 127–8
orbital motion:
De Motu Corporum in Gyrum (On the Motion of Bodies in Orbit) (Newton), 134, 138
of electrons, 257, 261
Hooke on, 135
innermost allowed orbits, 259
Kepler's laws, 136–7
moon's, 137–8
orbital radius of electrons, 259
of planets, 132–3
stationary states, 260
unobservable, 280
Oresme, Nicole, 95–7
organisms, classification of, 177
Ostwald, Friedrich Wilhelm, 217, 236, 240
Oxford University, *92*, 127, 131, 160, 209, 285
Merton College scholars, 91, 93–5, 158
Merton rule, *96*, 97
oxide of mercury (calx), 164–5, 168–9
oxygen, 152, 162–3, 165, 168, 170, 173–4, 178

Pais, Abraham, 240
Paleolithic era (Old Stone Age), 25; *see also* Neolithic era (New Stone Age)
Paleolithic humans, 25–6
Paracelsus, 129, 153, 155, *156*, 157–8
Pasteur, Louis, 187
patents, Simjian's, 176
Pauli, Wolfgang:
on Born's mathematics, 273
mentioned, 126, 272, 284, *287*
on Schrödinger's theory, 278
spin theory, 264
theory for deriving spectral lines of hydrogen, 274
Paul V, Pope, 110
Peebles, Jim, 237 and *n*
Penzias, Arno, 237*n*
Pepys, Samuel, 189
periodic table of elements, 177–82, *181*, 262
Petty, Sir William, 164
phenomena, 243, 257, 279–80
Philip II of Macedon, King, 71
Philosophiæ Naturalis Principia Mathematica (Mathematical Principles of Natural Philosophy), *see* Newton, Isaac; *Principia Mathematica* (Newton)
philosophy/philosophers, 51, 64, 66, 87, 217; *see also* thinkers (in general)

phlogiston, 165, 170
Phoenician alphabet, 52
photoelectric effect, 243, 249, 259
photons:
  Born's introduction of term, 264
  Einstein's theory of, 242–4, 248
  excitation of atomic energy by, 259
  orbital radius of, 259
  quantum of light theory, 128, 242, 244–6, 248–50, 261, 267
  reality of, 267
  see also light
physical laws, 60
physical principles, 69
physicists (in general), 68–9, 95, 103, 276
  application of Newtonian physics by, 216
  on Bohr's theories, 262–3
  competing theories among, 275
  on dark matter, 297–8
  definition/description of, 150
  discouragement among, 264
  as dreamers and technicians, 256
  experimental, 251
  fears of, 131
  first, 68
  insights/epiphanies of, 122
  issues for Jewish, 283–9
  mental pictures created by, 277
  modern, 103
  Nazi, 287
  passion of, 223, 248
  on Planck's work, 231
  promise of theory of everything, 298
  quantum, 275, 289, 295
  reactions to Heisenberg, 288–9
  on string theory, 297
  theoretical, 238, 251
  view of blackbody radiation by, 230
  view of their own field, 220
  see also specific physicists/theorists
physics:
  application of laws of, 80
  Aristotle's understanding of, 74–5, 79–80
  changing theories of, 158
  chemistry versus, 149–51
  classical, 216 (*see also* Newtonian [classical] physics)
  compared to art, 248
  effects of Einstein's work in, 239–40
  Heisenberg's approach to, 272
  Lavoisier's take on, 166
  laws of, 79 (*see also* laws of nature)
  mathematics as language of, 95
  Merton rule, 95–7, *96*
  nature of change in, 73
  Newtonian (*see* Newtonian [classical] physics)
  nineteenth-century, 220
  Planck's study of, 221–2
  progress in, 258
  qualitative analysis of, 77–8
  quantitative, 77
  quantum (*see* quantum physics)
  revisions of, 233
  statistical, 229
  theoretical, 237–8
  twentieth-century, 219
  see also Aristotelian physics; Newtonian (classical) physics; Newton's laws
pictograms, 49–51
Pinker, Steven, 46
pioneers/revolutionaries, 71, 165, 268, 270, *287*, 289
Planck, Max:
  atomic theory and, 221
  blackbody radiation research, 225, 227–8, 230–1, 242, 249, 259
  early life and education, 217–21
  on existence of atoms, 222–3
  formulas for describing radiation and high frequencies, 227–8
  interest in Boltzmann's research, 228–30
  mentioned, *224*, *229n*, 237, 286–7, *287*, 288
  Nazis and, 288
  origin of quantum concept, 230
  professorship at University of Berlin, 223
  quantum experiments, 250
Planck's constant, 230–2, 298
planets, *see* astronomy
Plato, 71, 72, 75
population growth, 125, 200
position, 271; *see also* motion (change of position)
Priestley, Joseph, 163–5, 167–8

Principia Mathematica (Newton), 122, 130, 134–5, 138, 142–4
printing press, 98
problem solving, 33–5
Protungulatum donnae, 11
Ptolemaic astronomy, 107
purpose:
  absence of, 59
  Aristotle's search for, 78–9
  as driver of events, 201, 233
  Kepler's assignment of, 58
  loss of common, 283
  Merton scholars' theory of motion and, 93
  in nature, 78, 114, 201, 233
  yearning for, 79
pyramids, 55, 66–7
Pythagoras/Pythagorean theorem, 69–71, 78, 109, 114, 129, 151

quantity of motion, 140
quantum of light theory, 128, 242–6, 248–50, 261, 267; *see also* photons
quantum mechanics, 245, 270, 283
quantum physics, 215–89
  applications of, 279
  Bohr's intuition about, 252, 267
  Born's thoughts on, 264, 286
  challenges of, 265
  creation of, 282–3
  deterministic world and, 280
  development of, 166
  development of competing theories, 275–9
  dismissal of, 261
  Einstein and, 128, 240–1, 242, 244, 245, 248
  electron orbits, 261
  evolution of, 230–3, 273–80
  Feynman's version of, 277
  as general theory of nature, 262
  Heisenberg's approach to, 273–9, 286, 289
  invention of, 222, 278, 295
  laws governing atoms, 215–16
  laws of, 148, 241
  Lindeman's views of, 266
  Nazis' view of, 284
  object properties in, 282
  philosophy behind, 216
  photon concept, 249–50, 252, 264
  probabilistic interpretation of, 282
  products of, 211
  properties of objects in, 279
  randomness in, 281
  Schrödinger on, 278
  Schrödinger's version of, 277, 285
  technologies derived from, 295
questions:
  act of questioning, 22
  asking scientific, 67–8
  children's, 21–2
  of freethinkers, 224–5

radar technology, 256
radial motion, 135
radiation:
  blackbody, 216, 225, 227–8, 242–3
  electromagnetic, 225
  infrared, 227
radioactive elements, 251
randomness, 201, 281
Rayleigh, Lord, 261
rays (alpha, beta, gamma), 250
reality:
  of atoms, 217, 240
  differing views of, 232, 239, 272, 275–9
  Heisenberg's theory, 273
  of invisible/intangible objects, 241, 267
  Newton's concept of, 269, 279
  number as essence of, 71
  objective, 273, 282
  of objects, 226
  philosophical questions about, 278
  representations of, 99, 222
reason and instinct, 11
reasoning, guidance of human, 69
Recorde, Robert, 54–5
Redi, Francesco, 187–8
relativity theory, 143–4, 235, 238–40
  general relativity, 95, 143, 248, 298
  special relativity, 143, 248
religion/religious issues:
  Bible, 28, 129, 197, 281
  as Boyle's motivation, 160
  Catholic Church (*see* Catholic Church)
  Christianity (*see* Christianity)
  Church of England, 144

religion/religious issues *(continued)*:
  conservative forces in medieval period, 87
  Darwin's loss of faith, 202
  distinction between church and state, 44
  Göbekli Tepe monument as church, 27–30
  heliocentric universe, 106
  idea of law and ethical codes in, 56–7
  indications of worship, 28
  madrassas (charitable trusts), 88
  Newton's beliefs, 129–30
  objections to Darwin's theories, 209
  opposition to science in, 204
  questioning traditions in, 22
  Torah, 57
  see also God; gods
religious/spiritual worldviews, 85–6
Renaissance, 87, 89–93, 97–9
reproduction, God's laws of, 199
resistance, to objects in free fall, 141
respiration experiments, 161–2
Revolutionibus Orbium Coelsestium (On the Revolutions of the Heavenly Spheres) (Copernicus), 158
Riccioli, Giovanni, 103
Rilke, Rainer Maria, 67
Ripoli Press, 98
Röentgen, Wilhelm, 250, *263n*
roentgenium, *263n*
rolling-ball experiment, 103–4
Roman conquests, 86–7
Royal Society of London, 132, 134, 142–5, 160, 174, 188, 192
Rubens, Heinrich, 228
Russell, Bertrand, 75
Rutherford, Ernest:
  alpha particle experiments, 253–5, *256*
  atomic model of, 256–9, 261
  on Bohr's work, 261–2, 269
  discoveries, 250–1, *251*
  energy loss from electrons, 259
  experiments on electric charge in atoms, 252–3
  force fields, 257
  gold foil experiment, 253–5, *254*, 257
  on Heisenberg's work, 275
  mentioned, 263*n*, 271, 284–5
  Nobel prize in Chemistry, 251
  orbiting electrons model, 257
  partnership with Bohr, 252
  predictions based on work of, 258
rutherfordium, 263

samarium, *263n*
Samarsky-Bykhovets, Vasili, *263n*
San people of Africa (Bushmen), 27
Sarpi, Paolo, 108
Sceptical Chymist, The (Boyle), 160–1
Scheele, Carl, 150
Schmidt, Klaus, 28
Schrödinger, Erwin, 275–8, 285, *287*
Schweppe, Johann Jacob, 164
science:
  Aristotle's approach to, 77–9
  attitudes toward, 5
  Darwin's view of, 201
  development of modern, 6–7
  disconnect between philosophy and, 78
  Einstein on, 9
  European revival of, 89–93
  framework of ancient versus modern, 63
  medieval, 91
  as metaphysical system, 86
  negative reactions to, 204–5
  philosophy versus, 217
  separation of arts and, 99
  societal support for, 87–8
  theoretical and experimental aspects of, 167–8
scientific knowledge, 42–3
scientific laws, 56–7, 59–60
scientific method, 127–8
scientific revolution, 89–93
scientific worldviews, 85–6
Seaborg, Glenn T., *263n*
seaborgium, *263n*
Searle, John, 91
Sedgewick, Adam, 205
self-awareness, 24
Shakespeare, William, 99
shapes, mathematics of, 54–5
sign language, 45–6
Simjian, Luther, 176
Sklodowska-Curie, Marie, *263n*
Socrates, 81, 294–5
solar system, *see* astronomy

Sommerfield, Arnold, 237 and *n*, 261, 266–7
sound, speed of, 140
space and time relationship, 238–9, 261
specialized professions, 42–3
special relativity, 143, 248
species differences, 15, 198–9, 211
spectral lines, 260, 271, 274
spectroscopy, 259–60
speech/spoken language, *see* language(s)
speed, 93–4
  defining, 124
  of matter versus light, 238–9
  measuring, 125
  Merton rule, 94–5, *96*
  objects in free fall, 103–4
  rolling-ball experiment, 103–4
  weight and, 102
spin theory, 264
spontaneous generation theory, 187
spyglasses, 107–8
squaring the circle, *266n*
Starry Messenger, The (Galileo), 109
stationary states, 260
statistical physics, 229
Stevens, Wallace, 277
Stokes, Pringle, 196–8
Stone Age, *see* Neolithic era (New Stone Age); Paleolithic era (Old Stone Age)
Stone Age humans, 26
string theory, 247, 297
substance, science of, *see* biological organisms; biology; chemistry
substances:
  Aristotle's theories of transformation of, 154
  breakdown/change of, 151–2
  combustion of, 152
  conversion of, 161
  decomposition of, 152
  discovery of elements, 151–2
  interaction of, 161
  nature of, 174
succession, law of, 198
Sumer, 47, 50–2
superiority of humans, ancient depictions of, 32
supernatural realm, 68, 91, 143
superstition, 66, 188
survival instinct, 11
symbiotic relationships, 49

tangential motion, 135
tax farms, 168–9
technology:
  applications of new, 48
  for chemistry, 162
  communication, 46
  evolution of, 5–6
  innovations of Renaissance period, 97–8
  late-nineteenth-century advances, 249–50
  photoelectric effect in, 243
  quantum, 295
  revolutionary advances in, 55–6
  see also innovation(s); inventions
telescopes, 107–8
television, 250
terminal velocity, 141–2
Thales, 66–7, 69
theology, *see* God; gods; religion/religious issues
theoretical physics, 237–8
theories (in general):
  development of, 269
  different but correct, 277
  discovery versus invention of, 277–8
  lack of support for, 223
  metaphysical questions about, 222
  proving, 248
  strength of, 281–2
  see also specific theory or theorist
theory of everything, 211, 296
thermodynamics, 221–4, 229; *see also* Einstein, Albert; Planck, Max
thinkers (in general):
  accomplishments of great, 34–5
  cultural ratcheting among, 36
  development of mathematical laws by, 55, 60
  discouragement of, 89
  discoveries of, 8
  first professional, 50
  logical thinking, 52
  struggles of pioneering, 71
  see also philosophy/philosophers; pioneers/revolutionaries; *specific theorists*

thinking:
  challenging processes, 96
  conventional, 76–7
  critical, 219
  logical, 52
  new modes of, 55–6
  tools for, 38
  unconventional, 234–5
Thomas, Mark, 37
Thomson, J. J., 243, 250–5, 258
3M Company, 96
tides, theory of, 107
time:
  arrow of time, 221
  measurement of, 93–4, 124
  space and time relationship, 238–9, 261
  water clock invention, 102–3
tools/tool use:
  ability to design, plan, use, 17–18
  by animals, 29, 35
  first use of, 15–16
  for thinking, 38
Torah, 57

uncertainty principle, 279 and *n*, 281–2
uncommon sense, 75
uniform motion, 105
universal gravitation theory, 122
universe:
  Aristotle on function of, 78
  causality in quantum, 216–17
  expansion of, 298
  God's plan for, 199
  Greek philosophers' views of, 64
  heliocentric view of, 106–7, 110
  modern understanding of, 295–6
  in Newtonian worldview, 280
  Pythagorean view of, 114
universities, 89–91
University of Göttingen, 261
uranium 235, 288
urbanization, 43, 49, 53–4
Urban VIII, Pope, 110–11
Uruk, 40–5, 48–50

Van De Mierop, Marc, 44–5
velocity, 97, 125, 141–2
Vestiges of the Natural History of Creation (Chambers), 203–5
villages, 30–3, 39–40
violations of laws, 57
violent change doctrine, 74

Wallace, Alfred Russel, 206–8
Waste Book (Newton), 123–7, 135
water clock, 102–3
Watt, James, 34
wave theory, 276
weight:
  atomic, 172, 174, 178–81, 262
  of mercury, 168
  relative, 173
weights and measures, 53
Westfall, Richard, 118, 120
why question study, 21–2
Wichmann, Eyvind, 286*n*
Wilson, Robert, 237*n*
wisdom, conventional, 76
workforce, organization of laborers, 43
worldviews:
  Aristotelian, 161
  Merton scholars', 93
  Newtonian, 232–4, 239–40, 249, 280, 289
  physicists', 248
  Renaissance, 99
  scientific versus religious bases of, 85–6
  see also beliefs/belief systems
worship, *see* religion/religious issues
Wren, Christopher, 133
writing systems/written word:
  ancient inventions of, 46–8
  earliest use of, 48–9
  Phoenician alphabet, 52
  symbols as representations of vocalizations, 51–2
  use in professions, 50
  see also communication; language(s)

Zilsel, Edgar, 58–9

Illustration Credits

11: From Maureen A. O'Leary et al., "The Placental Mammal Ancestor and the Post–K-Pg Radiation of Placentals," *Science* 339 (February 8, 2013): 662–7.
16: Courtesy of Nachosen/Wikimedia Commons
29: Courtesy of Teomancimit/Wikimedia Commons
30: Created by Derya Kadipasaoglu
53: U.S. Navy photo by Photographer's Mate First Class Arlo K. Abrahamson. Image released by the United States Navy with the ID 030529-N-5362A-001.
72: © Web Gallery of Art, created by Emil Krén and Daniel Marx, courtesy of Wikimedia Commons
92: Picture of the interior of the Merton College Library, from *The Charm of Oxford*, by J. Wells (London: Simpkin, Marshall, Hamilton, Kent & Co., 1920). Courtesy of fromoldbooks.org.
96: Created by Derya Kadipasaoglu
101: Courtesy of PD-art/Wikimedia Commons
123: Created by Derya Kadipasaoglu
136: Created by Derya Kadipasaoglu
141: Courtesy of Zhaladshar/Wikimedia Commons
145: (Left) Courtesy of Science Source®, a registered trademark of Photo Researchers, Inc., copyright © 2014 Photo Researchers, Inc. All rights reserved.
145: (Right) Courtesy of English School/Wikimedia Commons
156: Courtesy of Science Source®, a registered trademark of Photo Researchers, Inc., copyright © 2014 Photo Researchers, Inc. All rights reserved.
171: Courtesy of *Popular Science Monthly*, vol. 58/Wikimedia Commons
175: Courtesy of Science Source®, a registered trademark of Photo Researchers, Inc., copyright © 2014 Photo Researchers, Inc. All rights reserved.
181: Courtesy of Wikimedia Commons
189: Lister E 7, Pl. XXXIV, courtesy of The Bodleian Libraries, The University of Oxford
202: Courtesy of Duncharris/Wikimedia Commons
210: (Top) Courtesy of Richard Leakey and Roger Lewin/Wikimedia Commons
210: (Center) Courtesy of Maull and Polyblank/Wikimedia Commons
210: (Bottom) Courtesy of Robert Ashby Collection/Wikimedia Commons
218: (Above) Courtesy of Science Museum, London/Wikimedia Commons
218: (Below) Courtesy of Maximilien Brice (CERN)/Wikimedia Commons
224: Courtesy of Wikimedia Commons
231: Courtesy of The Dibner Library Portrait Collection—Smithsonian Institution/ Wikimedia Commons
234: Courtesy of Canton of Aargau, Switzerland/Wikimedia Commons

245: Courtesy of F. Schmutzer/Wikimedia Commons
251: Courtesy of Science Source®, a registered trademark of Photo Researchers, Inc., copyright © 2014 Photo Researchers, Inc. All rights reserved.
254: Created by Derya Kadipasaoglu
256: Created by Derya Kadipasaoglu
271: Created by Derya Kadipasaoglu
274: Photography by Paul Ehrenfest, Jr., courtesy of AIP Emilio Segre Visual Archives, Weisskopf Collection
287: Courtesy of Benjamin Couprie, Institut International de Physique de Solvay/ Wikimedia Commons

About the Author

Leonard Mlodinow received his Ph.D. in theoretical physics from the University of California, Berkeley, was an Alexander von Humboldt Fellow at the Max Planck Institute, and was on the faculty of the California Institute of Technology. His previous books include five *New York Times* best sellers: *Subliminal: How Your Unconscious Mind Rules Your Behavior* (winner of the 2013 PEN/E. O. Wilson Literary Science Writing Award), *War of the Worldviews* (with Deepak Chopra), *The Grand Design* and *A Briefer History of Time* (both with Stephen Hawking), and *The Drunkard's Walk: How Randomness Rules Our Lives* (a *New York Times* Notable Book), as well as *Feynman's Rainbow* and *Euclid's Window.* He also wrote for the television series *MacGyver* and *Star Trek: The Next Generation.*

A Note on the Type

This book was set in Janson, a typeface long thought to have been made by the Dutchman Anton Janson. However, it has been conclusively demonstrated that these types are actually the work of Nicholas Kis (1650–1702), a Hungarian, who most probably learned his trade from the master Dutch typefounder Dirk Voskens.

Composed by North Market Street Graphics, Lancaster, Pennsylvania

Printed and bound by Berryville Graphics, Berryville, Virginia

Designed by M. Kristen Bearse